SCIENCE PRESERVED
A directory of scientific instruments
in collections in the
United Kingdom and Eire

Frontispiece
Abacus, Russian, early 17th century
John Tradescant the elder (1567-1637) probably brought this Russian abacus, called a schety, to England in 1618. It was displayed at the Ashmolean Museum in Oxford when it opened in 1683, thereby having claim to be the first scientific instrument publicly displayed in a museum in the British Isles.
(Ashmolean Museum, Oxford. O2.16)

SCIENCE PRESERVED
A directory of scientific instruments in collections in the United Kingdom and Eire

Mary Holbrook
with additions and revisions by
R G W ANDERSON and D J BRYDEN

Trustees of the Science Museum
on behalf of the
British National Committee for the
History of Science, Technology and Medicine

LONDON: HMSO

© Copyright 1992 Trustees of the Science Museum
First published 1992
ISBN 0 11 290060 7

British Library Cataloguing in Publication Data

A CIP catalogue record for this book is available from the British Library

The editors are most grateful to Roderick and
Marjorie Webster of Chicago, staunch supporters of
the study of the history of scientific instruments,
who have provided financial assistance towards the
publication of this volume.

HMSO publications are available from:

HMSO Publications Centre
(Mail, fax and telephone orders only)
PO Box 276, London, SW8 5DT
Telephone orders 071-873 9090
General enquiries 071-873 0011
(queuing system in operation for both numbers)
Fax orders 071-873 8200

HMSO Bookshops
49 High Holborn, London, WC1V 6HB
071-873 0011   Fax 071-873 8200 (counter service only)
258 Broad Street, Birmingham, B1 2HE
021-643 3740   Fax 021-643 6510
Southey House, 33 Wine Street, Bristol, BS1 2BQ
0272 264306   Fax 0272 294515
9-21 Princess Street, Manchester, M60 8AS
061-834 7201   Fax 061-833 0634
16 Arthur Street, Belfast, BT1 4GD
0232 238451   Fax 0232 235401
71 Lothian Road, Edinburgh, EH3 9AZ
031-228 4181   Fax 031-229 2734

HMSO's Accredited Agents
(see Yellow Pages)

and through good booksellers

Set in 9.5/10.5 Palatino from WordPerfect and Pagemaker at the Science Museum
Printed in the UK for HMSO
Dd 291800 7/92 C15 531/3 12521

Contents

List of illustrations

Acknowledgments

Many institutions have graciously supplied illustrations of instruments in their collections. We gratefully acknowledge permission to use these photographs. Picture captions identify the source of each illustration.

Foreword

There is a remarkable wealth of historically important scientific instruments in this country. The question of how to make this wealth available to scholars was tackled from the first meeting under the chairmanship of Sir Harold Hartley of the British National Committee for the History of Science. The publication of this book is the result of one of Sir Harold's initiatives. Difficult decisions had to be made about the balance between detailed descriptions of selected instruments and a more or less complete inventory, and about the practical meaning of the words 'historical', 'important' and 'scientific'. It was also necessary to obtain funds for the work and to find a suitable scholar to do it. At last, in 1967, Sir Harold heard from Lord Murray of the Leverhulme Trust that its trustees could provide a generous grant for the work to proceed. This made it possible to appoint Dr Mary Holbrook, who was returning that year from similar work at the Historisches Museum in Frankfurt to take up a curatorship at Holburne of Menstrie Museum, Bath.

The critical early period during which the co-operation of museums, institutions and private collectors was secured, and press interest aroused, was under the supervision of Dr Francis Maddison, Curator of the Museum for the History of Science at Oxford (itself a tribute to Sir Harold's enterprise in earlier days). From the start, too, support came from successive Directors of the Science Museum: in the early stages of the project Sir David Follett, and then Dame Margaret Weston. To Dame Margaret we are indebted particularly for her advice and for her recommendation for publication by HMSO. The debt to the Science Museum extends further, to Dr R G W Anderson (now Director of the British Museum) and Mr D J Bryden (now Keeper of the Department of Science, Technology and Working Life in the National Museums of Scotland), who have done much both in revising and substantially adding to the text, and in catalysing the reaction pathway to publication.

It would have gladdened Sir Harold's heart to see this volume, and I would like to express most warmly the appreciation of the British National Committee to all those who have brought about its completion.

W D M Paton FRS
Chairman, British National Committee for the History of Science, Medicine and Technology, 1980-85

Preface

Towards a world inventory of scientific instruments

The concept of a world inventory of scientific instruments was formulated in 1946 by the late A Leveille, but not until ten years later did the Union Internationale d'Histoire et de Philosophie des Sciences, Division d'Histoire des Sciences, set up a group, the Commission pour l'Inventaire Mondial des Appareils Scientifiques d'Intérêt Historique, whose object would be to promote the compilation of such an inventory by encouraging the production of national inventories. The scheme was sponsored by Unesco through the Union. The background to the project was described by Henri Michel in a paper dated 1958,[1] who at the same time outlined its aim and its form, based on a standard 'fiche' to be completed for each object or group of objects.

Henri Michel defined the object of the inventory as follows: 'Il ne s'agit pas de cataloguer *tous* les objets scientifiques ayant une valeur archaeologique, artistique ou historique. L'inventaire mondial doit permettre de retrouver, s'ils existent, les principaux jalons de l'histoire des sciences. Entendons par là les instruments qui ont servi à des savants, ou encore qui marquent un progrès dans l'évolution des sciences'.[2]

The first three contributions to the world inventory, by Belgium,[3] Italy,[4] and France,[5] followed very closely the principles established by Michel, with 'fiches' being completed for only those items associated with distinguished scientists, or which could be considered of outstanding importance in the history of science. However, subsequent inventories tended to cover a wider range of instruments, not all of which were associated with distinguished scientists, or could necessarily be said to mark a major step forward in scientific progress, but were nonetheless intrinsically interesting and important in the study of the development of scientific instruments, or of scientific instrument making.

Since the publication of the first three inventories, contributions have been produced by Poland,[6] Czechoslovakia[7] and Russia.[8] The first two are unpublished, but can be seen on microstrip or microfilm in major libraries, including the Science Museum Library at South Kensington, London. Of all the inventories the Russian one, in which the 'fiche' had been abandoned in favour of a brief catalogue entry, was the only contribution to be published in book form, in 1968.

In more recent years responsibility for coordinating production of national inventories has passed to the Scientific Instrument Commission of the International Union of the History and Philosophy of Science, Division of the History of Science.

The British contribution

In the British Isles, responsibility for contributing to the world inventory was assumed by the British National Committee for the History of Science, Technology and Medicine, a sub-committee established by the Royal Society, originally under the chairmanship of the late Sir Harold Hartley. A generous grant from the Leverhulme Trust made it possible for someone to undertake the compilation of the British contribution. Since the compilation of this work, a detailed interim Irish inventory has been published,[9] sponsored by the Royal Dublin Society and the Irish Science and Technology Agency (EOLAS).

The British National Committee felt that in the case of the British Isles, where a large proportion of instruments were concentrated in six museums to which every scholar had easy access, a major contribution could be made by locating material outside these museums, much of which would be virtually, if not completely, unknown. In view of the formidable task of contacting so many potential owners of scientific instruments, it was clearly valuable in the completed work (it was intended that the British contribution should, if possible, be published) to throw the net over as wide a range of instruments as possible. The use of a 'fiche' for each instrument was thus quite impracticable, and it was decided, in the case of the smaller collections, that a system of brief entries should be adopted for each instrument. For the six major instrument collections (the Science Museum, the Wellcome Museum (now at the Science Museum) and the National Maritime Museum, all three in London, the Museum of the History of Science in Oxford, the Whipple Museum of the History of Science, Cambridge, and the National Museums of Scotland (a constituent of which is the former Royal Scottish Museum, Edinburgh)), it was decided that no attempt should be made to list their contents in detail, but rather the scope of the collections should be summarised. Thus the British contribution would take the form of a guide to scientific instruments in collections in the British Isles, which, it was felt, would be of great value to researchers and other interested people.

When the book was first proposed the exact quantity of material which would be located was difficult to estimate, but in retrospect the British National Committee's decision was fully justified. Some highly important individual items, and groups of items, were located, as well as a multitude of other instruments which are perhaps not individually of outstanding historical importance, but which are nonetheless of potential value as a resource in the investigation of the development of a particular instrument, of instrument making in general, or of any study of the output of a particular instrument-making workshop. Equally the book can be a useful tool for curators anxious to locate exhibition material. But perhaps most importantly, scientific instruments can be used as evidence in the history of science, alongside the more traditional resources of printed matter and manuscripts.

The object and scope of the book

The primary object behind the compilation of the book has been to locate and record scientific instruments in collections throughout the British Isles, thus making them

11

accessible for study. The book does not aim to provide a detailed catalogue entry for each item, but simply to indicate what may be found in a given location. Emphasis throughout is on material located outside the major collections of scientific instruments.

Instruments in private hands have not in general been included. At best such a listing would have been incomplete, for many owners of scientific instruments would have been reluctant for there to be public knowledge of their collection. A system of coding entries to preserve the anonymity of private collections was considered, but was finally abandoned since such coded entries, when unaccompanied by an address to which enquiries could be made for further details, would have been of little value. The only exceptions to this general rule have been the Egestorff collection in Dublin, which may be visited regularly, and some old established country house collections, of which instruments form part, which are regularly open to the public. Examples are Longleat, Chatsworth, Petworth and Burton Constable, where most of the instruments have been in the hands of the family for several generations.

The term 'scientific instrument' is used to cover both the artefacts normally constructed by a scientific instrument-maker and the relatively few surviving instruments actually designed by, used by, or constructed for distinguished scientists in their research.

In general instruments are not included which were made after the middle to later years of the nineteenth century. However, later material has been included where it is sufficiently important, or where it forms part of a sequence of instruments and its inclusion is warranted in order to give a rounded impression of the scope of a collection. Areas covered include surveying, navigation and astronomical instruments, mathematical instruments (sectors, slide rules, calculators), optical instruments (microscopes, telescopes, early spectroscopes, etc., but excluding most ophthalmological instruments), and philosophical, physical and chemical apparatus.

Certain groups of peripheral material have been excluded. Among these are the majority of barometers, most of which are purely domestic in nature. The same is true of clocks and watches, except for those clocks and a few chronometers used for scientific or astronomical purposes (normally forming part of the equipment of an observatory) and some examples of clocks incorporating an instrument such as an armillary sphere, astrolabe or orrery. Not every astronomical regulator has been sought out, but they are listed where they occur in collections including other scientific material. For the purposes of the book, sundials, normally produced by a scientific-instrument maker, have been considered to be scientific instruments. Clockmaker's wheel-cutting engines have been listed, but other items of clockmaking equipment, in common with most other machine tools, normally only receive a reference in the introductory paragraphs which precede the list of instruments in each collection.

Commercial weights and measures and trade standards are generally excluded, other than in the introduction to each collection. However chemical, assay and hydrostatic balances are listed. Goldsmith's and coin balances are normally omitted, although a particularly fine early goldsmith's balance such as the one at the City Museum and Art Gallery, Birmingham receives an entry. So too does a coin balance which forms part of a collection of philosophical instruments (see under Department of Natural Philosophy, University of Aberdeen).

With very few exceptions medical and surgical instruments have been excluded from the book. There can be hardly a museum in the British Isles which does not possess a few such instruments, in the main of late eighteenth- and nineteenth-century date. Where these occur in quantity in collections which receive entries, reference has been made to them, and some particularly important items have been listed.

Compilation procedure

A preliminary investigation of the locations of scientific instruments revealed museums, scientific societies, and the science departments of older universities as major sources, followed by private collections, company museums, old-established schools and some government departments. In 1967 a letter describing the proposed book and outlining its scope was prepared for circulation to potential owners or custodians of scientific instruments. At the same time the project was publicised in *Notes and Records of the Royal Society*, *The Times*, *Nature*, *New Scientist*, the *British Journal for the History of Science*, and the *Monthly Bulletin* of the Museums Association.

A list of museums was compiled from the Museums Association's *Museums Yearbook*, and from the publication *Museums and Galleries of the British Isles*. *The World of Learning* was the main source of information on colleges, university departments, learned societies and other organisations which might possibly possess relevant material.

Several thousand copies of the letter were distributed during the early stages of the project, and brought a good response. Replies varied greatly in detail, but nonetheless served to indicate whether a collection contained likely material. However, it soon became apparent that the letter presented a conscientious museum curator with a difficult problem in view of the curator's lack of familiarity with what was only a minor part of the collection. What should be considered a scientific instrument? Was an item used by a navigator, a land or mine surveyor or a brewer to be included in this category? Fortunately curators tended to err on the side of safety and included details of all items they thought might possibly be relevant.

Where no response was received to the first circular letter, a further letter was sent where there were grounds to believe that the collection might contain relevant material. From replies to the letter, from information from specialists in the field of scientific instruments, and from a search through photographs and archives in the Museum of the History of Science in Oxford, a list was assembled of museums and other locations containing likely material. Nearly all these collections were to be visited at a later stage. The programme of visits to collections was a time-consuming exercise, but one without which the book would have been worthless, since the majority of instruments were located in collections whose staff lacked any specialist knowledge in this field. Additional material of relevance to the survey was frequently observed during the course of a visit, although on other occasions the contrary was true, and apparently promising instruments proved to be of little or no interest. Every effort has been made to achieve completeness in the book, but nonetheless further instruments undoubtedly exist, even in collections visited by the author. It is hoped that the glossary of instruments, in conjunction with the illustrations, may assist in the process of identification.

In the course of a visit to a collection, photographs were taken where possible to aid identification and subsequent description, but storeroom conditions, shortage of time, and other logistical constraints did not always permit this. Entries are compiled in the main from an inspection of the instrument rather than from museum records. Inventory numbers or other identifying features are provided wherever possible. An effort was made to determine the provenance of important instruments.

In all more than 900 museums were contacted, as well as university departments, some schools and colleges, observatories and learned societies, and the owners of numerous country houses. The completed book contains entries for more than 200 collections and lists over 2800 instruments. Only in the case of a few collections was a visit not possible, often because information had been received too late for the collection to be included in an itinerary. In these cases the entries are based on information supplied, normally, by the curator.

The compilation of the book occupied a number of years, largely as a result of the extensive travel involved. During that period, from 1968 to 1976, many changes occurred in the museum world, partly as a consequence of local government reorganisation. Old-established museums changed their names, collections began to be relocated among museums within an enlarged local authority, and more museums began to develop a positive collecting policy with regard to scientific instruments. Hence it was deemed wise to send all collections represented a draft of their entry for comments and amendments. This took place during 1976, and thus entries are as far as possible corrected up to that date. Subsequently, up to 1986, further considerable additions and emendations were made by the editors. The final two entries were added only shortly before the date of publication. It is recognised that some data will have changed since they were compiled. Nevertheless there is considerable value in being provided with an overall picture of the recent past, even if a few dispersals, amalgamations and additions have changed matters slightly.

Organisation

The book falls into three distinct sections, comprising a glossary of instrument types (with which the illustrations are closely linked), inventory entries of collections containing scientific instruments, and the indexes.

1 Glossary

A glossary has been compiled which covers frequently occurring instruments as well as less familiar items. In general modern physical and electrical apparatus has not been included. It is hoped that the glossary may assist those using the book, and particularly those without a specialist knowledge of scientific instruments, to identify items which come into their possession.

In order to achieve this, the glossary entries are closely linked with the illustrations, which have been selected on two counts: to present a particular category of instrument included in the glossary; and where possible to illustrate interesting and unfamiliar material from outside the major collections.

2 Entries

Entries are arranged alphabetically according to location, and each collection is assigned an identification code: A 1, Aberdeen, Anthropological Museum; A 11, Aylesbury, Buckinghamshire County Museum; L 1, Lacock, Fox Talbot Museum, etc. Collections in London are identified by the codes Lo 1, 2, 3, etc., and follow the L entries. Collections in the Irish Republic are designated Eire 1, 2, 3, etc., and appear in a separate section after the entries for the United Kingdom. A list of the collections included in the book precedes the main body of the entries, together with the codes which have been assigned to them.

Entries for each collection follow a standard form. First there is a brief introductory section which outlines the history of the collection or institution, the nature (and source if known) of its collections, as well as including a reference to any peripheral material it may possess (for example clocks and watches, barometers, weights and measures, medical, surgical and ophthalmological instruments). This introduction is followed by brief entries for each scientific instrument in the collection.

The principle of brief individual entries is departed from in the case of the six largest collections mentioned above. Entries for these collections do not seek to list individual instruments. Indeed, this would be impossible, for they contain a very high proportion of all early instruments in the British Isles. For these institutions, the entries indicate the sources of the collection and the categories of instruments represented. This accords with the principle that the main function of the book should be to act as a finding list for material in less well-known locations.

An intermediate form of entry has been adopted for the Department of Medieval and Later Antiquities at the British Museum, where makers of individual instruments are specified as well as important instruments.

3 Indexes

Two indexes are provided, one of instrument makers and those associated with particular instruments, and the other of types of instrument.

Information for users

1 Authorship of entries

The majority of instruments were inspected by the authors. However, occasionally information about specific instruments was not forthcoming until after the collection had been visited, and so some entries have been submitted by curators, or by the editors of the text. More rarely, entries for an entire collection have been provided by a curator. This sometimes results from the fact that the remoteness of a collection containing only a small number of items was not thought to warrant the time and expense of a visit, or from the fact that information about the collection was received too late for a visit to be arranged.

2 *Dating of instruments*

While signed and dated instruments located during the preparation of this book have occasionally modified the known period of activity of a maker, in the majority of cases dates for signed instruments have been obtained from the following sources:

Bryden, D J, *Scottish Scientific Instrument Makers, 1600-1900* Royal Scottish Museum, Information Series, Technology, **I**, Edinburgh 1972

Goodison, N, *English Barometers 1680-1869*, London 1969, second edition 1977

Taylor, E G R, Mathematical Practitioners of Hanoverian England, Cambridge 1966

Taylor, E G R, *Mathematical Practitioners of Tudor and Stuart England*, Cambridge 1954, reprinted 1967

Webster, R S and M K, *An Index of Western Scientific Instrument Makers to 1859, A-B, C-F*, Winnetka, Illinois 1968, 1971

Two further works, published since the compilation was made, will be of great benefit in identifying Scottish and Irish makers:

Clarke, T N, Morrison-Low, A D and Simpson, A D C, *Brass and Glass: Scientific Instrument Making Workshops in Scotland*, Edinburgh 1989

Burnett, J E and Morrison-Low, A D, *Vulgar and Mechanick: the Scientific Instrument Trade in Ireland 1650-1921*, Edinburgh and Dublin 1989

Where an entry does not provide a date for a signed instrument, the following work may be found useful in tracing trade literature which may lead to a date:

Anderson, R G W, Burnett, J and Gee, B, *Handlist of Scientific Instrument Makers' Trade Catalogues 1600-1914*, Edinburgh 1990

When its results are published, the SIMON project, which derives data from guild records, will provide a great deal of further information about London instrument makers. The Project was devised by, and is under the supervision of, G L'E Turner.

3 *Dimensions*

Key dimensions are given in the case of those instruments of sufficient importance to warrant a more detailed entry, or where dimensions are critical to the description or dating of an instrument. Imperial measurements are given since most of the British-made instruments would have been constructed using these units.

4 *Inventory numbers*

Museum inventory numbers are quoted, when known, in parenthesis after the entry for each instrument.

5 *Literature*

Where a collection has published a handbook or guide which includes some description or mention of instruments, a keyword author entry is included in the collection entry. Where an individual instrument is known to have been published, this keyword reference is included, following the entry for the instrument concerned. Full citation is provided in the bibliography at the back of the book.

References

1 Michel, H. Note sur l'inventaire mondial des instruments scientifiques d'intérêt historique, *Archives internationales d'histoire des sciences*, 1958, **XI**, pp394-401.

2 We do not have to catalogue every scientific instrument of artistic, historic or archaeological value. The world inventory should allow us to find, if they exist, the main milestones in the history of science, those instruments which have been of use to scholars or which mark a step forward in the development of science.

3 Centre national d'histoire des sciences. *Inventaire des instruments scientifiques historiques conservés en Belgique*, Brussels: two parts, 1959, 1960.

4 Union internationale d'histoire et philosophie des sciences . . . Commission pour l'inventaire mondial des appareils scientifiques d'intérêt historique. *Inventaire des instruments conservés en Italie*, Milan: five volumes 1963.

5 Comité national français d'histoire et de philosophie des sciences. *Inventaire des instruments scientifiques historiques conservés en France*, Paris: Centre de documentation d'histoire des techniques 1964.

6 Union internationale d'histoire et philosophie des sciences . . . Commission pour l'inventaire mondial des appareils scientifiques d'intérêt historique. *Polish national inventory of scientific instruments*, microfilm in the Science Museum Library, London, no date.

7 Union internationale d'histoire et philosophie des sciences . . . Commission pour l'inventaire mondial des appareils scientifiques d'intérêt historique. *Contribution tchechslovaque à l'inventaire . . .*, microstrip in the Science Museum Library, London 1970.

8 Maistrov, Leonid Efimovich. *Nauchnye pribory. Redaktor-sostavitel' L E Maistrov. Pribory i instrumenty istoricheskogo znacheniya*, Moscow: Akademiya Nauk SSSR Institut Istorii Estestvoznaniya i Tekhniki 1968.

9 Mollan, Charles. *Irish national inventory of historic scientific instruments: interim report 1990*, Dublin: Royal Dublin Society 1990. (There is also a 1989 edition of this work.)

Introduction

Collections of scientific instruments in the British Isles

The collections of the British Isles possess a great wealth of early scientific instruments, both of British and of European and Eastern origin. A large proportion of these, perhaps three quarters of the total holding, are to be found in six collections: the Science Museum, the Wellcome Museum and the National Maritime Museum in London, the Whipple Museum of the History of Science in Cambridge, the Museum of the History of Science in Oxford, and the Royal Museum of Scotland (formerly the Royal Scottish Museum) in Edinburgh. However, despite this concentration of instruments in a few centres, much important and interesting material is distributed in collections throughout the country.

Of all museums of the British Isles, the Science Museum in London possesses not only the largest but also the widest ranging collection. Originally part of the South Kensington Museum, encompassing both the decorative arts and science and technology, it was established as an independent museum in 1909. Among its many treasures must be counted the King George III cabinet of philosophical instruments, and extensive and important collections of microscopes, surveying, astronomical and time measurement instruments.

The Royal Museum of Scotland, Chambers Street, Edinburgh, was established as the Industrial Museum of Scotland in 1854, and presents a clear parallel to the South Kensington Museum. One important group of instruments, the Playfair collection of chemical and philosophical material, was presented shortly after the Museum's foundation. In common with the Science Museum in London, the Royal Museum of Scotland received material from the Admiralty and from the Fox Talbot collection, while other items came from the Scottish Meteorological Office and from departments of Heriot Watt or of Edinburgh universities. The collection is thus particularly rich in instruments of eighteenth- and nineteenth-century date. More recently there has been increasing emphasis on material of Scottish origin, in part acquired from the collection of Mr Arthur Frank.

The Museum of the History of Science in Oxford, although possessing very important eighteenth- and early nineteenth-century material from Oxford colleges, such as the Orrery and Daubeny collections and the instruments from the University Observatory, is remarkable for its outstanding collection of medieval and renaissance instruments, including more than one hundred astrolabes, several thousand sundials, and a wide range of early surveying and astronomical instruments. The Museum itself is of relatively recent foundation, having been established in 1935 after the gift to the University of the Lewis Evans collection of instruments and books. The building in which the Museum is located, originally erected in 1683 to house the Ashmolean Museum, the School of Natural History and a chemical laboratory for the University, provides a worthy setting for this unique collection.

Although smaller, the Whipple Museum of the History of Science in Cambridge has in many respects undergone a similar development. It is based on the collection of R T S Whipple which was presented to the University of Cambridge in 1944, and has been augmented by gifts and loans from Cambridge colleges, departments and university museums.

The National Maritime Museum, too, is a recent foundation, having been established by Act of Parliament in 1944. Starting out from an important nucleus of instruments from the Old Royal Observatory, it has extended its collections by gift, purchase and loan, and is now one of the most extensive instrument collections in the country, specialising in navigation and astronomy. The Museum owes much to one man, Sir James Caird, who presented important items and whose legacy formed the basis of a purchase fund.

The Wellcome Museum, the last of the six major instrument collections, was established by the late Sir Henry Wellcome, who died in 1936. His wide-ranging collection (nearly 100,000 items) centres around medicine, but also includes ethnographic, archaeological and anthropological material. The Museum was until recently maintained by the Wellcome Trustees, but since 1977 has been established at the Science Museum as a distinct department, the Wellcome Museum of the History of Medicine. The range of the collection, however, is such that certain sections are on loan to several other departments of the Science Museum, while others have been lent to the Museum of Mankind in Burlington Gardens and many other museums in Great Britain. The sheer size of the Wellcome collection is daunting, including for instance, more than 1,200 microscopes and 25,000 surgical and medical instruments.

A very important collection of early instruments is to be found at the British Museum in the Department of British and Medieval Antiquities, although some important material is located in other departments of the Museum. The collection is specialised, consisting in the main of sundials, many of sixteenth- to seventeenth-century date, but there are also early surveying and astronomical instruments, nearly all of which are of interest both from a scientific and from a decorative standpoint.

In addition to the collections so far discussed, London, with its great wealth of museums, colleges and institutions, possesses several instrument collections of moderate size, as well as many single items in unexpected locations. Of the collections the largest, and certainly the most important, is in the Royal Institution of Great Britain. The Institution has a long history, dating back to the penultimate year of the eighteenth century, and was one of several such foundations whose objectives were to extend the benefits of science to the lower classes. While its original function, of educating the poorer classes, was soon abandoned, it has always carried on a dual role, combining the diffusion of knowledge to the general public with the pursuit of research. In common with the Natural Philosophy departments of the Scottish universities, much early demonstration apparatus has survived, as well as large quantities of experimental equipment and apparatus associated with distinguished past professors, including Humphry Davy, Michael Faraday, John Tyndall and Lord Rayleigh.

Also in London, the Royal Geographical Society, the Royal Institute of British Architects, the Institute of Ophthalmology, the Pharmaceutical Society of Great Britain, and the Royal Botanic Gardens have collections of instruments associated with, or presented by, members of the institute or society concerned. Of these, the collections at the RGS and RIBA are the largest. The latter owns a fine architectural sector constructed by George Adams, and an early seventeenth-century sector by Daniel Chores, author of an early treatise on this instrument. The Society of Antiquaries owns an important fourteenth-century Hispano-Moorish astrolabe, while the Royal Society retains a reflecting telescope associated with Sir Isaac Newton and two chronometers by Arnold, most other surviving material from this source being deposited with the Science Museum.

An important, but little known, collection of moderate size is housed in the Museum of Artillery at the Rotunda in Woolwich. It is highly specialised, since the majority of instruments are associated with gunnery, and of some antiquity, having been established in 1788 by Captain (later Sir William) Congreve as an aid to instructing officers of the Royal Artillery in the handling of guns. Several instruments are associated with a Colonel Borgard, an early member of the Royal Artillery, and the origins of other important and sometimes unique instruments signed by such makers as John Rowley, George Adams or Francis Morgan, although uncertain, are presumed to be similar.

Several London museums possess small or moderate-sized collections of instruments, notably the Victoria and Albert Museum, the Museum of the Clockmaker's Company (at the Guildhall Library), the National Army Museum, the Museum of London, and the Horniman Museum. Each contains important early instruments.

Moving away from London, substantial collections of instruments are relatively scarce. The rapid growth of technological collections in major provincial museums has rarely been paralleled by a similar development in collections of scientific instruments, a fact which is in part a reflection of the scarcity and high cost of instruments in recent years when compared with technological items, but is also probably due to lack of expertise and understanding of them.

Among the major provincial collections, that at the Liverpool Museum, part of the National Museums and Galleries on Merseyside, is undoubtedly one of the most interesting and important. It includes a medieval equatorium, Renaissance sundials, and eighteenth- and nineteenth-century astronomical instruments. The basis of the Museum's collection was formed by Joseph Mayer, and was presented to Liverpool Corporation in 1867, and it is from this source that several important early scientific instruments are derived.

Many local authority museums, such as the Museum and Art Gallery at Kelvingrove, Glasgow, and the Museum of Science and Industry in Birmingham (established in 1950 as an offshoot of the Museum and Art Gallery) have surprisingly weak collections. In Manchester, the Museum of Science and Industry, another recent foundation, is assembling a collection of instruments, in the main with

regional or local connotations, such as material connected with John Dalton, instruments by J B Dancer, an important Manchester-based maker of the nineteenth century, and apparatus from the University of Manchester.

Moving away from collections in industrial centres, the Castle Museum at York possesses an extensive range of instruments, including many microscopes and telescopes. Several of these formed part of the collection assembled by Dr J L Kirk, founder of the Museum. In its extraordinary range, covering all aspects of folklife and bygones, the collection as a whole is an impressive monument to the collecting energy of one man.

The collection at Snowshill Manor, near Broadway, Gloucestershire, now in the ownership of the National Trust, resembles in many ways the Kirk collection. Assembled by the late Charles P Wade, it includes all manner of bygones: agricultural implements, equipment associated with rural industries, furniture, toys, costume and a large number of scientific instruments, mainly eighteenth-century in date.

The Castle Museum, Norwich, a major provincial museum, possesses many instruments, including several unusual early microscopes, of which the most important must be a rare example of a Leeuwenhoek aquatic microscope signed by John Yarwell and stamped *E.C.* (probably Edmund Culpeper). In addition there are interesting horary quadrants and other sundials of seventeenth- and eighteenth-century date.

The Museum of Science and Engineering in Newcastle upon Tyne contains a substantial, interesting group of instruments, mostly connected with navigation, surveying and mining, but including an example of an Armstrong hydro-electric machine and a very rare adjustable bow for drawing arcs. Still in the North-East, Sunderland Museum has an instrument collection consisting mainly of navigational instruments, and material associated with Sir Joseph Wilson Swan, a native of Sunderland and inventor of the Swan incandescent filament lamp. Only a few miles away, at South Shields, the Marine and Technical College has a sizeable display of navigational instruments.

Further south at Whitby, an important centre of the whaling industry in the late eighteenth and early nineteenth centuries, the Museum of the Literary and Philosophical Society contains a remarkable collection of instruments, again mainly nautical in character. The most interesting group are however concerned with magnetism, and are associated with the Arctic explorer William Scoresby who carried out magnetic studies in the Arctic regions during the first decades of the nineteenth century. Instruments in this group, bequeathed by Scoresby to the Museum on his death, include dip circles, compasses and compass needles and magnetic batteries.

The Town Docks Museum at Hull possesses another interesting nautical collection. As well as material related to whaling and shipping, there are early scientific instruments, mostly navigational in character and dating from the eighteenth and nineteenth centuries, and several earlier items.

On the opposite side of the country, Bristol, another port possessing a major provincial museum, has a collection of instruments which includes some items by local makers. During the eighteenth and nineteenth centuries Bristol was a regional centre of instrument making, but, despite its importance as a port in past centuries, navigation instruments, although present in the collection, do not form a major part.

One of the most ancient instrument collections is to be found in the Museum of the Gentlemen's Society of Spalding. Its chief treasure is a fine sixteenth-century astrolabe by the Flemish maker Regnerus Arsenius, presented to the Gentlemen's Society only three years after its foundation in 1720 by one of its early members. In 1761 the Society purchased four new instruments from the London maker Benjamin Martin, and one of these, a Gregorian reflecting telescope, survives today. The Museum contains many treasures, accumulated during the course of the Society's long history, the instruments forming only a small, but important, part of its collection.

At Bury St Edmunds may be found a somewhat larger but specialised collection of time measurement devices: clocks, watches, sundials and related instruments. These were presented to the Corporation of Bury St Edmunds by the late Frederick Gershom Parkington in memory of his son, who was killed in the Second World War, and are admirably exhibited in a Queen Anne house belonging to the National Trust.

During recent years there has been a tendency for scientific instruments, formerly in the hands of universities, to find their way into museums, either as gifts or as loans – a tendency which is particularly pronounced in the cases of Oxford, Cambridge, London and Edinburgh. Outside these centres, however, university departments may be a rich source of scientific instruments, and this is especially true of some old-established Scottish universities. Aberdeen, St Andrews, Glasgow and Strathclyde universities are all important in this respect. Instruments from the University of Edinburgh and Heriot Watt University, also in Edinburgh, are now mainly in the care of the Royal Museum of Scotland. The city of Glasgow is particularly richly endowed as regards university collections, since the departments of Natural Philosophy at its two universities, the University of Glasgow and the University of Strathclyde, possess quantities of important eighteenth- and early nineteenth-century material. At the University of Glasgow the first Professor of Natural Philosophy was appointed in 1727 and inventories of equipment survive from 1727 and 1760, the latter being a copy of an earlier inventory dated 1756, so that the full range of early equipment is known. Only fragments of this equipment actually survive, but among them are some items figuring in the 1727 inventory as well as considerable quantities of material from the second half of the eighteenth and from the nineteenth centuries, including items associated with James Joule, Lord Kelvin and John Kerr.

The University of Strathclyde, as such a relatively new foundation, traces its origins to the end of the eighteenth century when John Anderson, Professor of Natural Philosophy at the University of Glasgow, provided in his will for the establishment of 'Anderson's Institution' with objectives similar to those of the Royal Institution in London. A few of the surviving instruments antedate the foundation of Anderson's Institution, and may have belonged to Anderson himself. There are considerable quantities of late eighteenth- and nineteenth-century demonstration and teaching apparatus, which formed an integral part of the teaching of Natural Philosophy. The collection includes instruments by such makers as George Adams, John Bird, and James Crichton of Glasgow, who was active during the early years of the nineteenth century.

The Department of Natural Philosophy of the University of Aberdeen also possesses much eighteenth- and early nineteenth-century equipment, almost all of which can be identified with the help of an inventory compiled in 1823, since it formed part of the collection of teaching and demonstration apparatus at Marischal College (one of two ancient colleges amalgamated in 1860 to form the University of Aberdeen). Patrick Copland, Professor of Natural Philosophy from 1775 to 1822, was responsible for setting up an observatory, some of whose equipment survives in the Department of Natural Philosophy.

Although the collections at Aberdeen and Glasgow are the more extensive, that in the Department of Physics at the University of St Andrews contains early instruments of outstanding importance, including a large (two feet diameter) sixteenth-century astrolabe, a rare mariner's astrolabe, an armillary instrument of late sixteenth-century date, and a splendid eighteenth-century orrery. Several of these instruments may have formed part of the collection assembled by James Gregory for the equipment of an observatory. Gregory, Professor of Mathematics and Astronomy from 1668 to 1674, purchased several instruments in 1673, and these may very well have included the great astrolabe by the late sixteenth-century London maker Humphrey Cole, which possesses a later plate for latitude 56° 25', is signed by John Marke and dates from the 1660s or 1670s. Not all items connected with the Observatory are in the Department of Physics: the University Library possesses three clocks by Joseph Knibb and an incomplete late seventeenth-century telescope, all of which can be identified in the extant 1679 inventory.

The University of St Andrews is remarkable in a further respect – the possession of an important collection of eighteenth- and nineteenth-century chemical glassware. Because of its bulk and fragility, chemical glassware rarely survives in quantity (however, there are such items in the Playfair collection in the Royal Museum of Scotland, Edinburgh, the Wellcome Museum of the History of Medicine, and at the Museum of the History of Science in Oxford). This collection at St Andrews owes its preservation at least in part to the fact that it lay undiscovered in the tower at St Salvator's Chapel until 1925. It is thought that it may have been acquired from Thomas Thomson, an Edinburgh chemical practitioner who abandoned his private chemistry course there in 1811. In addition to the glassware, the Department of Chemistry also possesses a fine long beam balance of mid-eighteenth century date, signed by the London maker George Adams.

In the Republic of Ireland, Trinity College Dublin possesses fragments of what may once have been an extensive collection of philosophical apparatus (air-pumps, a telescope, and among other items a very splendid lodestone from the first half of the eighteenth century), while the Department of Physics at University College Dublin also possesses a good collection of teaching apparatus. However, the most important and extensive collection is at St Patrick's College, Maynooth, a recognised College of the National University of Ireland and a major seminary for the training of priests, where the Revd Nicholas Callan,

Professor of Natural Philosophy in the 1830s, worked in the field of electromagnetics, as well as pioneering high tension electricity. The College Museum houses a collection of ecclesiastical material as well as quantities of apparatus used by or constructed for Callan himself. More recently, a collection of instruments by Irish makers has been assembled, and this, together with the Egestorff collection in Dublin, is the most important source for the study of Irish instruments.

Also in Dublin, the National Museum of Ireland possesses a small collection of instruments, some of them by Irish makers, but its most important single item in this field is a fine astrolabe by Erasmus Habermel, who, at the end of the sixteenth century, was instrument maker to Emperor Rudolph II in Prague.

Most of the contents of the observatories which flourished around the country in the eighteenth and nineteenth centuries have been dispersed. The equipment from the Old Royal Observatory at Greenwich, London, still largely survives, while the instruments from the University Observatory at Oxford are now in the Museum of the History of Science. Equipment from the Observatory of St John's College, Cambridge and from the University Observatory is to be found at the Whipple Museum of the History of Science. In the Irish Republic the Dunsink Observatory, founded in 1783 as part of Trinity College, Dublin, still possesses a pair of regulators, apparently acquired in 1787, and a number of nineteenth-century instruments, including a telescope purchased by Sir James South for his private observatory at Camden Hill, Kensington, and later presented to Trinity College, Dublin.

Armagh Observatory in Northern Ireland possesses a part of its original equipment, and, indeed, the Troughton equatorial, installed in 1796, still rests on its original mountings. As well as two Earnshaw clocks there are several instruments from the early nineteenth century, and a group of instruments from the King George III collection, transferred to the Observatory in 1840. These and other instruments are now housed in the modern Planetarium.

The Calton Hill Observatory in Edinburgh was established in 1776 and later became known as the Royal Observatory, a title granted in 1822 and held until 1894 when it became the seat of the Astronomical Society of Edinburgh. It retains astronomical clocks and transit telescopes, although the six-foot mural circle by Troughton & Simms, installed in 1833, is on loan to the Royal Museum of Scotland. The Royal Observatory was transferred in 1894 from its Calton Hill premises to the present site on Blackford Hill. It also possesses some instruments, though much material, including the direct standards made for Charles Piazzi Smyth after his measurements of the Great Pyramid in 1866, has been transferred to the Royal Museum of Scotland.

Reference has already been made to the Observatory at St Andrews University which was equipped by James Gregory c1673. At Aberdeen an observatory was established at Castle Hill in 1781 (it was demolished in 1795) and was subsequently moved to Marischal College. Some equipment from this observatory is in the Department of Natural Philosophy at King's College. A second observatory was installed in the Cromwell Tower at King's College in 1790. It seems to have been re-equipped

inexpensively in the mid-nineteenth century and equipment from it is also housed in the Department of Natural Philosophy.

Throughout the country there were numerous local astronomical societies, many of which had their own small observatory. One such which still survives is at Dumfries, where the modern museum is housed in the observatory building opened in 1836 for the Dumfries and Maxwelltown Astronomical Society. Some of its equipment, including telescopes and a large camera obscura installed in 1836 by a local maker, is still located there.

Throughout the British Isles, often in small local museums, may be found astronomical equipment associated either with a local society or with an individual local astronomer. The universal equatorial instrument constructed by Abraham Sharp, Flamsteed's assistant at the Royal Observatory Greenwich from 1684 to 1690, which was originally presented to the Yorkshire Philosophical Society in 1836 and long housed in the hall of the Yorkshire Museum, may now be seen at Greenwich. However, other smaller instruments constructed by Sharp, some of them experimental in character, may be seen at Bolling Hall, Bradford. At Talgarth, Brecon, Hawick, Dumfries, Dundee and Darlington, to name only a few instances, may be seen equipment constructed by, or for, local astronomers.

Equally interesting are the relatively rare surveying instruments known to have been used by a particular surveyor, or on the occasion of a particular survey. Apart from instruments in the Science Museum used in connection with the triangulation of Great Britain and Ireland such as Ramsden's three-foot theodolite and Colby's compensation bars, or the boning telescope and steel chain made by Ramsden and used by General Roy to measure the Hounslow Heath baseline in 1784, the Department of Geology at the National Museum of Wales in Cardiff possesses a quadrant owned by the geologist Henry de la Beche. Leicester Museum possesses instruments used in connection with surveys for the Leicester Swannington railway in the 1820s and 30s, while the Rydale Folk Museum at Hutton le Hole in Yorkshire houses three surveying instruments used by Joseph Ford, a local surveyor, c1745 for designing water courses to supply the villages of the southern edge of the Yorkshire Moors. Keswick Museum and the Dean and Chapter Estate Office of Durham Cathedral possess other examples of surveying instruments whose former ownership and use is recorded. The items at Keswick are particularly interesting since they appear to be in part home constructed, and belonged to a local geologist/cartographer/surveyor.

Collections of instruments remaining in the great country houses of England are seemingly rare, although few cannot possess a waywiser, pair of globes, garden sundial or other such items. Of the many owners of houses contacted relatively few were able to provide details of instruments in their possession. The most notable collections are those at Burton Constable, Longleat, Petworth, Chatsworth and Kedleston. Many of the great collections of the eighteenth century were dispersed, often in the same century, and while the collection of instruments at Longleat, a small but self-contained group, may be relatively complete, those at Chatsworth, Kedleston, and Burton Constable represent only fragments of what there must have been in the eighteenth century. Single instruments, or small groups, are found at Luton and Waddesdon, but both represent more recent acquisitions of the nineteenth and twentieth centuries.

At Longleat the collection of instruments, consisting of telescopes, microscopes, globes, and a small group of navigation instruments, may be divided into three groups, of *c*1700, 1720-40 and late eighteenth/early nineteenth century. At Chatsworth several of the instruments are likely to have belonged to the scientist Henry Cavendish (1731-1810), although a few highly decorative items are of earlier date. The items at Kedleston include tantalising and incomplete fragments of Marshall, Watkins and Martin drum microscopes, as well as a surveying instrument by John Bird and a fine carriage waywiser by Richard Glynne. These date from the first quarter and the middle of the eighteenth century. No details are available regarding their acquisition. The case of the Petworth instruments is particularly fascinating, since they were purchased for the private laboratory of Elizabeth Ilive, mistress of the third Earl of Egremont.

Burton Constable possesses quite extensive remnants of a collection which in its day must have been very impressive. Surviving items include electrical machines, air pumps, chemical glassware and other items acquired by William Constable (1721-91). As well as complete instruments there are numerous fragments, plus receipts which serve to give an indication of the original extent of the collection. William Constable was elected FRS at the suggestion of Sir Joseph Banks, President of the Royal Society. Constable shared Banks' botanical interests, and was also an active patron of artists, architects and craftsmen. In its day this collection must have been one of several. That of Charles Boyle (1676-1731), fourth Earl of Orrery, is now in the Museum of the History of Science in Oxford, while the later collection of philosophical instruments assembled by John Stuart, the third Earl of Bute (1713-92) was dispersed by auction in the year after his death. The Burton Constable collection represents a rare survival, for it is still located in the house in which it was assembled.

A recent interesting new source of early scientific instruments has been from archaeological work. In particular, small dials have been found in archaeological excavations and in wrecks. Examples may be seen in the Museum of London, Castle Museum, Norwich, and the Mary Rose Exhibition, Portsmouth. With the increasing interest in urban and underwater archaeology, these frequently datable artifacts will become of increasing value to historical studies.

Such a brief introduction to the collections included can do no more than whet the appetite of the user by drawing attention to the many and varied collections in the British Isles. Throughout the book will be found both collections and individual instruments of interest and importance, often in most unexpected locations. Perhaps, having searched the pages of this book, the reader may be encouraged to locate further material, which has either been added to collections since this survey was undertaken, or is still unreported and unrecognised in the many collections of instruments in the British Isles.

Material may be reported to:

The Head of Science, Collections Management Division, The Science Museum, London SW7 2DD (for England and Wales)

The Keeper, Department of Science, Technology and Working Life, Royal Museum of Scotland, Edinburgh EH1 1JF (for Scotland)

The Science Officer, The Royal Dublin Society, Ballsbridge, Dublin 4 (for Ireland).

List of collections, with identification codes

United Kingdom

A1	Aberdeen, Anthropological Museum, University of Aberdeen
A2	Aberdeen, Cromwell Tower Observatory, University of Aberdeen
A3	Aberdeen, Department of Anatomy, University of Aberdeen
A4	Aberdeen, Department of Physics, University of Aberdeen
A5	Aberystwyth, Department of Physics, University College of Wales, Aberystwyth
A6	Airdrie, Airdrie Museum
A7	Alloway, Burns' Cottage and Museum
A8	Alton, Curtis Museum
A9	Anstruther, Scottish Fisheries Museum
A10	Armagh, The Observatory
A11	Aylesbury, Buckinghamshire County Museum

B1	Banff, Banff Museum
B2	Barnard Castle, Bowes Museum
B3	Barrow-in-Furness, The Furness Museum
B4	Basingstoke, Willis Museum
B5	Bath, The American Museum in Britain
B6	Bath, Pump Room
B7	Bath, Royal Photographic Society National Centre of Photography
B8	Bedford, The Bedford Museum
B9	Belfast, Ulster Museum
B10	Birmingham, Assay Office
B11	Birmingham, Birmingham Museum and Art Gallery
B12	Birmingham, Dollond and Aitchison Museum
B13	Birmingham, Birmingham Museum of Science and Industry
B14	Blair Atholl, Blair Castle
B15	Blantyre, The David Livingstone Centre
B16	Bournemouth, Natural Science Society
B17	Bournemouth, Russell-Cotes Museum
B18	Bracknell, The Meteorological Office
B19	Bradford, Bolling Hall
B20	Bradford, National Museum of Photography, Film and Television
B21	Brecon, Breconshire War Memorial Hospital
B22	Brecon, The Brecknock Museum
B23	Bridgwater, Admiral Blake Museum
B24	Bristol, City of Bristol Museum and Art Gallery
B25	Bristol, University of Bristol Library, Special Collections
B26	Broadheath, The Elgar Birthplace
B27	Broadway, Snowshill Manor
B28	Buckler's Hard, Maritime Museum
B29	Burton Constable, Burton Constable Hall
B30	Bury St Edmunds, The Gershom Parkington Collection

C1	Cambridge, Cavendish Laboratory
C2	Cambridge, Gonville and Caius College
C3	Cambridge, Department of Earth Sciences
C4	Cambridge, Queen's College
C5	Cambridge, The Library, Trinity College
C6	Cambridge, Whipple Museum of the History of Science
C7	Canterbury, Dean and Chapter of Canterbury Cathedral
C8	Cardiff, National Museum of Wales
C9	Carlisle, Tullie House Museum and Art Gallery
C10	Chatham, Royal Engineers Museum
C11	Chatsworth, Devonshire Collection
C12	Cheltenham, Cheltenham College Laboratories
C13	Cheltenham, Cheltenham Art Gallery and Museum
C14	Chester, Grosvenor Museum
C15	Chipping Camden, Woolstapler's Hall Museum
C16	Colchester, Colchester and Essex Museum
C17	Coventry, Herbert Art Gallery and Museum

D1	Darlington, Darlington Museum
D2	Dartford, Dartford Borough Museum
D3	Derby, Derby Museum and Art Gallery
D4	Devizes, Museum of the Wiltshire Archaeological and Natural History Society
D5	Dorchester, Dorset County Museum
D6	Douglas, Manx Museum
D7	Dover, Dover Museum
D8	Dumfries, Dumfries Museum
D9	Dundee, Dundee Art Galleries and Museums
D10	Durham, Dean and Chapter Library

E1	Edinburgh, Astronomical Society of Edinburgh
E2	Edinburgh, Department of Chemistry, University of Edinburgh
E3	Edinburgh, Heriot Watt University
E4	Edinburgh, Huntly House
E5	Edinburgh, Lauriston Castle
E6	Edinburgh, Museum of the Royal College of Surgeons of Edinburgh
E7	Edinburgh, Royal Observatory
E8	Edinburgh, Royal Museum of Scotland (Chambers Street)
E9	Edinburgh, Royal Museum of Scotland (Queen Street)
E10	Edinburgh, Scottish United Services Museum
E11	Elgin, Elgin Museum
E12	Exeter, Royal Albert Memorial Museum

F1	Forfar, Forfar Museum
F2	Fort William, West Highland Museum

G1	Glasgow, Glasgow Museums and Art Galleries
G2	Glasgow, Department of Astronomy, University of Glasgow
G3	Glasgow, Department of Natural Philosophy, University of Glasgow

G4   Glasgow, Department of Natural Philosophy, University of Strathclyde
G5   Glasgow, Hunterian Museum
G6   Gloucester, City Museum and Art Gallery
G7   Greenock, The McLean Museum

H1   Hawick Museum and the Scott Gallery
H2   Helston, Helston Folk Museum
H3   Hertford, Hertford Museum
H4   Holywood, Ulster Folk and Transport Museum
H5   Huddersfield, Tolson Memorial Museum
H6   Hull, Town Docks Museum
H7   Hutton-le-Hole, Rydale Folk Museum

I1   Inverness, Inverness Museum and Art Gallery
I2   Inverurie, Inverurie Museum
I3   Ipswich, Ipswich Museum

K1   Kedleston, Kedleston Hall
K2   Kendal Museum
K3   Keswick, Keswick Museum and Art Gallery
K4   Kilmarnock, Dick Institute
K5   King's Lynn, Lynn Museum

L1   Lacock, Fox Talbot Museum of Photography
L2   Leamington Spa, Warwick District Council Art Gallery and Museum
L3   Leeds, City Museum
L4   Leicester, Leicestershire Museum and Art Gallery
L5   Letchworth, Letchworth Museum and Art Gallery
L6   Lincoln, Museum of Lincolnshire Life
L7   Liverpool, Department of Physics, University of Liverpool
L8   Liverpool, Liverpool Museum
L9   Longleat, Longleat House
L10   Ludlow, Ludlow Museum
L11   Luton, Werhner Collection

Lo1   London, Royal Armouries, HM Tower of London
Lo2   London, British Museum
Lo3   London, British Museum (Natural History)
Lo4   London, British Optical Association
Lo5   London, Bruce Castle Museum
Lo6   London, Cuming Museum
Lo7   London, Galton Laboratory, University of London
Lo8   London, Gunnersbury Park Museum
Lo9   London, Horniman Museum
Lo10   London, Institute of Ophthalmology, University of London
Lo11   London, Museum of Artillery in the Rotunda
Lo12   London, Clockmakers' Company Collection, Guildhall Library
Lo13   London, Museum of London
Lo14   London, Museum of the Pharmaceutical Society of Great Britain
Lo15   London, National Army Museum
Lo16   London, National Maritime Museum
Lo17   London, Royal Botanic Gardens, Kew
Lo18   London, Royal Geographical Society
Lo19   London, Royal Institute of British Architects
Lo20   London, Royal Institution of Great Britain
Lo21   London, Royal Society

Lo22   London, Science Museum
Lo23   London, Sir John Soane's Museum
Lo24   London, Society of Antiquaries of London
Lo25   London, Victoria and Albert Museum
Lo26   London, Wallace Collection
Lo27   London, Wellcome Museum of the History of Medicine, Science Museum
Lo28   London, Wesley's House and Museum

M1   Maidstone, Maidstone Museum and Art Gallery
M2   Manchester, Manchester Literary and Philosophical Society
M3   Manchester, Museum of Science and Industry
M4   Monmouth, Nelson Collection and Local History Centre

N1   Newcastle-upon-Tyne, Museum of the Department of Mining Engineering, University of Newcastle
N2   Newcastle-upon-Tyne, Museum of Science and Engineering
N3   Northampton, Museum of Leathercraft, Central Museum and Art Gallery
N4   Norwich, Castle Museum
N5   Nottingham, Industrial Museum
N6   Nottingham, Department of Physics, University of Nottingham

O1   Olney, Cowper and Newton Museum
O2   Oxford, Ashmolean Museum, University of Oxford
O3   Oxford, Merton College, University of Oxford
O4   Oxford, Museum of the History of Science, University of Oxford
O5   Oxford, Pitt Rivers Museum, University of Oxford

P1   Peterhead, Peterhead Arbuthnot Museum
P2   Petworth, Petworth House
P3   Plymouth, Plymouth City Museum and Art Gallery
P4   Portsmouth, The Mary Rose Exhibition
P5   Portsmouth, Portsmouth City Museum and Art Gallery
P6   Portsmouth, Royal Naval Museum
P7   Port Sunlight, The Lady Lever Art Gallery

R1   Reading, Reading Museum and Art Gallery
R2   Rochester, Guildhall Museum
R3   Rothbury, Cragside
R4   Rotherham, Rotherham Museum

S1   St Andrews, Department of Chemistry, University of St Andrews
S2   St Andrews, Department of Physics, University of St Andrews
S3   St Andrews, University Library
S4   St Fagans, Welsh Folk Museum
S5   St Helens, St Helens Museum and Art Gallery
S6   St Helens, Pilkington Glass Museum
S7   St Mary's, Isles of Scilly Museum
S8   Salford, Salford Mining Museum
S9   Salisbury, Salisbury and South Wiltshire Museum

S10	Saltcoats, North Ayrshire Museum
S11	Scunthorpe, Scunthorpe Museum and Art Gallery
S12	Shaftesbury, Shaftesbury Local History Museum
S13	Sheffield, Sheffield City Museum
S14	Shugborough, Staffordshire County Museum
S15	Sidmouth, Sidmouth Museum
S16	Skipton, Craven Museum
S17	Southampton, City Museums and Art Gallery
S18	South Shields, Marine and Technical College
S19	Spalding, Spalding Gentlemen's Society Museum
S20	Stirling, Smith Art Gallery and Museum
S21	Stockport, Stockport Museum
S22	Stoke-on-Trent, Wedgwood Museum
S23	Stranraer, Wigtown District Museum
S24	Stromness, Stromness Museum
S25	Sunderland, Sunderland Museum and Art Gallery
S26	Swansea, University College of Swansea and Royal Institution of South Wales Museum
S27	Swindon, Swindon Museum and Art Gallery
T1	Talgarth, Howell Harris Museum
T2	Taunton, Somerset County Museum
T3	Teddington, National Physical Laboratory Museum
T4	Torquay, Torquay Museum
T5	Truro, Royal Institution of Cornwall, Royal Cornwall Museum
W1	Waddesdon, Waddesdon Manor
W2	Warley, Avery Historical Museum
W3	Whitby, Museum of the Whitby Literary and Philosophical Society
W4	Winchester, Hampshire County Museum Service

W5	Wisbech, Wisbech and Fenland Museum
W6	Worthing, Worthing Museum and Art Gallery
Y1	York, Castle Museum
Y2	York, National Railway Museum
Y3	York, Vickers Instruments
Y4	York, Yorkshire Museum

Eire

Eire 1	Birr, Birr Observatory and Museum Trust
Eire 2	Carlow, College Library, Carlow Regional Technical College
Eire 3	Castleknock, Dunsink Observatory
Eire 4	Cork, Experimental Physics Department, University College, Cork
Eire 5	Dublin, Chester Beatty Library and Gallery of Oriental Art
Eire 6	Dublin, Dublin Civic Museum
Eire 7	Dublin, Egestorff Collection
Eire 8	Dublin, National Botanic Gardens
Eire 9	Dublin, National Museum of Ireland
Eire 10	Dublin, Royal College of Surgeons of Ireland
Eire 11	Dublin, Royal Dublin Society
Eire 12	Dublin, Chemistry Laboratory, Trinity College
Eire 13	Dublin, Physical Laboratory, Trinity College
Eire 14	Dublin, School of Engineering, Trinity College
Eire 15	Dublin, Department of Physics, University College Dublin
Eire 16	Galway, Physics Department, University College Galway
Eire 17	Maynooth, St Patrick's College (Maynooth College)

Glossary of instrument types

Abacus (frontispiece)

The term abacus can be used as a generic term to describe a wide variety of calculating devices which use counters on boards or beads on wires. They have been developed over thousands of years and in many cultures. There are frequent contemporary references to the use of pebbles for reckoning in classical Greece and Rome. Special 'jettons' with boards and cloths were developed in medieval times. In India, the chukkrum board was used.

There is a variety of bead-on-wire types. From China comes the suan-pan, from Japan the soroban and from Russia the schety. In 1618, John Tradescant the Elder brought a schety to England. Sixty-five years later it was displayed in the newly opened Ashmolean Museum in Oxford, and it could be considered to be the first scientific instrument viewable by the public in a museum in Great Britain (Ryan (1972); MacGregor (1983)).

Lit: BARNARD (1916); PULLAN (1970)

Air pump or vacuum pump (figs 1-3)

Otto von Guericke, *c*1650, used an air pump for producing a vacuum in order to demonstrate the effects of air pressure. It was subsequently improved by Christiaan Huygens, Robert Boyle, Denis Papin, Robert Hooke and Francis Hauksbee, and became a fundamental part of any collection of philosophical instruments during the 18th and 19th centuries.

The vacuum was formed beneath a glass dome seated on a brass plate, the joint being made air-tight by the use of grease. The air inside the dome was evacuated by the action of one or more pistons operated by means of a rack and pinion, or by means of a stirrup. In England the rack-operated air pump with two barrels (figs 1 and 2) was generally used, while on the Continent the stirrup-operated pump with a single barrel was preferred. The pumps were originally free-standing, but by the mid-18th century table-top versions were being produced.

With the air-pump were associated numerous accessories for performing experiments – Magdeburg hemispheres, guinea and feather apparatus, cohesion discs, etc. By the mid-18th century air pumps were being constructed which could, at will, exhaust or compress the air beneath the glass dome.

A simple compression pump apparatus was developed in the late 18th century for experiments at enhanced pressures (fig 3).

Fig 1
Air pump, English, post-1717
A double-barrel air pump with rack-operated pistons, to evacuate a receiver. This type of pump was originally described by Francis Hauksbee in 1709, but continued in use for several decades. This particular example was made to a later adaptation described by William Vream in 1717. (Royal Museum of Scotland (Chambers Street), Edinburgh. E8.I.1)

Fig 2
Air pump, English, early 19th century
Small double-barrel air pump of mahogany and brass. The pistons are activated by two racks working against a pinion concealed inside the curved wooden pediment, operated by means of a crank handle. (Museum and Art Gallery, Maidstone. M1.20)

Fig 3
Air-pressure vessel, by Bate, London, first half of the 19th century
Single-barrel air pump for creating pressures above atmospheric. It consists of a receiver clamped between a brass lid and stand by means of a mahogany press. The lid is equipped with a hook from which various objects can be suspended for experiments in the jar.
(The National Museums and Galleries on Merseyside, Liverpool. L8.77)

Fig 4
Armillary sphere, by C Vinchx, Naples, 1601
An armillary sphere of brass demonstrating the Ptolemaic cosmological system. The stand is engraved with the name and arms of Camillo Gonzago of Novellara.
(Museum of the History of Science, Oxford. O4.II.2)

Lit: For the early history, see WILSON (1849), ANDRADE (1929 and 1957), SHAPIN (1984) SHAPIN/SCHAFFER (1985); for French versions, see DAUMAS (1972) pp217-8; for Dutch instruments see TURNER/LEVERE (1973) pp229-231 and STROUP (1981).

Armillary sphere (fig 4)

An instrument for demonstrating and teaching cosmological theory. It consists of a series of rings representing the main circles of the celestial sphere – meridian, ecliptic, equator, tropics, etc., centred around a small globe or ball. The majority of armillary spheres are constructed according to the Ptolemaic (geocentric) system, although from the late 16th century onwards some spheres based on the Copernican (heliocentric) system were produced, as well as occasionally ones representing other cosmological systems (e.g. Heraclidean). Normally the armillary sphere is mounted on a stand, although the earliest ones were often hand-held. As well as the Earth or Sun, the Moon and stars were frequently represented. The armillary sphere was primarily a teaching instrument,

although it did permit some astronomical problems to be solved without calculation.

Armillary spheres survive from the late medieval period onwards, but existed at least as early as the 13th century. During the 17th and 18th centuries they were sometimes constructed in pairs in order to contrast the Ptolemaic and Copernican cosmologies. Pasteboard examples were made in France in the 18th and 19th centuries.
Lit: NOLTE (1922); PRICE (1954); TURNER (1973A); MADDISON (1969)

Artificial horizon (fig 5)

A device which enables the altitude of the Sun, Moon, or a star to be measured with a sextant, octant, etc. when the natural horizon is obscure. It consists either of a bubble level attached to an instrument, or of an independent device such as a mercury trough or an optically-parallel plate of glass floated on a mercury surface, or of spirits of wine, or a glass plate with adjustable screw feet equipped with a spirit level.
Lit: COTTER (1968) p91ff; LONDON (1971) section 2-1, MAY (1973) pp148-51; COTTER (1983) pp188-206

Fig 5
Artificial horizon, English, 19th century
When the horizon was not visible, navigators needed a device to provide
the horizontal datum when measuring the elevation of the Sun or a star.
In this example, a rectangular trough of mercury was used as a reflecting
surface. When not in use, the mercury was stored in the wooden bottle.
(Science Museum, London. Lo22.10)

Fig 6
**Hispano-Moorish astrolabe, by Muhammed b as-Sâffar, Cordova,
AH417 (AD1026/27)**
A brass astrolabe made in Moorish Spain and inscribed in Kufic script.
This is the earliest known scientific instrument, constructed in Europe,
which is exactly dated.
(Royal Museum of Scotland (Chambers Street), Edinburgh, E8.II.1)

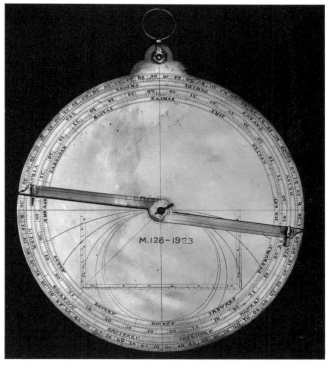

Fig 7A & B
Astrolabe, French, c1410
The usual form of planispheric astrolabe, based on a stereographic
projection of the celestial sphere onto the plane of the equator, and
requiring separate engraved plates for each latitude (side A). The rete of
this example, which may be dated to c1410, is of a characteristic French
type associated with Jean Fusoris (? 1365-1436). Side B includes a
shadow square, used in surveying.
(Victoria and Albert Museum, London. Lo25.38)

Fig 8A & B
Astrolabe, by E Habermal, Prague, c1600
*A gilt copper astrolabe by Erasmus Habermal, instrument-maker in
Prague to Emperor Rudolph II. The astrolabe is engraved on side B with
the coat of arms of Franciscus de Padoanis, the Emperor's physician.
(National Museum of Ireland, Dublin. Ei9.1)*

Astrolabe (figs 6-11)

An astronomical and astrological computing instrument
which could be used for time measurement and for
surveying. The instrument normally consists of a thick
plate (*mater*) with a cavity in one side into which fit one or
more plates (*tympana*) engraved with stereographic
projections of the celestial sphere from the pole onto the
plane of the equator, each projection constructed for use in
one particular latitude. Over the tympana is mounted a
rotatable star map (*rete*). The tympana, rete and an alidade
or sighting rule are held in position by means of a pin and
wedge. A bracket and ring allows the instrument to be
suspended when in use. One type of astrolabe, of which
only one complete example survives (fig 11), is the spherical
astrolabe, described in Islamic literature of the 13th century
(*Lit:* MADDISON (1962)). More bulky than the planispheric
astrolabe described above, it also had the drawback that it
could only be used in the latitude for which it was
constructed. The weight of the normal planispherical
astrolabe was increased by the need to provide it with
several tympana, permitting it to be used in several
latitudes. Several astrolabes were designed for use in any
latitude. The two most common universal astrolabes are:

Fig 10
Universal astrolabe, by E Dantes, Perugia, c1580
*The astrolabe has no loose plates. The mater is engraved for latitude 43°
40' (? Florence). The reverse, shown here, is engraved with the Roias
universal astrolabe projection.*
(Museum of the History of Science, Oxford. O4.II.3)

Fig 11
Spherical astrolabe, by Musa, AH855 (AD1480/81)
*Brass, damascened and laminated (on the ecliptic and equatorial circles)
in silver; silver suspension ring. Of eastern Islamic origin.*
(Museum of the History of Science, Oxford. O4.II.3)

1 The *sapharea*, described by Ibn az-Zarquellu (13th century)
and revived by Gemma Frisius of Louvain in the 16th
century. It is based on a stereographic projection of the
celestial sphere from the vernal point on to the colura of
the solstices. (fig 9)
Lit: MADDISON (1969)

2 *Roias projection*, first described in a treatise by Juan de
Roias on the astrolabe, 1550. It consists of an orthographic
projection of the celestial sphere on the colura of the
solstices, and occurs mainly during the second half of
the 16th century. (fig 10)
Lit: MADDISON (1966)

The earliest surviving astrolabe is Islamic and dates
from the late 9th century. European astrolabes survive
from the 11th century onwards; the earliest are Moorish or
Hispano-Moorish (fig 6). In Europe few instruments can
be dated later than the second half of the 17th century, but
in Islamic lands the tradition continued until much more
recent times. There are many fake astrolabes in existence,
some with meaningless markings and others where the
rete has degenerated into performing no function other
than decoration. See also the *mariner's astrolabe*.
Lit: GUNTHER (1932); MAYER (1956); PRICE (1955);
GIBBS/HENDERSON/PRICE (1973); HARTNER (1932);
NORTH (1974); SAUNDERS (1984); WEBSTER (1974);
JENKIN (1925)

Astronomical compendium (fig 12)

This term is applied to an elaborate combination instrument
much loved by wealthy amateurs of science in the 16th and
early 17th centuries. The compendium is normally of gilt
brass, and of rectangular, circular, octagonal or hexagonal
shape. It invariably includes at least one, and often several,
sundials – horizontal, vertical or equinoctial dials
constructed for any of a number of different hour systems.
A nocturnal, a windrose, tables of latitude, a compass, a
lunar volvelle and scales for the conversion of the Gregorian
and Julian calendars frequently occur. Some compendia
incorporate astrolabes in different projections, or maps
and projections of the terrestrial hemisphere which permit
the solution of problems concerning local time in other
parts of the world. Other compendia include surveying
instruments – geometric squares, sinical quadrants and
circumferentors. The example illustrated, constructed in
1588 by Tobias Volckmer in Munich, incorporates a
clinometer on a cord with reel which could be used for
mine surveying and levelling.

Compendia were constructed most frequently by makers centred in southern Germany (Augsburg, Munich, Bamberg), Saxony (Dresden) and Austria (Vienna).

Lit: MADDISON (1957) pp74-77; ZINNER (1956)

Fig 12 A & B
Astronomical compendium, by T Volckmer, Munich, 1588
A gilt brass compendium, which contains a sundial, and a compass (A), and a stereographic astrolabe projection, as well as a clinometer, cord, and reel for use in surveying (B).
(British Museum, London. Lo2.I.4)

Backstaff or Davis's quadrant (fig 13)

Invented by John Davis c1594. It replaced the *cross staff* as an instrument for measuring the Sun's meridian altitude as a means of establishing latitude at sea. It had the great advantage over the cross staff that the observer was no longer forced to look directly at the Sun. By the 1630s it was an essential part of the navigator's equipment, and gradually replaced the cross staff completely. The backstaff itself was only superseded by the *octant* towards the middle of the 18th century. Most surviving are of 18th-century date.

The instrument consists of a large and a small arc with a horizon vane at their common centre, a shadow vane on

the smaller arc and an eye vane on the larger one. The observer adjusted the shadow vane until it was in a suitable position, then slid the eye vane up and down its arc until the shadow of the edge of the shadow vane was thrown on to the horizon vane and coincided with the line of the horizon. The sum of the angles indicated by the two vanes on their respective arcs gave the zenith distance.

Around 1680 the backstaff was improved by the invention, by John Flamsteed, the first Astronomer Royal, of a shadow vane incorporating a glass lens (known as a Flamsteed glass). This made it possible for the backstaff to be used in hazy sunshine.

Lit: HILL/PAGET TOMLINSON (1958) pp10-13; WATERS (1958) pp205-6, 302-6; TAYLOR/RICHEY (1962) pp49-51; COTTER (1968) pp70-73; MAY (1973) p125ff.

Fig 13
Backstaff, English, early 18th century
This example is equipped with a vane set with a glass lens (Flamsteed glass) on the small (horizon) arc. This was used to determine the angle of the Sun in hazy weather, instead of the shadow vane.
(Science Museum, London. Lo22.10)

Balance (figs 14-16)

The balance has been used for thousands of years. The first type to be developed for a specialised scientific process was probably the assayer's balance, used in the determination of the precious metal content of alloys. By the 16th century the thin, solid beam had developed swan-

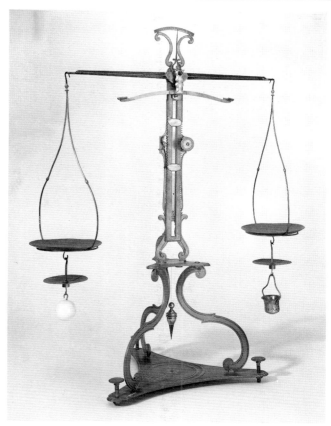

Fig 14
Hydrostatic balance, French, late 18th century
The ratchet device enables the ivory ball to be lowered into a liquid whose specific gravity is being determined, or the bucket into water to measure the density of a solid contained in it.
(Science Museum, London. Lo22.3)

neck ends which could be used to adjust the balance point. The assayers balance underwent relatively little change subsequently.

Hydrostatic balances were used in the 17th century to measure specific gravities of solids and liquids – one was described by Robert Boyle in 1690. They were constructed shortly afterwards by Francis Hauksbee (1666-1713) in London, and though very rare, a few of these survive. Rather elaborate versions were constructed in Paris later in the century (fig 14). Perhaps the greatest technical advance at the time was made by Jesse Ramsden (1735-1800), who was commissioned by the Royal Society in 1789 to make a balance for accurate measurement of specific gravity of alcohol/water mixtures (the British Government needed this data for taxation of spirit liquors). Ramsden solved the problem of minimising the flexing of the beam (when loaded) by constructing it of a double cone of brass. However the massive beam had a high moment of inertia, hence a long periodicity of swing, and readings took a long time to take. This actual balance survives in the Science Museum, London. In Paris, Nicholas Fortin (1750-1831) used a more traditional solid beam design for Lavoisier's balance, though the level of craftsmanship was extremely high. The sole example of this type in the British

Isles can be seen at the Whipple Museum of the History of Science, Cambridge.

The problem of the beam was solved in London by the instrument maker TC Robinson (1792-1841), who in the 1820s designed beams in the form of brass frameworks which were rigid yet light. For precision suspension points, Robinson used steel knives on agate planes. These were also used by Ludwig Oertling, who came from Berlin to London in the 1840s, and whose firm became the world's largest supplier of balances. For even greater accuracy, Oertling introduced a small weight which could be adjusted along a graduated beam (the 'rider'). For much of the 19th century it was believed that long beam balances (of, say, 10in in length) were more sensitive than shorter.

In Germany, Paul Bunge showed that length was not of the greatest importance, and he and Florenz Sartorius developed high successful, sensitive short beam (about 5½in) balances from the 1870s which became universal. At around this time semi-automatic balances were introduced in which weight could be loaded and unloaded from outside the glazed cases. Albert Ruprecht and Ludwig Seyss, both of Vienna, perhaps produced the earliest designs.

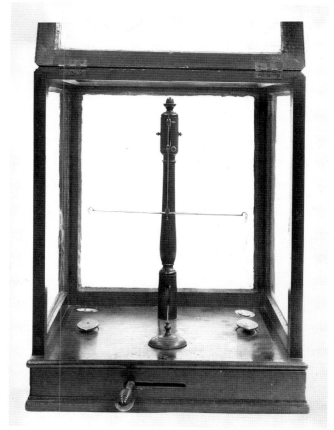

Fig 15
Assay balance, English, c1773
A steel balance beam with brass pans and movable rests for pans, housed in a glazed mahogany case. The balance was used in the Assay Office, Birmingham, from 1773 until about 1860.
(The Assay Office, Birmingham. B10.1)

Most non-chemical balances (such as coin, bullion and letter balances) are outside the scope of the inventory. Specialised types which have been included are *dotchins*, or Chinese opium scales, which are miniature steelyards with single pans and graduated ivory beams, and *chondrometers*, developed to determine the quality of grain. A cylindrical brass cup is filled with a specific type of grain, and a sliding counter weight on the calibrated beam reads off the weight per unit volume (for example, pounds per bushel).

Lit: STOCK (1969, 1973); JENEMAN (1977, 1979); BUCHANAN (1982)

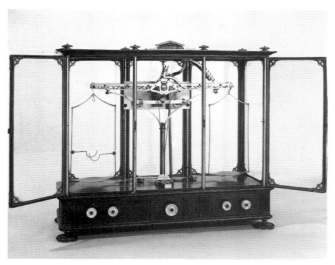

Fig 16
Chemical balance, by L Seyss & Co., Vienna, c1880
An early semi-automatic balance, weights being lowered on to the beam.
(Science Museum, London. Lo22.3)

Camera lucida (fig 17)

This instrument was invented by WH Wollaston before 1806 (British Patent No. 2993) in order to facilitate the drawing of objects in perspective. It consists of a small four-sided prism mounted on a stand which can be clamped to a book or a table holding the drawing paper. The image of the object to be copied is directed by means of the prism on to the paper, the user viewing through an aperture between the prism and his eye.

Wollaston's original model for the camera lucida, in a wooden frame, is in the Whipple Museum, Cambridge (*Lit:* GUNTHER (1937A) p109) and bears a label with the date 1786, which may mark the date of its invention, although the camera lucida was not patented until 1806. Early examples are found with the initials WHW and a serial number scratched on the prism. An improved and simpler version was developed later by Amici.

Attempts to invent instruments and machines which would facilitate the production of drawings in perspective date back to the 16th century, when Albrecht Dürer in his *Underweysung der Messung* (Nuremberg 1525) illustrated two pieces of apparatus designed for this purpose.

Lit: WOLLASTON (1807); HAMMOND/AUSTIN (1987)

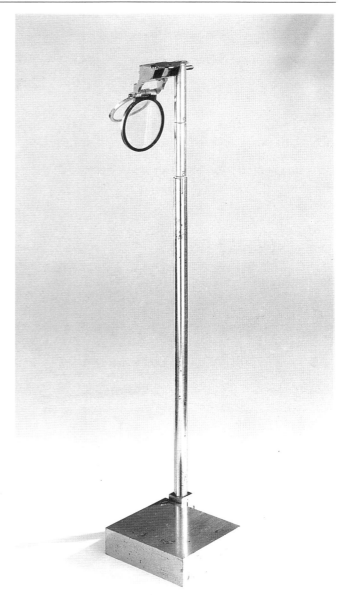

Fig 17
Camera lucida, English, c1820
Wollaston camera lucida of brass. The camera lucida was invented by W H Wollaston before 1806 to facilitate the drawing of objects in perspective. (Science Museum, London. Lo22.12c)

Camera obscura (fig 18)

A device which enables images of distant objects to be projected into a darkened room, or, in a portable form, on to a ground glass plate or white screen in a box. In room form the camera obscura was used for observing sunspots as well as for more amateur pastimes. Old camera obscuras for observing sunspots are at the Royal Observatory Edinburgh and at Herstmonceux, but the earliest still in operation is that erected for the Dumfries and Maxwelltown Astronomical Society in 1836, now an integral part of the Dumfries Museum. In portable form the camera obscura permitted the reduction of a landscape to a size where it could be copied directly from its image on a sheet of paper. Occasionally in the 18th century the camera obscura appears in book form, and could be adapted so that it also served as a print viewer (example in the Victoria and Albert Museum, London, Department of Prints and Drawings).

In either form, the image is transmitted by means of a double convex lens, and often by means of a mirror which directs the image on to a convenient horizontal surface.

Although the camera obscura existed at a much earlier date, surviving instruments are mostly dated to the latter half of the 18th or the 19th century.

Lit: GERNSHEIM (1969) pp17-29; HAMMOND (1981)

Fig 18
Camera obscura, by W Storer, London, post-1788
This version of a camera obscura, entitled 'Royal Deliniator' by its inventor William Storer, was patented in 1788. On the label pasted inside the lid, Storer states that his instrument is intended as a drawing aid, and is superior to the ordinary camera obscura because it can function in both sunlight and artificial (candle) light.
(Hertford Museum, Hertford. H3.1)

Chemical glassware (fig 19)

Vessels developed specifically for chemical purposes date from at least 2000 years ago, and probably further back than that. The earliest surviving unambiguous apparatus for distillation is ceramic ware from the Indus Valley made at the beginning of the Christian era. Glass alembics, in which the vapours were condensed in a still, were probably used in Hellenistic Alexandria. The earliest surviving vessels of this type are from the Near East and are of the 9th-12th century AD (*Lit:* ANDERSON (1983)). Stylistic changes were minimal over very long periods – alembics to be found in 19th-century dealers' catalogues are closely similar to their Hellenistic ancestors. However the range and specificity of glassware increased from medieval times onward. For example, the Woulfe bottle was developed in late 18th-century France to remove unwanted constituents of gas mixture by passing it through appropriate liquids or solids.

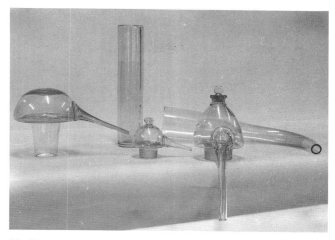

Fig 19
Chemical glassware, early 19th century
Illustrated are a funnel-shaped tube, a tall cylindrical vessel, and three alembic heads, all of clear white glass. Two of the alembic heads have ground joints for connecting with the body, one is also equipped with a neck and ground stopper. The alembic head formed part of a distillation apparatus, and was used to close a heated vessel. The distillate condensed inside the head and ran out through a narrow tube into a receiver.
(Department of Chemistry, University of St Andrews. S1.2h, i & o)

From the late 18th and early 19th century, glassware was developed for volumetric procedures, in particular the burette and the pipette. Prominent in their evolution was the French pharmacist FAH Descroizilles (1751-1825). PJ Kipp invented 'Kipp's apparatus' for generating hydrogen sulphide, used in qualitative analysis; this may have been based on *Nooth's apparatus*. More effective condensing columns were produced, from the simple condenser (not invented by Justus von Liebig) onwards.

Soda or potash glass was used for chemical ware until the early 20th century. 'Pyrex' glass became available *c*1916.

Lit: CHILD (1940) pp93-111

Chondrometer
See *balance*.

Circle, reflecting (figs 20-1)

The circle serves the same purpose as the *sextant*, but the limb is extended to 360°. Angles are either measured and remeasured, as in Mayer's repeating circle, *c*1752, or Borda's repeating circle, *c*1772 (fig 21), or measured from three separate verniers, as with Troughton's reflecting circle, *c*1796 (fig 20), which allows three distinct but simultaneous readings. Either process allows the elimination of errors of construction, graduation and reading.

Reflecting circles were used extensively by the French and German navies, but English seamen preferred the sextant. The circle was also used by astronomers, surveyors and by mariners on making a landfall. For this purpose it was slightly modified and mounted on a stand with counter-weight. Edward Troughton's design of 1796 increased the popularity of the reflecting circle in England.

Lit: LONDON (1971) part 8; DAUMAS (1972) pp183-6

Fig 20
Reflecting circle, by Troughton and Simms, London, c1830
*An example of the non-repeating but triple-verniered circle invented by
Edward Troughton in 1796. Equipped with a counterpoised stand, the
circle could also be used by surveyors and astronomers, as well as by
navigators seeking an accurate fix of latitude and longitude on making
a landfall.*
(Vickers Ltd, York. Y3.6)

Fig 22
Circumferentor, Dutch, early 17th century
*A surveyor's circle of gilt brass with inset compass, double shadow-
square, and wind-rose. There are fixed sights at the four cardinal points,
as well as an alidade with sights.*
(Department of Physics, University of St Andrews. S2.16)

Fig 21
Repeating circle, by J Cox, London, early 19th century
*In Borda's 1772 design more accurate angular measurements are made
as two sets of readings are taken.*
(Science Museum, London. Lo22.10)

Fig 23
Circumferentor, by E Nairne, London, c1750-75
*This instrument can be set up as illustrated here, or as a theodolite, for
which see Fig 132.*
(Smith Art Gallery and Museum, Stirling. S20.1)

Circumferentor (figs 22-3)

A surveying instrument consisting of a graduated horizontal circle with inset compass and two fixed sights on the north-south axis. A second pair of sights are mounted on an alidade. Another variation, the Holland circle, is equipped with sights on the N-S and on the S-W axis as well as on an alidade. Such instruments normally date from the 17th and 18th centuries.

In Ireland, from the late 17th century onwards, a variation of the circular rentor seems to have been developed which closely resembles the simplest form of miner's dial. It consists of a compass with an attachment for a tripod. On the N-S axis of the compass two small lugs project to which can be screwed two arms carrying slit and hair sights. See also *graphometer*.

Lit: BROWN (1982A); KIELY (1947); RICHESON (1966)

Compass, azimuth (fig 24)

An instrument used to observe magnetic variation by comparing the bearing of the Sun or star with the calculated bearing. The most usual form of azimuth compass during the latter part of the 17th and the early 18th century consisted of a compass mounted in gimbals, over the top of whose bowl rotated a broad flat ring. An L-shaped index equipped with an inclined thread was pivoted at a point on the outer end of the ring, and traversed across a graduated scale on the other side of the ring. After correctly aligning the cross wires on the ring to the N-S axis of the compass card, the index was turned until the thread threw its shadow onto a line marked on the index, and the variation could be read off the scale. Another type of azimuth compass was equipped with an alidade with slit and hair sights (Kenneth McCulloch's azimuth compass, British Patent 1663 (1788)).

Ralph Walker's meridional compass, an improvement of the azimuth compass, was designed in 1793. Intended for finding longitude by means of an accurate determina-tion of the magnetic variation, it was in general used as an improved azimuth compass. It consists of a compass in gimbals on which a universal equinoctial dial is mounted.

Charles Schmalcalder mounted a prism by the near sight, permitting the bearing to be viewed and read off simultaneously (British Patent 3545 (1812)).

Lit: TAYLOR/RICHEY (1962) pp31-33; MAY (1973) pp83-92

Compass, mariner's (fig 25)

A compass, normally in a brass bowl, suspended inside a box by means of gimbals, so that it always hangs in a horizontal position. While the earliest compasses were not hung on gimbals, a compass of late 16th-century date in the National Maritime Museum, Greenwich is mounted in a gimbals in a circular box. In contrast to the surveyor's compass where the needle is pivoted over the compass card, compasses for nautical use generally mount the compass card over the needle so that the two pivot together.

The use of a liquid to reduce oscillation of the compass was investigated during the late 18th and early 19th century, and is now common.

Lit: TAYLOR/RICHEY (1962) p45ff; MAY (1973) pp18-21

Fig 25
Mariner's compass, by J F Hervouet, Vannes, 1764
A dry card compass mounted on gimbals in a rectangular box. (Smith Art Gallery and Museum, Stirling. S20.2)

Fig 24
Azimuth compass, by J Fowler, London, c1720
Compass card graduated to 32 points, mounted in gimbals in a wooden box. Compass fitted with alidade and diagonal scale. The compass card, signed: H Gregory, Near the India House London, is a replacement of c1760.
(National Maritime Museum, London. Lo16.II.10)

Cross staff (fig 26)

An instrument used for measuring the altitude of the Sun and Pole Star at sea, so that a seaman could find his latitude. It probably originated in the 14th century as an astronomical instrument, but was only adopted by seamen in the 16th. It was ousted by the *back staff* or Davis's quadrant in the 17th century. The instrument consists of a square-sectioned staff about 30in long with scales on each side. On it can be fitted one of a set of three or four cross pieces. The staff was pointed in the direction of the Sun with the end resting by the eye of the observer. One of the cross-pieces was slid along the staff until the bottom edge coincided with the horizon and the top with the Sun, after which the altitude of the Sun could be read off the staff.

Very few examples have survived and most lack their original vanes.

Lit: HILL/PAGET TOMLINSON (1958) pp8-10; COTTER (1968) p64ff; MADDISON (1969) p46ff; STIMSON/DANIELL (1977); ROCHE (1981)

Fig 26
Cross staff, by T Tuttell, London, c1700
A presentation instrument in ivory.
(National Maritime Museum, London. Lo16.II.12)

Dial

An instrument for measuring the time by means of (a) variation of the Sun's altitude during the day or (b) the variation of the Sun's azimuth along the equinoctial circle. Most dials have to be orientated by means of a compass and can be adjusted for use in more than one latitude. The earliest surviving English sundial, of Saxon date, is in the possession of the Dean and Chapter of Canterbury Cathedral, but sundials were known at the latest by 500 BC. Medieval portable sundials were primarily in quadrant form (see *quadrant, horary*) and the function of a time-measuring device was combined with other astronomical functions. By the second quarter of the 15th century a whole range of differing types of sundials was being developed, and this extended greatly during the Renaissance. With the improvement in the mechanical clock some characteristics of the former were transferred to the sundial (hour dial, hands) as in the mechanical equinoctial dial (below). During the 18th century small portable dials still enjoyed great popularity, notably the Augsburg-made universal equinoctial dial (below), the French Bloud dial (below), and the equinoctial ring dial (below). Side by side with these, produced in enormous quantities, a number of very complex instruments were constructed for a more expensive and limited market. The vertical or horizontal fixed dial for use in the garden or on

the facade of a building was also popular.

By the 19th century the heyday of the sundial was over, although many instruments were still being produced at that date.

Lit: HIGGINS (1957) for a useful classification scheme; JOSTEN (1955); MADDISON (1957); ROHR (1970); COUSINS (1969); BOBINGER(1966); ZINNER (1956); HERBERT (1967); GOUK (1988) for Nuremberg dials

Analemmatic dial (fig 27)

A self-orientating dial, the design of which was published by Thomas Tuttell in his *Description and Use of a New Contriv'd Double Dial* (London 1698). It consists of two dials combined in one instrument – an ordinary *horizontal dial* and an *azimuth dial* which consists of an orthographic projection of the equinoctial on to the plane of the horizon.

Fig 27
Analemmatic dial, by E Baradelle, Paris, 1750-75
Dial of brass, for use in a single latitude, 46° 18' N.
(National Maritime Museum, London. Lo16.II.29)

When the instrument is placed in a horizontal position and rotated until both dials give the same reading, that indicates the correct time.

Azimuth dial

A horizontal dial in which concentric hour-scales are graduated in the solar azimuth angles for each month of the year.

A modification of the azimuth dial is the *magnetic azimuth* dial. Here a compass needle is mounted on a pivot at the centre of the dial, and sights attached to the 12 o'clock line. When the sights are directed towards the Sun, the compass needle indicates the time.

Bloud dial (fig 28)

This is a variation of the magnetic azimuth dial and is associated with Charles Bloud and other Dieppe sundial makers around 1660. The series of circular hour scales is replaced by an elliptical hour ring adjustable against a calendar scale to allow for the time of the year. When this adjustment has been made, and the dial turned so that the shadow of the upper leaf falls exactly over the lower, the position of the compass needle over the hour ring indicates the time. Bloud dials normally incorporate *polar, horizontal*

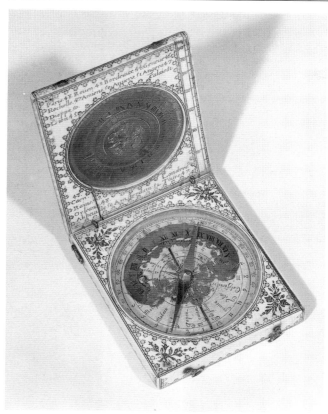

and *string-gnomon* dials. They continued to be produced in quantity until well into the 18th century.

Butterfield dial (fig 29)
Horizontal dial of distinctive design which derives its name from Michael Butterfield, an English mathematical instrument-maker who worked in Paris at the end of the 17th century. The dial is usually octagonal, sometimes oval, and sometimes constructed of silver rather than brass. Its characteristic feature is a hinged style supported by thin folding plates shaped in the form of a bird. The style is engraved with a degree scale for a range of latitudes (normally between 40 and 54°), and the bird's beak acts as pointer on the scale. Butterfield dials long retained their popularity and were still being constructed around the middle of the 18th century.

Diptych dial (fig 30)
A sundial, consisting of two tablets of ivory, wood or brass which close flat, and can be opened so that they stand at right angles to one another. It incorporates a compass for

Fig 28
Magnetic azimuth dial, by C Bloud, Dieppe, late 17th century
Ivory dial with typical Dieppe decoration. The volvelle in the upper leaf, and the analemmatic dial scale in the lower leaf are made of silvered pewter.
(Science Museum, London. Lo22.16)

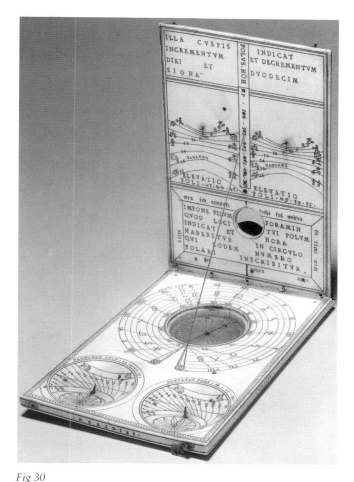

Fig 29
'Butterfield' dial, by E Baradelle, Paris, mid-18th century
The gnomon on this silver dial can be adjusted for any latitude between 40° and 60°. There are four hour scales, calibrated for 40°, 45°, 49°, and 52°. The compass is calibrated to allow fine adjustment for magnetic variation. On the reverse is a list of 27 European towns with their latitudes.
(Science Museum, London. Lo22.16)

Fig 30
Diptych dial, by H Tucher, Nuremberg, 1589
Ivory dial. There is a range of attachment points for the string style, and a corresponding series of scales for the horizontal dial on the lower leaf. In addition there are vertical and horizontal pin-gnomon dials, only accurate at their design latitude.
(Science Museum, London. Lo22.16)

correct orientation of the instrument. A thread stretched taut between the two tablets serves as gnomon for a horizontal (hour scale on the base tablet) and a vertical (hour scale on the vertical tablet) dial. In addition the instrument usually incorporates several subsidiary pin-gnomon dials reading in differing time measurement systems – Babylonian, Italian, planetary hours, etc. Usually the diptych dial incorporates a list of towns and their latitudes and some means of adjusting the angle of elevation of the gnomon so that the instrument can be used in a range of locations.

Universal equinoctial (or equatorial) dial (fig 31-2)
In this type of dial a gnomon placed parallel to the Earth's axis casts a shadow on to an hour-scale parallel to the equator. The instrument has to be correctly orientated prior to use, and portable examples are invariably equipped with a compass. The hour ring is adjusted to the latitude of use by means of a degree scale engraved on a quadrant or folding arc.

Equinoctial dials frequently occur as part of an *astronomical compendium* and on some diptych dials. They achieved great popularity during the 17th and 18th centuries. In the late 17th century, under the influence of clock design, forms of mechanical equinoctial dials were designed where the time was measured from the Sun's position, either by means of a gnomon, or a radial pointer with sights, connected by means of gearing with hour and minute hands (fig 32). Such sundials were particularly popular in Central Europe, the earliest known example being made by Michael Bergauer of Innsbruck in 1671. The design was adapted by some English makers, notably John Rowley, who introduced it as a modification to the standing universal equinoctial ring dial in or before 1715.

Universal equinoctial ring dial (fig 33)
A self-orientating universal equinoctial dial which originated in the early 17th century, probably as a simplification of the astronomical ring designed by Gemma Frisius in or before 1554. It enjoyed a great popularity for two hundred years, for it could be used in any latitude; the only pre-requisites being a knowledge of the day of the month and the latitude of use. The dial consists of two

Fig 31
Universal equinoctial dial, by E Allen, London, c1630
The hour ring is attached to a quadrant with a degree scale, one of whose radii is extended as a gnomon. Thus the hour ring can be inclined until the degree scale indicates that the latitude of the place of observation is reached.
(Science Museum, London. Lo22.16)

Fig 32
Universal mechanical equinoctial dial, German, second half 18th century
Gilt brass, with pewter hour ring and glazed minute dial. With contemporary blind-stamped leather case.
(City Museum and Art Gallery, Bristol. B24.19)

Fig 33
Standing universal equinoctial ring dial, by G Adams, London, late 18th century
When the instrument is adjusted for latitude, the hour ring lies parallel to the equator. The pin-hole gnomon is set for the date by sliding it along the diametrical bar. The dial is rotated until the hour scale is illuminated by the gnomon, giving the time.
(Science Museum, London. Lo22.16)

rings which fold inside one another when not in use. The outer ring is provided with a sliding shackle and suspension ring and engraved with a degree scale against which the shackle can be set in order to allow for latitude when held by the outer, and carries an engraved scale. A bar fixed as a diameter to the meridian ring, thus corresponding to the Earth's axis, is engraved with a scale of months and zodiac, and a small slide bearing a pin hole can be set against this scale according to the time of the year. The instrument is held by its suspension ring so that a ray of Sun, passing through the pin-hole lights on the hour ring, indicates the time of day. A version set on a stand with a compass in the base was produced by English makers from the early 18th century; as the dial is self-orientating, this instrument could be used to establish the magnetic variation.

Horizontal dial (fig 34-5)

The horizontal dial was the most popular form of garden sundial in the 18th century. Being in a fixed position it was constructed for one latitude only.

Portable horizontal dials were orientated by means of a compass. The style, the shadow-casting surface or edge, is mounted so that its edge is parallel to the Earth's axis. It consists either of a triangular brass plate mounted at right angles to the base-plate of the dial, or a thread stretched taut between the base-plate and a support, or between the two leaves of a diptych dial. Many horizontal sundials could be used in several latitudes (cf. Butterfield dial, diptych dial) but only for a limited range, since a different hour-scale was required for each latitude.

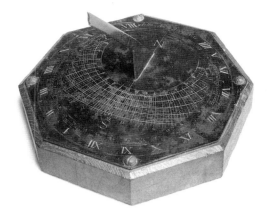

Fig 35
'Double' horizontal dial, by J Allen, London, 1632
Within a standard horizontal dial is the projection published by William Oughtred in his Description and Use of the Double Horizontal Dyall (London 1632).
(Museum and Art Gallery, Maidstone. M1.11)

Fig 34
Horizontal dial, by J Rowley, London, early 18th century
The dial is constructed for use at 44° 35', the latitude of Turin. It is engraved with a monogram, beneath a ducal crown, probably that of Amadeus II, Duke of Savoy.
(City Museum and Art Gallery, Gloucester. G6.1)

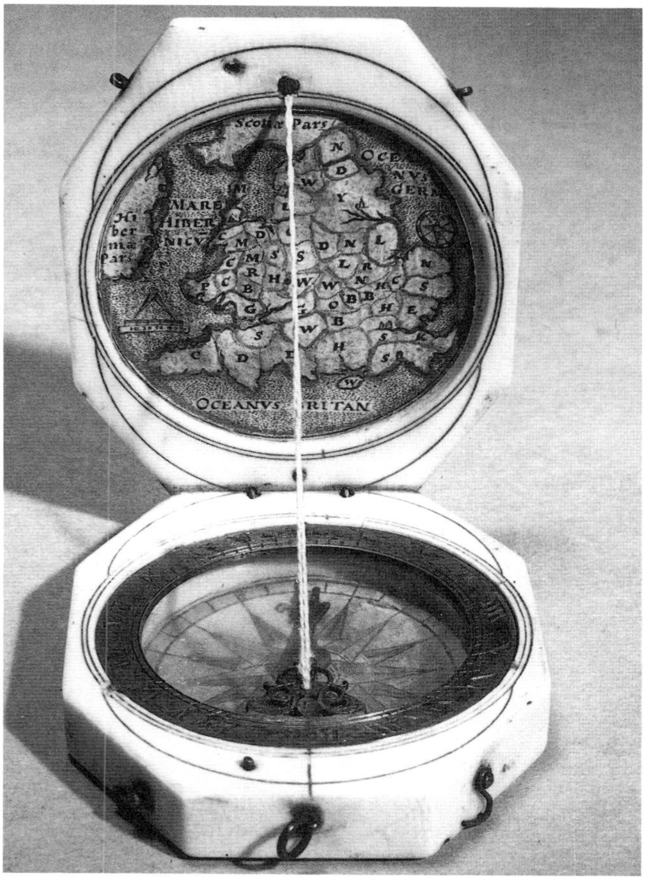

Fig 36
Horizontal string-style dial, by ? Ralph Greatorex, London, mid-17th century
Ivory dial with silvered hour ring. The compass card and the county map of England are printed from an engraved plate and hand-coloured.
(Horniman Museum, London. Lo9.4)

Horizontal string-style dial (fig 36)

A form of horizontal sundial popular in the 16th and 17th centuries. A taut thread, held in position by a support hinged to lie flat when not in use, replaces the metal style of the usual horizontal sundial. Often the dial is of diptych form, and in this case the lid of the dial, in its open position, serves as the support for the style.

Inclining dial (fig 37)

An ordinary horizontal sundial could be used in one latitude only. The inclining dial is a horizontal sundial which, by means of a hinge at its north point, can be inclined against a quadrant with a scale of latitudes. By adjusting the angle of the plate the dial can be brought parallel to the horizontal plane of the original latitude and thus be used in a series of latitudes. Such dials were common during the period 1750-1850 particularly, although not exclusively, in England, and in Central Europe.

Fig 38
Pillar dial, ? French, 18th century
Ivory with brass gnomon.
(Museum of the History of Science, Oxford. O4.II.27)

Fig 37
Inclining dial, by J Morgan, St Petersburg, 1775-1800
Brass, part silvered. The signature and the list of latitudes of cities in Russia is engraved in Russian script. The maker, a London-trained craftsman, worked in St Petersburg from c1772 to 1804.
(Museum of the History of Science, Oxford. O4.II.27)

Pillar dial (fig 38)

This form of altitude dial is described in medieval texts, although the type was still being made in the 19th century, for example for the use in peasant communities, notably the Pyrenees. Late examples are usually of wood; earlier ones may be of wood, brass or ivory. They are also known as *cylinder dials*. The dial consists of a hollow pillar with a removable top with attached gnomon which can be folded away when not in use. The pillar is marked with a calendar scale and with curving hour lines. The instrument is prepared for use by bringing the gnomon in line with the appropriate month on the calendar scale. The whole instrument is then turned until the shadow from the gnomon falls vertically beneath it. The time is read from the position of the tip of the shadow cast by the gnomon. Such sundials can only be used in the latitude for which they were designed.

Polyhedral dial (fig 39)

A form of multiple dial which became popular during the 16th century. Each face of a polygon – frequently a cube or a dodecahedron – was occupied by a sundial. In the latter part of the 18th century portable cube dials became popular, normally consisting of a wooden cube, on each of whose faces were pasted printed sundial scales (such sundials were constructed, among others, by Nuremberg dial makers such as David Beringer). On the Continent and in Scotland, stone polyhedral sundials of very complex shapes

Fig 39
Cube dial, by D Beringer, Nuremberg, late 18th century
Wood-covered with engraved hand-coloured paper scales, brass styles. The dial is orientated so that the compass points north. The time can then be read from more than one of the dial faces.
(The National Museums and Galleries on Merseyside, Liverpool. L8.51)

Fig 40
Ring dial, by M F Poppell, Passau, 1696
Gilt brass, constructed for latitude 49°. There are separate pin-hole gnomons for summer and winter.
(National Maritime Museum, London. Lo16.II.29)

were frequent features in gardens. Portable polyhedral dials require a compass for correct orientation, and this was usually incorporated in the base, or in a horizontal face. Occasionally such multiple dials were constructed on the outside of caskets.

Ring dial (fig 40)
An altitude dial where the time is ascertained from a ray of sunlight passing through a small hole in a suspended ring and striking an hour scale engraved on the inside of the ring. Some ring dials were provided with two holes, one for use during the summer, the other during the winter solstice. More elaborate dials were provided with a sliding ring permitting an adjustment to be made according to the Sun's declination.

The ring dial was known in medieval times but most surviving examples are probably of 18th century date, although their very simplicity makes an accurate dating difficult. Ring dials occur frequently in English collections. They are also known as *poke* dials.

Dip circle or dipping needle (fig 41)

An instrument for measuring magnetic inclination or dip, i.e. the deviation of the compass needle from the horizontal. The earliest such instrument was described by Robert Norman in *The New Attractive* (London 1581). In its simplest form the dip circle consists of a compass needle mounted on a horizontal axis through its centre of gravity in the centre of a graduated circle. The vertical circle is mounted in turn on a horizontal circle and can be revolved until it lies in the plan of the magnetic meridian. The position of the needle against the graduations of the vertical circle indicates the magnetic dip. The Department of Physics, St Andrew's University, owns a dip needle of 17th-century date. The majority of surviving instruments, however, date from the late 18th or 19th century. By the third quarter of the 18th century the axle of the compass needle rested on agate edges or was mounted on frictionless bearings as devised by the Revd William Mitchell in 1772. To avoid air disturbance affecting the compass needle the vertical circle was enclosed in a glass case. Around 1840 the accuracy of the instrument was further improved by the addition of microscopes to facilitate accurate readings from the graduations.

Lit: WATERS (1958) for the early development; LONDON (1971) section 14; TURNER/LEVERE (1973) p195; McCONNELL (1980) pp12-21; McCONNELL (1986A)

Fig 41
Dip circle, by J Sisson, London, c1780
Brass dip circle mounted in gimbals. With inscription marking its presentation to Marischal College in 1795.
(Department of Natural Philosophy, University of Aberdeen. A4.13)

Drawing instruments (figs 42-45, 52)

One of the fruits of the scientific Renaissance was the establishment of a formal and theoretical mathematical basis for such arts as gunnery, surveying, navigation, shipbuilding, architecture and engineering. The large instruments used by master craftsmen to lay out work were refined for use by the new mathematical practitioners to record and plan their work on paper. Whilst the basic drawing instruments (scriber, rule, square, and compasses) were in use in antiquity, the majority of draughting instruments have a history that can be traced to the late Renaissance. Since the draughtsman needed a range of instruments to carry out his work, these were frequently collected in sets, whose extent depended on both the needs of the user and his wealth (figs 42-44). Larger sets might well include more specialist instruments, such as the *curve bow* (fig 45), *ellipsograph* (fig 52) and *proportional compasses* (fig 110), which would also be supplied and kept separately with other specialist draughting devices like the *pantograph* (fig 107). In England the *sector* (fig 119) was almost

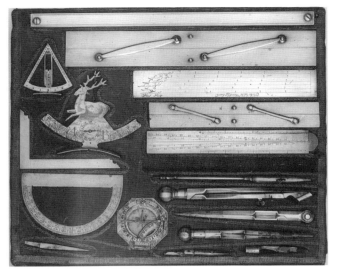

Fig 42
Set of drawing and gunnery instruments, European, 16th to 18th centuries
The leather case appears to be of German origin, of early or mid-18th century date, and made for this collection of instruments. The earliest piece is the rule signed: MARCVS PVRMAN FECIT MONACHIO 1589. The sector, by D Chorez of Paris, is early 17th century. The universal equinoctial sundial, made in Augsburg, was added later to the set, and the box adapted to take it.
(Royal Institute of British Architects, Drawings Collection, London. Lo19.1)

Fig 43
Case of instruments, by P Galland, Rome, c1700
The instruments, housed in a gold-stamped leather case, include a simple microscope, a levelling square, a compass, dividers with interchangeable points, and a set square/protractor. The microscope is of a type described by Zahn in 1702.
(Victoria and Albert Museum, London. Lo25.64)

universally included in sets of drawing instruments, even into the early decades of the present century, long after its effective demise as a calculating aid.

Lit: DICKINSON (1956); HAMBLY (1982); FELDHAUS (1959); NEDOLUHA (1957-9)

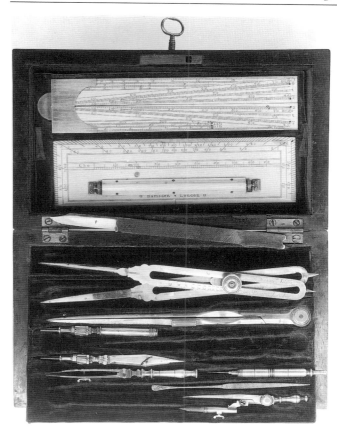

Fig 44
Box of drawing instruments, by J Ramsden, London, 1775-1800
Included in this set are a sector, a combined protractor and rolling parallel rule, and a proportional compass.
(Science Museum, London. Lo22.8)

Electrical machine (figs 46-51)

What is frequently called the first frictional electrical machine was described by Otto von Guericke in 1672. His machine consisted of a sulphur sphere mounted on an iron axle, and resting on two supports within which the sphere could be rotated. A hand applied to the sphere acted as a rubber to create an electric charge, though at that date neither von Guericke nor his contemporaries appreciated the nature of the phenomenon. The sulphur ball was later replaced by one of glass and a rubber consisting of a wool, silk or leather pad replaced the hand. A conductor was later added and the *Leyden jar* was used to store the charge.

Fig 45
Curve bow, English, c1700
Made of boxwood and brass. An adjustable bow for drawing large radius arcs. The curvature of the rule is altered by the five brass screws.
(Museum of the History of Science, Oxford. O4.II.9)

Fig 46
Globe electrical machine, by E Nairne, London, c1775
A table model described by Nairne in his Directions for using the Electrical Machine as made and sold by Edward Nairne (London 1773). The box contains gearing operated by means of a crank handle (now missing).
(Museum and Art Gallery, Glasgow. G1.21)

The earliest electrical machines were of globe form (figs 46 and 47). They were largely replaced by the cylinder machines of Cavallo and Nairne (fig 48). The earliest plate electrical machines were constructed shortly after the middle of the 18th century and proved particularly popular on the continent (fig 49). The electrical machine was only replaced for experimental purposes by the induction of influence machine *c*1865.

Chronology of the electrical machine
The dates given are those in which the machines were first published or are known to have been invented, and include only the main types.

*c*1663 von Guericke's sulphur globe
1709 Hauksbee's spherical electrical machine with a glass globe. Hauksbee also developed an electrical machine employing a glass cylinder.
1750 Wilson's electrical machine, a table model with a glass cylinder.
1760s Development of the plate electrical machine. Constructed by Ingenhousz and Ramsden in England, although Planta and Sigaud de la Fond on the Continent have been regarded as being the originators, at some time during the late 1750s or very early 1760s.
*c*1767 Lane's combined generator, Leyden jar and electrometer (constructed by J Read).
pre-1773 Edward Nairne's spherical electrical machine with a clamp for fixing to a table (fig 46).
pre-1777 Cavallo's cylinder electrical machine
pre-1782 Nairne's patent medico-electrical machine, Winter electrical machine.

Induction or influence machines:
1843 Armstrong's hydro-electric machine (fig 50)
1865 Holtz machine
*c*1870 Carré machine, Topler machine (modification of Holtz machine with addition of tin foil on glass disc)
*c*1880 Voss machine
1883 Wimshurst machine (fig 51)

Lit: DIBNER (1957); LAVEN/VAN CITTERT-EYMERS (1967); HACKMANN (1973, 1978A); and for measuring devices, HACKMANN (1978B); HEILBRON (1979)

Fig 47
Globe electrical machine, ? Scottish, mid-18th century
This machine is mounted between high H-shaped supports on a wooden base. A pinion on the axis of the globe connects with a gear wheel mounted on the axis of the handle.
(Anthropological Museum, Marischal College, University of Aberdeen. A1.11)

Fig 48
Cylinder electrical machine, by G Adams, London, c1780
Machines such as this example were used to produce electrical charges for medical therapy. Among the accessories shown are Leyden jars and an insulating stool.
(Science Museum, London. Lo22.7)

Fig 49
Plate electrical machine, English, c1770
This plate or disc machine has four pairs of friction pads creating the charge. The micrometer screw on the left is incorporated in the discharging electrometer, so regulating the strength of the discharge. This instrument forms part of the King George III Collection.
(Science Museum, London. Lo22.5,7)

Hydro-electric machine (fig 50)
Invented 1841-3 by WG Armstrong as an improvement on the ordinary electrical machine, it consists of an insulated wrought-iron boiler from which steam escapes through a specially designed hardwood-lined nozzle. The action depends on the friction of the particles of water in the partially condensed steam on the wooden lining of the nozzle. The electricity resulting from the use of pure water is positively charged. If turpentine is added the charge is negative. Collecting combs in front of the nozzle are connected to the prime conductor as with an ordinary frictional electrical machine. Armstrong's machine was very much more powerful than its predecessors, but was more difficult to use.

   Lit: ARMSTRONG (1840, 1843)

Fig 51 (above)
Wimshurst electrical machine, by J Wimshurst, London, 1885
James Wimshurst's design for an induction machine dates from 1883. It was generally thought to be the best of the various induction electrical machines available in the second part of the 19th century. The standard Wimshurst machine has two plates, each with sectors of tinfoil. The plates rotate in opposite directions and a charge is induced between them. Larger machines were made with more than one pair of plates.
(Science Museum, London. Lo22.7)

Fig 50
Hydro-electric machine, by H Watson, Newcastle upon Tyne, 1842
The machine generates electric charge by the friction of wet steam passing through a nozzle. The design was due to W G Armstrong (later Lord Armstrong), who presented this example to Michael Faraday.
(Science Museum, London. Lo22.7)

Fig 52 (right)
Ellipsograph, English, London, c1815
John Farey's design for an instrument to draw small ellipses was published in the Transactions of the Royal Society of Arts. Farey himself was a draughtsman, not an instrument-maker. In the first instance, William Harris of London made and sold the ellipsograph. That firm probably made this example, which is inscribed: Farey London and numbered: No. 22. Later, other London instrument-makers sold the device.
(Royal Institute of British Architects (Drawings Collection), London. Lo19.8)

Equatorial instrument (fig 53)

An astronomical telescope mounted on an axis at right angles to the axis of the celestial pole. Thus the telescope is parallel to the polar axis and to the Earth's axis, and once the star has been brought in line with the telescope objective, it can be kept in view for a long period of time by means of a single motion. Such a mounting also permitted the direct measurement of the right ascension and declination of a star (equatorial co-ordinates). Large fixed equatorials formed an important part of the equipment of astronomical observatories from the late 18th century onwards. Somewhat earlier small equatorial instruments were designed in England, by Jonathan Sisson, James Short, Edward Nairne and Jesse Ramsden. Frequently they were *universal equatorial instruments* which could also measure zenith and altitude, and the universal nature of the design led to the name 'portable observatory'.

Lit: REPSOLD (1908); KING (1955)

Fig 53
Universal equatorial instrument, by J Ramsden, London, c1775
This instrument could be used to measure equatorial co-ordinates (right ascension and declination), zenith, and altitude. The design follows that of Jesse Ramsden's patent of December 1775 for an astronomical equatorial instrument.
(Science Museum, London. Lo22.12b)

Equatorium (fig 54)

An instrument developed in the late Middle Ages to permit, by mechanical means, the determination of planetary positions according to the Ptolemaic mathematical theory without the necessity for lengthy calculations. Few such instruments survive in metal although they occur relatively frequently in paper or vellum. In the latter materials, separate volvelles were constructed for each planet. Instruments of brass were more complex since they were designed so that a single instrument served all planets.

Accurate planetary positions were calculated by reference to astronomical tables and manual computation. The equatorium, because of its small size, could not achieve great accuracy, and must have been used primarily for demonstration and teaching purposes, providing rapid approximate calculations.

Lit: POULLE (1969); NORTH (1969); PEDERSEN/PHIL (1974) pp258-61

Fig 54
Equatorium, French, c1600
Unsigned brass instrument for the rapid approximate calculation of planetary positions.
(The National Museums and Galleries on Merseyside, Liverpool. L8.30)

Globe (figs 55-57)

The globe, or more frequently the pair of globes – one terrestrial, the other celestial – was one of the commonest symbols of the increase in knowledge brought about through direct study of natural phenomena in positive attempts to discover the unknown. As navigators ventured into uncharted seas, so the surface of the Earth became more fully mapped. As astronomers searched the heavens with improved telescopes or from new vantage points in the southern hemisphere, so fainter stars and new

Fig 56
Terrestrial and celestial globes, by J & W Cary, London, 1815
The plates from which the gores of this 'pair' of 12in diameter globes were printed, were originally issued at different dates. The celestial globe is dated 1799, the terrestrial 1815, the latter having been updated as a result of recent geographical discoveries.
(Science Museum, London. Lo22.2)

Fig 55
Globe goblet, by A Gessner, Zurich, 1571
An elegant example of the work of an eminent Swiss goldsmith; presented to Sir Francis Drake by Queen Elizabeth I in 1582. The terrestrial globe is constructed in two halves which join at the equator. The northern hemisphere, topped with an armillary sphere, can be removed, and the lower half used as a goblet.
(City Museum and Art Gallery, Plymouth. P3.1)

Fig 57
Pocket terrestrial globe, by J Miller, Edinburgh, 1793
In common with many pocket globes, this 3in diameter terrestrial globe has a case lined with celestial gores.
(Royal Museum of Scotland (Chambers Street), Edinburgh. E8.II.2)

constellations were added to the heavens. Primarily the globe was a teaching and demonstrative device.

   Lit: STEVENSON (1921). For Islamicate celestial globes, SAVAGE-SMITH (1985)

Graphometer (figs 58-59)

A surveying instrument, similar in use to the *circumferentor*, but consisting of an alidade with sights mounted over a semicircle with fixed sights on the diameter. A compass is mounted either at the swivel point of the alidade, or inset in the main radius of the semicircle. The instrument originated at the end of the 17th century in France or Germany, and its instrument (Philippe Danfrie, *Declaration de l'usage du graphometre* (Paris 1597)). It continued to be constructed in both countries until the late 18th century. In

Fig 58
Graphometer, French, second half 17th century
An example of the early form of graphometer on the lines of Philippe Danfrie's instrument. There are two fixed sights on the extended diameter, and two movable ones, mounted on the alidade.
(Town Docks Museum, Hull. H6.64)

Fig 59
Graphometer by Lennel, Paris, 1774
The decorative pierced scroll work is typical of Paris-made instruments of the period. In contrast, the telescopic sights and the tangent-screw slow motion of the alidade are unusual innovative design features.
(Royal Museum of Scotland (Chambers Street), Edinburgh. E8.II.19)

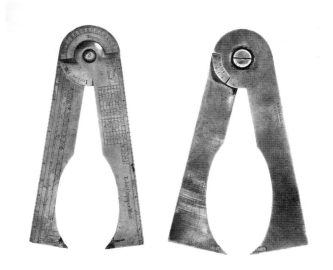

Fig 60A & B
A Gunner's callipers, by E Culpeper, London, 1700-25
B Gunner's callipers, North American, 19th century
Both instruments are made from brass, with steel insets at the tips of each limb. The North American example was used at the United States arsenal founded in 1812 at Watervliet, in New York State.
(Royal Armouries, HM Tower of London. Lo1.3 & 5)

Fig 61
Gunner's level, by C Treschler (the elder), Dresden, 1617
Gilded brass, with decoration typical of continental work of the period.
(Victoria and Albert Museum, London. Lo25.66)

England the *circumferentor* was normally preferred. During the second half of the 18th century some French makers constructed graphometers with telescopic sights in place of the slit and window sights of the earlier design.

Gunnery instruments (figs 60-62; see also fig 42)

From the Renaissance onwards, instruments designed to serve gunners formed an important specialised group. They range from small quadrants with a plumb bob and rod to be inserted in the mouth of the cannon or equipped with a curved base sighting device, to one or more clinometers or levels. Other instruments used by gunners were rules engraved with scales which permitted the estimation of weight of shot of varying materials from its diameter. Callipers permitted the measurement of shot or the bore of the cannon and were equipped with similar scales. Such callipers were constructed well into the 19th century.

Lit: KIELY (1947) for some gunnery instruments of the Renaissance; BION (1972) for the 18th century.

Fig 62
Gunner's quadrant, German, 1585
Made in the form of an axe, whose 'blade' is calibrated to give the elevation of the ordnance in whose mouth the 'handle' is inserted. Made for Duke Julius of Braunschweig and Lüneberg.
(Royal Armouries, HM Tower of London. Lo1.1)

Holland circle
see **Circumferentor**

Hydrometer (figs 63-66)

An instrument used for measuring the specific gravity or density of a liquid. The hydrometer was described by Andeas Libavius, *c*1600. Early hydrometers were made of glass, ivory, silver, copper or wood, and took the form of a bulb surmounted by a stem marked with graduations or a reference point. The bulb was filled with mercury or lead shot until it would float in liquid with the stem in an erect position. The extent to which the stem sank into the liquid indicated the greater or lesser specific gravity of one liquid when compared with another (fig 63). In Britain in the second half of the 18th century the hydrometer was considerably improved in order to meet the needs of the Excise for reliable methods for determining the specific gravity of alcohol/water mixtures. Most hydrometers could be adjusted with weights, either to bring a zero mark

level with the surface of the liquid or to read the specific gravity directly from the graduated stem above the weighted bulb. By means of weights and graduations on the stem, such hydrometers could be used for a wide range of specific gravities. In the 19th century the range of uses to which hydrometers were put increased widely.

The more important types of hydrometers are:

1 Fahrenheit's: differed from the earliest hydrometers in the addition of a tray to take weights. The stem was ungraduated but bore a single reference mark.
2 Clark's: *c*1730. Weights screw on to the bottom of the instrument.
3 Dicas': Patented 1780 (British Patent no. 1259). Brass with ovoid float. Lower rod with counter weight, upper rod graduated from 1-10, and equipped with a point to take weighted tokens. A slide rule supplied with the hydrometer permitted readings to be normalised to a standard temperature.
4 Nicholson's: Described 1789. Pans above and below a spherical or pear-shaped body. Weights added to bring zero point level with the surface of the liquid.
5 Guyton's gravimeter: A French design similar to Nicholson's hydrometer. The lower pans take the weights, the globe is of glass. (fig 64)
6 Sikes': Combined use of weights with a graduated stem to cover a wide range of specific gravities. Adopted as a standard by Parliament in 1816 and 1818. A slide rule or a book of tables was used for normalisation. Continued in official use until the 1970s.
7 Twaddel's: Used mainly in Britain. Invented during the early part of the 19th century. Several hydrometers are provided to give a range of use.

Fig 64
Guyton's gravimeter, ? French, c1800
Made to the design of Guyton de Morveau. In contrast to Nicholson's hydrometer of 1789, the flotation bowl is glass rather than brass. Weights are added to the pan below the bowl to set the fiducial point on the scale level with the surface of the liquid.
(Department of Natural Philosophy, University of Aberdeen. A4.33)

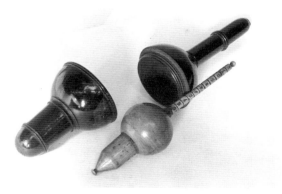

Fig 63
Hydrometer, ? English, first half 18th century
Wooden hydrometer with case. The stem is graduated, apparently to an arbitrary scale. The instrument is counter-weighted with lead below the flotation bowl, and floats higher or lower according to the density of the liquid in which it is placed.
(Department of Natural Philosophy, University of Strathclyde, Glasgow. G4.44)

During the late 18th and early 19th century numerous other variations of these main types were produced (examples can be found in *Science Museum, London*). Later 19th-century hydrometers became standardised as a hollow glass instrument weighted with mercury, and with stem graduated with scale devised by Gay-Lussac, Baume, Cartier, Beck, Tralles, etc. Related instruments were developed for special functions, for example the lactodensimeter, urinometer, acetometer (for measuring the strength of acetic acid) and saccharometer. (fig 65)

8 Specific gravity, or philosophical beads: (fig 66) Specific gravity could also be measured by means of glass bubbles or beads. These were supplied in sets which covered a range of specific gravities at a particular temperature. They were reputedly invented about 1757 by Alexander Wilson (1714-86) of Glasgow.

Lit: SCARISBRICK (1898); TATE (1930)

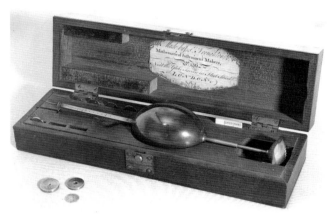

Fig 65
Saccharometer, Richardson's type, by J & E Troughton, London, c1790
A hydrometer dedicated to ascertaining sugar content, and made to the design published in John Richardson's Philosophical Principles of the Science of Brewing (York 1788). Weights are added to the top of the stem to make the instrument float at the fiducial point.
(Science Museum, London. Lo22.3)

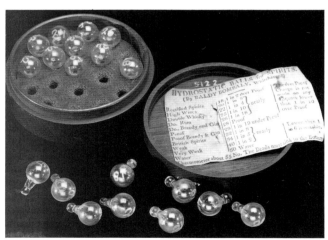

Fig 66
Specific gravity beads, by B Bombaly, Whitehaven, 1800-25
Each individual glass bead has been adjusted to a particular density. For example, that marked '29' should 'swim' in proof brandy, neither sinking to the bottom, nor floating on the surface.
(Science Museum (Wellcome Museum of the History of Medicine), London. Lo22.3)

Hydrostatic balance
see **Balance**

Level, surveyor's (**fig 67-68**)

Variations on the plumb-bob were the most frequently used levelling instruments during the Renaissance (see *Gunnery instruments*). However, at the same time, instruments of considerable complexity were developed, notably by Walter Ryff (1547), whose level consisted of a long wooden bar in which a trough for water had been cut, equipped at each end with sights and plumb bobs. A further plumb bob was mounted above the central points of the alidade (illd. KIELY (1947), fig 52). One of the oldest surviving instruments incorporating a water level is in the *Museum of the History of Science, Oxford*, and was made by Erasmus Habermel, *c*1600 (JOSTEN (1955) pl. XV).

1 Spirit level: The spirit level, making use of the bubble, is ascribed to the Frenchman Thevenot, and is described in a pamphlet dated 1666.
2 Level with telescopic sights: Introduced shortly after the middle of the 17th century, and initially combined with a plumb bob. Variations were described by Huygens (1660) and Picard (1669). The chief problem in the construction of levels with telescopic sights lay in the difficulty of constructing good telescopes of small size.
3 Sisson's level: An instrument combining a spirit level and a telescopic sight. Invented before 1725.
4 Troughton's improved level: Developed before 1812 by Edward Troughton. It was easier to use than the levels of Adams and Ramsden which preceded it.
5 Dumpy level or Gravett's level: Introduced *c*1828. Constructed initially by Troughton and Simms after a design by W Gravatt (1806-1866).

Lit: KIELY (1947) pp21, 54, 129; RICHESON (1966) p135; DAUMAS (1972) p55

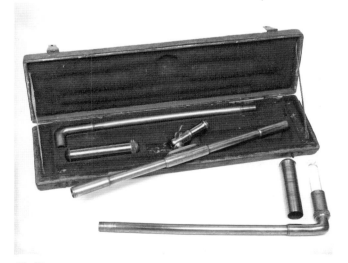

Fig 67
Water level, by Lennel, Paris, 1774
A late example of the surveyor's water level, which was by this date largely superseded by the bubble level.
(Royal Museum of Scotland (Chambers Street) Edinburgh. E8.II.19)

Fig 68
Surveyor's level, by B Cole (junior), London, third quarter 18th century
The sighting telescope is hinged to the base plate and set horizontal against the bubble level by adjusting the levelling screw.
(Herbert Art Gallery and Museum, Coventry. C17.2)

Lodestone (figs 69-70)

A piece of magnetite, or magnetic iron ore. The mineral is usually shaped, and equipped with soft iron pole pieces and a keeper. It is normally mounted in brass, but sometimes in silver. The lodestone was used by scientists, seamen and surveyors for magnetising the needles of pocket compasses and sundials. Most surviving examples are 17th or 18th century in date. The use of the lodestone declined rapidly once the art of magnetising steel compass needles became widespread at the end of the 18th century.

With the exception of very elaborately decorated specimens, where the decoration permits a reasonably close dating, many lodestones are difficult to date with any accuracy.

Fig 69
Lodestone, ? Irish, c1735
An oval-section lodestone with decorated brass mounts. Inscribed: The Gift of his Excely Thomas Lord Wyndham, Baron of Finlas, Lord Chancellor and one of the Lord Justices of Ireland, to Trinity College near Dublin.
(Physical Laboratory, Trinity College, Dublin, Eire. Ei13.1)

Terrella

A spherical lodestone, so named by William Gilbert in the 16th century, since it reproduces in miniature the magnetic properties of the Earth. Sometimes terrellae were stored in a fish-skin case; on other occasions they were mounted in brass or silver with end caps which gave the terrella an oval shape.

Lit: MADDISON (1969) pp17, 40, figs 12, 13, 37; LONDON (1971) section 18

Fig 70
Terella, by E Amory, ? London, second half 18th century
Magnetite ore worked into a spherical shape, and so able to demonstrate the magnetic properties of the earth. Steel pole pieces, with silver mounting and suspension ring. Said to have belonged to Henry Cavendish.
(Devonshire Collections, Chatsworth. C11.5)

Mariner's astrolabe (fig 71)

A simple instrument used by seamen for taking the altitude of the Sun or stars as a means of finding latitude. Astrolabes were first recorded as being used at sea during the last decades of the 15th century: both Vasco da Gama and Bartholomeo Diaz took 'astrolabes' on their voyages, which probably consisted of brass or wooden circles graduated at the rim and equipped with an alidade. The earliest dated mariner's astrolabe to survive to modern times (it was, however, destroyed in the Second World War) was of 1540.

Mariner's astrolabes are massive in construction, provided with a suspension ring, graduated in degrees at the rim, and cut away inside the degree scale in order to lower wind resistance. Though the instrument continued in use throughout the remainder of the 16th and the 17th century, relatively few have survived; however the number is increasing as a result of underwater archaeology.

Lit: WATERS (1966); MADDISON (1969); ANDERSON (1972)

Fig 71
Mariner's astrolabe, ? Portuguese (? Lopo Homem), 1555
The scale of this instrument is graduated in zenith distances, possibly indicating Portuguese manufacture. On the reverse is a later ownership inscription of a Dundee shipmaster: Androw Smyton 1688.
(City Museum and Art Galleries, Dundee. D9.1)

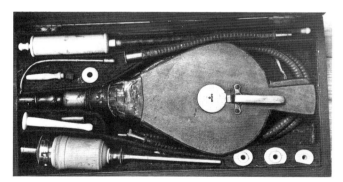

Fig 72
Resuscitation apparatus, by Evans, London, early 19th century
Bellows and related accessories for inflating the lungs of those apparently drowned.
(City Museum and Art Gallery, Plymouth. P3.14)

Microscope (figs 73-89)

Microscopes, both *simple* (with a single lens) and *compound* (with an eyepiece and an objective lens) originated in the early 17th century. However, only a very small proportion of surviving instruments were made before the opening of the 18th century. Some of the more important designs for early microscopes are described below.

Lit: CLAY/COURT (1928); DISNEY/HILL/WATSON-BAKER (1928); BRADBURY (1967, 1968); TURNER (1969, 1972, 1981); BRACEGIRDLE (1978, 1983).

Simple microscope (figs 73-77)
This term is used to cover all microscopes which have a single magnifying lens unit. Figure 73 shows a *'flea glass'* in which the object to be viewed was impaled on a spike.

Fig 73
'Flea glass', English, early 18th century
Turned cylinder of ebony, with a lens mounted in ivory and a specimen spike. The domed lid screws on to protect the lens and point.
(Castle Museum, Norwich. N4.1)

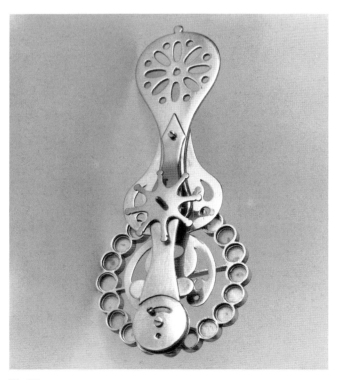

Fig 74
Simple microscope, ? German, early 18th century
Simple microscope of brass, with a rotating wheel of 20 objects and 2 objective lenses mounted on a swivel plate.
(Institute of Ophthalmology, University of London. Lo10.1)

Simple microscopes were often provided with an object holder containing a number of different objects and two or more lenses of differing magnifications (fig 74) (see also fig 43). Special forms of the simple microscope were used for botanical work and dissection during the late 18th and early 19th centuries.

Aquatic microscope. This is a simple microscope intended for wet preparations and living objects in water. A glass tube held the water sample with its living organisms, or a brass plate received a living fish for the study of the circulation of the blood in its tail. An early aquatic microscope was made and used by the Dutch microscopist Anton van Leeuwenhoek (c1689). An English instrument of a similar type is in the *Castle Museum, Norwich* (fig 75). Later aquatic microscopes consisted of a simple microscope mounted so that the lens could be traversed over a live box on the stage and the behaviour of the organisms in the water readily studied.

Screw-barrel microscope. This is a simple microscope invented c1694 by Nicolaas Hartsoeker and improved by the English optical instrument-maker James Wilson, c1700. The adjustment of the position of the objective relative to the object is accomplished by screwing one tube inside another. The microscope is equipped with a condensing lens. The object slide may be held between two metal plates secured by a spring (fig 76B). The plates are frequently shaped to take a glass tube in which a live specimen could be placed. Early screw-barrel microscopes were frequently made of ivory, later of brass with an ebony or ivory handle, and with fittings to view opaque objects.

Compass microscope. This was specifically intended for the study of opaque objects and was made by James Wilson (among others) in the early 18th century, as well as by continental makers. From the 1740s it was normally

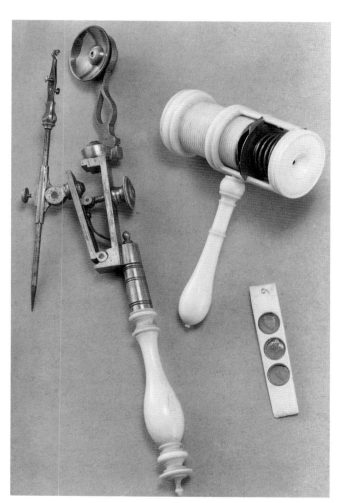

Fig 75
'Aquatic' microscope, by J Yarwell (and E Culpeper), London, c1700
A simple microscope similar in design to Leeuwenhoek's 'aquatic' instrument. The cylindrical specimen tube is intended to hold a small fish; there is a viewing panel so that the circulation of the blood in the tail can be examined.
(Castle Museum, Norwich. N4.6)

Fig 76
A Compass microscope, by B Martin, London, c1760
B Screw-barrel microscope, London, c1720
The simple lens of the compass microscope is mounted within a silvered reflecting mirror or Lieberkuhn used to increase the illumination on the specimen. The single lens ivory screw-barrel microscope follows the design first popularized in England by James Wilson. A slide is inserted between the spring-loaded stage plates, and brought into focus by screwing the threaded 'barrel' in or out.
(Science Museum, London. Lo22.12a)

fitted with a silvered reflector ('Lieberkuhn') to concentrate light onto the specimen. The object is held in forceps or attached to a small stage. The distance between it and the lens is adjusted by a screw working against a spring, or in the simpler earlier models, by direct adjustment of the compass-joint, from which the modern name of the instrument is taken. (fig 76A)

Pocket microscope. During the 18th century makers attempted to produce small pocket microscopes that incorporated a whole range of mechanical 'refinements'. The major design parameter was not optical performance but miniaturization and mechanical ingenuity. (fig 77)

Variations:

Early form. Culpeper-type *c*1725-30. Legs in two discontinuous stages. (fig 78)
Scarlett form, *c*1730-40. Continuous straight legs.
Loft form, *c*1740s. Legs curve outwards below the stage, 1750s onwards: most such instruments constructed of brass. Rack focusing introduced by John Cuff, the all-brass instrument continued to be constructed well into the 19th century when it was called the *three pillar microscope.* (fig 79)

Fig 77
Pocket microscope, by J Clarke, Edinburgh, c1750
A simple microscope in silver, to the design advertised by John Clarke in 1749 as being made for sale on a subscription basis. The instrument and its accessories all pack into a small shagreen-covered case.
(Royal Museum of Scotland (Chambers Street), Edinburgh. E8.II.12)

Compound microscope (figs 78-89)
The compound microscope mounted on a tripod stand above a stage illuminated from beneath, superseded the compound microscope mounted on a pillar which had remained in production throughout the first two decades of the 18th century. A round or rectangular base (the latter with a drawer to house accessories) carries the turned brass legs which support the stage. Three further legs connect the stage with a brass or wooden ring which supports the outer microscope tube. A second tube, sliding within the outer sleeve, carries the eyepiece and objective, and permits focusing.

Fig 78
Compound microscope, by E Culpeper, London, c1720
An early form of the genre now frequently called 'Culpeper-type'. The body tube is covered in tooled leather. In later examples polished and stained shagreen is used, whilst the stage and body are supported by three continuous legs.
(Science Museum, London. Lo22.12a)

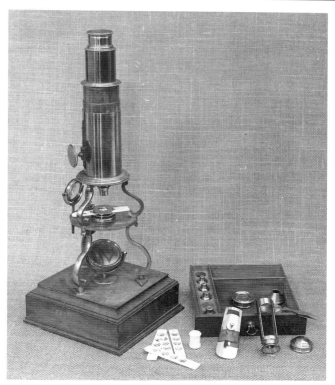

Fig 79
Compound microscope, by J Long, London, c1780
*A late 'Culpeper-type' microscope, in brass. The optical body and the
stage are supported on scroll legs. The sliding focus of the earlier design
has been replaced by a mechanical system using a rack and pinion.
(Museum of Science and Engineering, Newcastle-upon-Tyne. N2.1)*

Drum microscope. This was developed in England by
Benjamin Martin shortly after 1738. His instrument, constructed of pasteboard, consisted of one tube sliding inside
another which served as the base, cut away on one side to
permit light to fall on a mirror mounted at the lower end.
The outer tube was covered in ray-skin. Few early drum
microscopes survive; there are examples in the *Science
Museum, London* and the *Museum of the History of Science,
Oxford*. Brass became the preferred material of construction and the drum microscope continued to be made well
into the 19th century (fig 80). On the continent Oberhauser
and Nachet produced distinctive mid-19th century versions of the design.

Copies of English 18th-century microscopes – tripod,
solar and drum – were produced in a very simplified form
in Germany at the end of the 18th and in the early years of
the 19th century. They were normally made of pearwood
and cardboard, with paper-covered tubes. The branded
initials of makers (IM – Junker of Magdeburg – and JFF) are
sometimes found under the base. Such instruments have
frequently, and probably incorrectly, been said to have
been constructed by the toymakers of Nuremberg, hence
the name *Nuremberg microscope*. (fig 81)

In the 18th century the compound microscope, like the
simple microscope, developed mechanically rather than
optically. There were a number of design variations. In
England, one of the most enduring began with the *'varia-*

Fig 80
Drum microscope, by A Abraham, Bath, c1830
*A late, all-brass example of the drum microscope, with rack focusing. The
forerunner of this design is the pocket microscope designed by Benjamin
Martin in 1738, made from paste-board, with the outer body covered in
polished shagreen.
(Museum of the Pharmaceutical Society of Great Britain, London.
Lo14.6)*

ble' microscope designed by George Adams the elder (c1770)
(fig 82). The unwieldy rack and pinion inclination of the
optical unit was replaced by a compass joint in the *'most
improved' microscope* (fig 83). This instrument was widely
made in the first four decades of the 19th century until the
vastly superior performance of the achromatic microscope led to its demise. It is frequently found as the major
part of a large boxed set, with many accessories, including,
for example, a compass microscope. A small and essential-
ly portable compound microscope was developed by the
instrument-maker William Cary in the 1820s. For a compound microscope the *Cary type* (fig 84) is very small,
approximately 6in in height when screwed into the lid or
lockplate of its box.

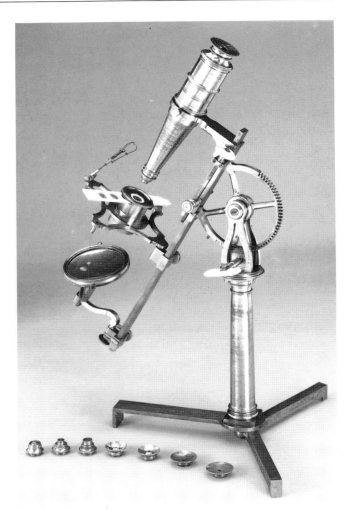

Fig 82
'Variable' microscope, by G Adams (elder), London, c1770
The pillar on which the optical elements and the stage are mounted can be inclined by the impressive but unwieldy rack and pinion mechanism, the design of which was attributed to the Third Earl of Bute.
(Science Museum, London. Lo22.12a)

Fig 81
'Nuremberg' microscope, German, early 19th century
Compound microscope of pearwood and cardboard. The outer body tube is covered in embossed paper.
(Museum of Science and Industry, Manchester. M3.III.2)

Solar microscope. This is a form of projection microscope, developed c1740 from the camera obscura, it permits the image produced by a microscope to be thrown upon a wall or screen, using the Sun as the light source and a screw-barrel microscope as the projection unit (fig 85). Mid-18th century solar microscopes were constructed of wood and brass, later ones entirely of brass. In the *lucernal microscope* (fig 86) an artificial light source was used to project the image on to the ground glass viewing screen.

Reflecting microscope. This was first constructed in 1737, on lines similar to a Gregorian reflecting telescope. In the second decade of the 19th century interest in the reflecting microscope revived as a result of the failure to produce satisfactory high-power achromatic lenses for microscopes. Giovanni-Baptista Amici of Modena produced (c1827) a successful reflecting microscope (fig 87), while in England a similar instrument was developed by John Cuthbert. The success of this design was shortlived; rapid improvements in the design of the achromatic lens system caused it to pass into disuse.

Achromatic microscope. The use of the achromatic objective for telescopes became widespread following John Dollond's patent of 1759. Attempts to apply the principle to the much smaller microscope objective lens did not yield any real results until the early decades of the 19th century. Once empirical achievement was married to JJ Lister's theoretical work on high-power achromatic doublet and triplet objectives for microscopes (1830), a major step forward was made. Objectives corrected for achromatic and spherical aberration gave significant increases in image quality and magnification. To match the improved optical performance, the design and construction of the microscope was also changed; it became sturdier, and was fitted with fine focus adjustment and a mechanical stage (fig 88). The binocular microscope, using the optical system developed by Wenham in 1860, gave an excellent stereoscopic effect when used with relatively low-power objectives. The single Wenham prism used as a beam splitter was, however, readily removable, so that the binocular instrument could be used as a monocular at higher magnifications. Subsequent binocular designs for use at high magnifications had no stereoscopic effect, but reduced fatigue on the microscopist. (fig 89)

Fig 84
'Cary type' microscope, by W Cary, London, c1825
A portable compound microscope incorporating a rack focusing stage. William Cary's design has provided a generic name, though others, notably Charles Gould and Henry Coddington, designed similar instruments.
(Science Museum, London. Lo22.12a)

Fig 85
Solar microscope, by J Cuff, London, c1750
An all-brass version of the design introduced into England in 1740 by the German physician Johann Lieberkühn.
(Science Museum, London. Lo22.12a)

Fig 83
'Most Improved' microscope, by Dollond, London, c1800
The stage and optical parts of the 'most improved' design of microscope are set on a rectangular sectioned limb, incorporating a rack and pinion motion for the stage. The whole can be inclined to the stand to which it is attached by a compass joint. Most instruments survive with their mahogany case and set of accessories that allow setting up as a 'single, compound, opake and aquatic microscope'.
(Science Museum, London. Lo22.12a)

Fig 86
Lucernal microscope, London, late 18th century
Made to the design illustrated by George Adams in his Essays on the Microscope (London 1787), but with an oil lamp rather than an Argand lamp to provide the illumination. By projecting the brightly lit image on to a ground-glass screen at the end of the mahogany body, the specimen could be examined by several observers.
(Science Museum, London. Lo22.12a)

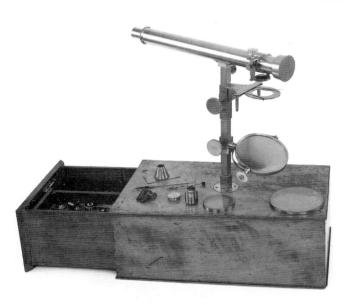

Fig 87
Reflecting microscope, by G-B Amici, Modena, c1830
By using a metal speculum as the magnifying element, Amici aimed to avoid the chromatic distortions then inherent in high magnification compound microscopes. This instrument was used by W H Fox Talbot, the photographic pioneer.
(Royal Museum of Scotland (Chambers Street), Edinburgh. E8.II.12)

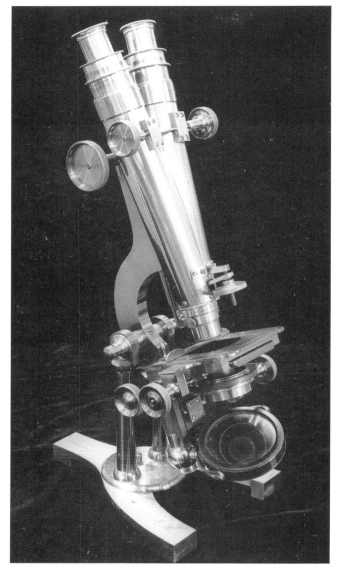

Fig 89
Achromatic binocular microscope, ? by J B Dancer, Manchester, c1870
Though signed by Dancer, this instrument was probably made by the London firm of Smith, Beck and Beck. The basic design incorporates a typical 'Lister' limb following the pattern introduced by Ross in 1839, apparently following J J Lister's suggestions. The binocular ocular system has the advantage of lessening fatigue during long periods of observation. However, not all binocular optical systems gave good stereoscopic vision at high magnifications.
(Museum of Science and Industry, Manchester. M3.14)

Fig 88 (left)
Achromatic microscope, by Powell & Lealand, London, 1846
Made to the design first published in 1843, the Powell & Lealand number three stand is particularly characteristic of the work of the partnership, and generally of the 'bar-limb' construction. Typical of the best practice of the era, the radically improved achromatic optical system is set on a sturdy tripod base, slung from trunnions. There is a mechanical stage, and to complement the well-engineered coarse focus of the whole optical section, there is a micrometer-screw fine focus of the objective.
(Science Museum, London. Lo22.12a)

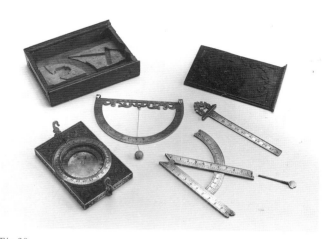

Fig 90
Mining instruments, Tyrolean, 1748
Pearwood, ivory, and brass. Inscribed: DAP VPH. The instruments, in a contemporary wooden case, consist of a rectangular compass with suspension hooks, a clinometer, a jointed square with divided arc, and a divided rule, which, used in conjunction with the jointed square, was used to set up a triangle similar to the one being surveyed.
(Royal Museum of Scotland (Chambers Street), Edinburgh. E8.II.13)

Mine surveying instruments (figs 90, 91)

The earliest mine surveying instruments to survive are from the Central European mining areas of the Tyrol. Initially the miner's compass consisted simply of a magnetic compass with a characteristic division into 24. By the end of the 16th century miner's compasses had been evolved which were equipped with suspension hooks, and are often associated with other accessories similar to those illustrated in fig 90. These accessories remained standard from the early 17th until the middle of the 18th century, and the set of instruments was normally contained in a simple wooden carrying box.

By the 19th century mining instruments ranged from a simple circumferentor with folding open sights and a spirit level, to a compass with vertical semicircle over which moves an alidade with sights and bubble level, Lean's dial (fig 91).

Lit: MICHEL (1956); KUHNELT (1962); SCOTT (1899-1985)

Models (figs 92-97)

Models were frequently used in lecture demonstrations in the 18th and 19th century. One of the most popular early

Fig 91
Lean's dial, by E T Newton, St Day, mid-19th century
A form of miner's dial in frequent use during the middle decades of the 19th century.
(Egestorff Collection, Dublin, Eire. Ei7.78)

Fig 92
Model of Murray's optical telegraph, late 18th century
The optical telegraph designed by the Revd Lord George Murray was adopted by the Admiralty in 1796.
(Scottish United Services Museum, Edinburgh. E10.1)

texts in natural philosophy, JT Desaguliers' *Course of Experimental Philosophy* (1734), devotes a great deal of space to the description of useful machines. Models are to be found in cabinets formed by wealthy amateurs and used at universities and other educational institutions. The King George III Collection at the *Science Museum, London* is the most extensive group to survive in the British Isles.

The range of models is very wide. Models may represent real systems, such as that of Murray's optical telegraph (fig 92), or of the steam engine (fig 93 and 94), or be used to demonstrate basic principles such as those of the pulley and lever (figs 95 and 96). Optical models, such as the model eye (fig 97), are less usual; examples incorporating ray diagrams are known from the post-1850 period.

For astronomical models see *Orrery* (figs 104-106).

Fig 94
Model beam engine, by W & S Jones, London, 1825-50
A working model in brass, with a mahogany casing. Intended to demonstrate the mechanical function of the parts of a rotative beam engine.
(Museum and Art Gallery, St Helens. S5.1)

Fig 93
Model Newcomen engine, by J Sisson, London, mid-18th century
A scale model, unsigned but supplied to the Department of Natural Philosophy in the University of Glasgow by the instrument-maker Jeremiah Sisson. In 1763, as a result of attempting to put this model into working order, James Watt first took an active interest in the design and manufacture of steam engines.
(Hunterian Museum, University of Glasgow. G5.1)

Fig 95
Pulley frame, by W & S Jones, London, early 19th century
Mahogany frame with pulleys and weights of brass. There are several arrangements of pulley blocks to demonstrate the different mechanical advantages that can be obtained.
(Department of Natural Philosophy, University of Aberdeen, A4.18)

Fig 96 (left)
Model to demonstrate the properties of levers, by W & S Jones, London, early 19th century
Brass levers and weights, mounted on a mahogany frame. The sovereign balance is a later addition.
(Department of Natural Philosophy, University of Aberdeen, A4.18)

Fig 97 (below left)
Model eye, by W & S Jones, c1794-5
Brass with glass lenses. The curvature of the 'eye' lens is such that rays of light from an object placed in front of the model are focused on the retina. The two additional lenses, one convex and the other concave, mounted below the eye, demonstrate the corrections which spectacles can provide to optical defects of the eye.
(Department of Natural Philosophy, University of Aberdeen, A4.25)

Fig 98
Napierian rods, ? English, late 17th century
Set of ivory rods, with tabulet, and wooden carrying case. The use of ivory led to the appellation 'Napier's bones'.
(Museum of the History of Science, Oxford. O4.II.7)

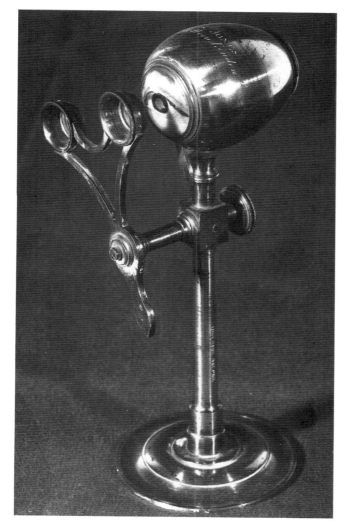

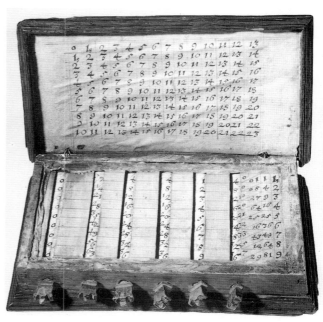

Fig 99
Napierian rods on cylinders, ? Scottish, c1700
Made to the mid-17th century design of Gaspar Schott. The 'rods' are calibrated in ink on paper. Note the addition table in the lid.
(Royal Museum of Scotland (Queen Street) Edinburgh. E9.3)

Napierian rods or Napier's bones (figs 98-99)

This calculating device was invented by John Napier of Merchiston, the inventor of logarithms, and described in his *Rabdologiae*, published in 1617. The instrument consists of 10 small four-sided wooden or ivory rods or 'bones', numbered from 0 to 9, and contained in a flat case. Each rod is divided by lines into nine equal parts, containing two columns of numbers, separated by a diagonal line, which correspond to the first nine multiples of the number at the head of the rod. By laying side by side the rods whose numbers correspond to the digits of the number to be divided or multiplied, it is possible to obtain the answer by means of simple addition (fig 98)

A cylindrical, later form of Napierian rods was made (fig 99). In this case the cylindrical rods are mounted side by side in a box, and can be revolved by means of small projecting knobs.

Most surviving sets of rods date from the 18th century.
Lit: MICHEL (1967) p48; WILLIAMS (1983)

Nocturnal (fig 100)

An instrument developed in the 16th century for determining the time at night by the observation of the apparent rotation of the Great Bear, and sometimes of the Little Bear, about the Pole. In its commonest form it consists of a disc with a handle, at whose centre a smaller disc and an index are pivoted. The small disc is set so that its pointer corresponds to the date of observation on the calendar scale marked on the larger disc. The Pole Star is then sighted through the hole in the centre of the disc, and the index arm is moved until it cuts a particular star in the Great Bear or Little Bear. Where the index arm passes over the hour scale, the correct time can be read off.

Nocturnals are common from the mid-16th until the 18th century. (See also fig 115B for planispheric nocturnal.)
Lit: TAYLOR/RICHEY (1962) pp60-63; MADDISON (1969) pp30-35

Nooth's apparatus (fig 101)

The chemist Joseph Priestley (1728-1804) invented an apparatus to dissolve carbon dioxide gas in water, and published a description of it in 1772. It was difficult to use, and an improved version consisting of three glass vessels mounted vertically was produced by John Mervin Nooth (1737-1828). The gas was produced by reacting marble chips with oil of vitriol (sulphuric acid) in the lowest vessel, water or a solution of mineral salts into which the carbon dissolved was contained above a valve in the middle vessel, and the uppermost bulb provided a head of water so the gas was under slight pressure.

Fig 100
Nocturnal, ? German, 1531
Brass nocturnal with lunar volvelle and aspectarium.
(Whipple Museum of the History of Science, University of Cambridge. C6.II.14)

Fig 101
Nooth's apparatus, English, late 18th century
From the collection of apparatus formed for King George III. Though unsigned, this apparatus was probably made and sold by W Parker, whose cut-glass manufactory was in Fleet Street, London.
(Science Museum, London. Lo22.3, 5)

Such apparatus was used to produce simulated spa waters. It was found in many homes in the late 18th and early 19th centuries, and was manufactured in large numbers especially, it seems, by the firm of Parker of Fleet Street, London. Few now survive.

Lit: ZUCK (1978)

Octant (figs 102-103)

The octant, or Hadley's quadrant, was named after its inventor, John Hadley, who patented the design in 1734. Within about twenty years it largely replaced the *backstaff* for measurement of the altitude of the Sun or stars in navigation. Angles between two objects were measured by bringing the double reflection of the stellar object in the index and horizon mirror into coincidence with the sighted image of the horizon. Because of the double reflection, the arc of the quadrant, though only 45°, is graduated into 90° (i.e. an eighth of a circle, hence *octant*). Early octants were made of mahogany, with a boxwood arc subdivided with transversals. The index arm was partly or entirely made of wood. During the period 1760-1780 both mahogany and ebony were in use. Most instruments after 1800 were constructed of ebony with inlaid ivory scales. There are few surviving octants with metal frames before 1850. The vernier, which superseded the diagonal scale, can be a useful guide to date. A vernier with the zero at the central point was used until about 1780. Thereafter a vernier with the zero to the right was used.

As a general rule the early octant tends to be of greater radius than the later specimens, but as late as the end of the 18th century some of the cheaper octants were constructed of mahogany, with a diagonal scale, and a radius of 13 or 14in.

The octant was superseded by the *sextant*, and finally went out of use at the end of the 19th century.

Lit: HILL/PAGET-TOMLINSON (1958) p13ff; COTTER (1968) p77ff; LONDON (1971) section 22; MAY (1973) p140; COTTER (1983) pp109-158

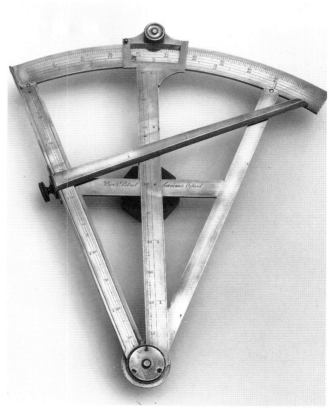

Fig 103
'Crow's Patent Seaman's Octant', English, post-1832
This brass device provides an instrumental solution to common trigonometrical problems encountered in navigation. It was patented by Francis Crow in 1832.
(Guildhall Museum, Rochester. R2.3)

Orrery (figs 104-106)

The orrery is both the commonest and the most complex of a range of astronomical models used to demonstrate astronomical phenomena. It is a hand- or clockwork-driven mechanism for demonstrating the motion of the heavenly bodies – the planets around the Sun, and the movements of the satellites around the planets. Orreries were particularly popular in England during the 18th and early 19th centuries. The first orrery seems to have been made by George Graham and Thomas Tompion *c*1709, but the name derives from a similar instrument made for Lord Orrery by John Rowley, *c*1712. The early orreries were costly and elaborate, but by the end of the 18th century

Fig 102
Octant, by Watkins, London, late 18th century
Ebony frame with ivory scale and brass index arm with vernier and slow-motion screw. Typical of later models, there is a back sight.
(Science Museum, London. Lo22.10)

relatively cheap instruments were being produced, either all of brass or with brass mechanisms, while the base was of wood covered with engraved paper. Later instruments, as well as a *planetarium* demonstrating the movements of the planets, also included a *lunarium* (demonstrating in detail the movements of the Moon) and a *tellurium* (demonstrating only the annual and diurnal motions of the Earth). Both were introduced by Benjamin Martin, *c*1760-70.

With the exceptions of the *terrestrial* and *celestial globe* and the *armillary sphere*, astronomical models other than the orrery are quite rare; surviving examples are found in collections such as those of George III in the *Science Museum, London*.

The *cometarium* demonstrates how the speed of a comet varies in its orbit according to Kepler's law, sweeping out equal areas in equal time periods. The example illustrated (fig 106) was formerly part of the apparatus of the Natural Philosophy Department of the University of Edinburgh.

Lit: MADDISON (1958); MILLBURN (1972-73); TURNER (1973A); KING/MILLBURN (1978) is the definitive work.

Fig 104
Grand orrery, London, mid-18th century
The design is typical of the grand orreries made by Thomas Wright, Benjamin Cole, and George Adams.
(Science Museum, London. Lo22.2)

Fig 105
Orrery, by G Adams (younger), London, c1790
This style of orrery was introduced by Benjamin Martin. On this example the outermost planet is Uranus, discovered by William Herschel in 1781, and named by him 'Georgium Sidus'. It is shown with two satellites, first seen by Herschel in 1787. Three separate demonstrations are possible; as here the planetarium illustrates the relative motions of the planets of the solar system. These fittings can be replaced with a tellurian to show the rotation of the Earth on its inclined axis, or with a lunarium to demonstrate the motion of the Moon round the Earth.
(Science Museum, London. Lo22.2)

Fig 106
Cometarium, by J Miller, Edinburgh, late 18th century
This instrument was originally part of the teaching apparatus of the Department of Natural Philosophy, University of Edinburgh.
(Royal Museum of Scotland (Chambers Street) Edinburgh. E8.II.2)

Pantograph (fig 107)

A drawing instrument used for enlarging or reducing maps or drawings. Early examples were often constructed of mahogany, later ones of brass. Surviving instruments are mostly of late 18th century or early 19th-century date, although the instrument was invented in the early 17th century by Christoph Scheiner; it was also known as a parallelogram.

During the 19th century several attempts were made to improve the pantograph. They included John Dunn's 'improved' pantograph, and the 'eidograph' invented by Professor William Wallace of the University of Edinburgh.

Lit: WALLACE (1836); DICKINSON (1956)

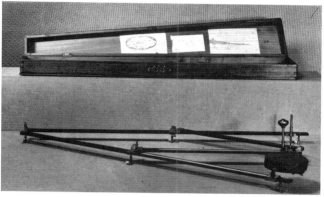

Fig 107
Pantograph, by B Cole, London, mid-18th century
The limbs are made of ebony, with scales inlaid on ivory. In later examples the whole instrument is made of brass. Note the maker's trade label in the lid of the mahogany case.
(City Museum and Art Gallery, Birmingham. B11.16)

Fig 108
Pedometer, by Fraser, London, 1788/9
A typical pedometer, in watch form. The case is hallmarked silver, the dial enamel.
(Museum of the Clockmaker's Company, Guildhall Library, London. Lo12.19)

Pedometer (fig 108)

An instrument for measuring walking distances. A dial records the revolutions of the mechanism which are actuated by a cord connected to the foot; i.e. the number of steps, rather than the distance walked. The pedometer achieved its most elaborate and decorative form in the German Renaissance, but also occurs frequently in watch form, operated by an internal pendulum, in the late 18th and early 19th centuries, when it was constructed by makers such as Spencer and Perkins, Fraser, and B Gray.

Perpetual calendar (fig 109)

The perpetual calendar embodies in a convenient form a body of pre-recorded calendrical information which can be read directly without the need for any computation. Information can include lists of Saint's days, and dates of birth of monarchs; but more usually it is restricted to astronomically related data – day, date and month and the related time of sunrise and sunset, position of the Sun in the zodiac. The perpetual almanac, in contrast, reduces cyclic calendrical information to a minimal tabulation, which by following certain rules allows key dates to be computed. Perpetual calendars and almanacs are often found associated with portable sundials – especially those made in Germany.

Fig 109
Perpetual calendar, ? German, 18th century
Fire-gilt metal. The small knobs permit discs set on the plate to be revolved. Thus the position of the Sun in the zodiac for any particular day can be read off, also the phase of the Moon, length of day and night, and the time of sunrise and sunset.
(The National Museums and Galleries on Merseyside, Liverpool. L8.54)

Proportional compass (fig 110)

The proportional compass is used to make reduced or enlarged copies of drawings. Fixed proportional compasses were known in classical times (i.e. compasses with sets of points at both ends of the arms, the distances between the two sets being in a fixed ratio to one another). In the second half of the 16th century attempts were made to devise a proportional compass where the ratio between the two sets of points could be adjusted. Jacques Besson produced one such instrument, but the form in which it has survived was devised by Jost Burgi and described by Levinus Hulsius in 1604.

In Burgi's instrument the two arms are pivoted by means of a clamping screw moving in an elongated slit between the legs, permitting any proportion between the ends to be attained. The arms are usually provided with scales of proportion for lines, solids, cubes and circles.

Lit: DICKINSON (1956) p77; SCHNEIDER (1970); ROSE (1968)

Fig 110
Proportional Compass, by N Witham, early 18th century
Typical adjustable proportional compass in brass with steel points.
(Art Gallery and Museum, Leamington Spa. L2.2)

Pyrometer (fig 111)

A pyrometer measures temperatures by the expansion(and contraction) of a solid. In practice, before electrical means of recording, this was difficult to achieve with any accuracy, and most 'pyrometers' should more strictly be called dilatometers, a device for demonstrating the property of expansion of one or more solids, usually in the form of metal rods.

Quantitative experiments on the expansion of metals were first carried out in the early 18th century; among the early workers were George Graham (1673-1751) and John Harrison (1693-1776), the latter known for his temperature-compensating gridiron pendulum which uses the property of differential expansion of two metals. Petrus van Musschenbroek (1692-1761) introduced the term 'pyrometer' in 1731. In 1735, Cromwell Mortimer (? 1698-1752) attempted to use the property of thermal expansion of metals to measure temperature.

The potter Josiah Wedgwood developed an instrument, described in 1781, which used the property of contraction of ceramic pieces to measure the temperature of furnaces. The pieces were pushed along wedge-shaped graduated grooves; the point where they stuck was an indication of temperature.

Lit: CHALDECOTT (1969, 1976)

Fig 111
Pyrometer, or dilatometer, by J Dunn, Edinburgh, c1830
A gas burner along the length of the instrument heats the metal rod, the expansion being indicated on the scale.
(Royal Museum of Scotland (Chambers Street), Edinburgh. E8.I.1)

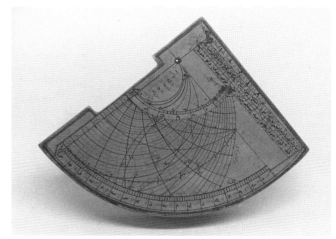

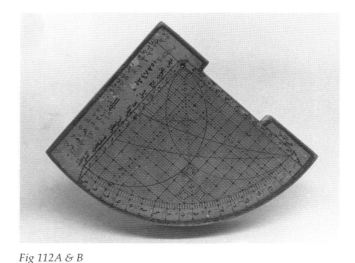

Fig 112A & B
Horary quadrant, Turkish, 18th century
Manuscript scales on paper, lacquered and pasted on wood. Side A is marked with Prophatius' astrolabe quadrant, side B with a sinical quadrant.
(National Museum of Wales, Cardiff. C8.5)

Quadrant (figs 112-117)

Several widely differing instruments are grouped under this name. Their main common feature is their shape, since they consist of a flat plate of wood or metal in the form of a quarter circle. In its simplest form the arc of the quadrant is graduated into 90° and the instrument used for the measurement of altitude (*altitude* or *mariner's quadrant*). More commonly the quadrant was designed to perform other functions as well. The most common distinct types are:

Astrolabe quadrant

The reduction to quarter circle of the lines of the normal stereographic astrolabe (cf. *Gunter's quadrant* below). It was described by Prophatius in the 13th century and was used for time measurement and the solution of the astronomical problems. The Prophatius quadrant, with a sinical quadrant on the reverse, is a characteristic Islamicate quadrant. (figs 112A and B)

Horary quadrant

An altitude sundial. The quadrant is equipped with sights on one radius and a plumb-line suspended from the apex. A declination scale permits allowance to be made for the Sun's declination at the time of the year at which the quadrant is being used. Some quadrants could be used in one latitude only (figs 113, 114A and B, 115A and B), others were provided with an adjustable latitude scale, and thus became universal instruments. The quadrant, together with the astrolabe, was the main portable astronomical and time-measurement instrument of the Middle Ages, but retained its popularity in later centuries, when a number of ingenious designs were evolved.

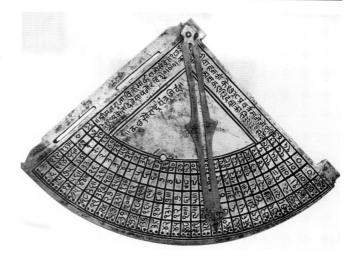

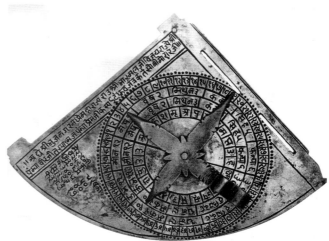

Fig 114A & B
Horary quadrant and nocturnal, Indian, 1814
Known as dhruvabhrama yantra, this Sanskrit instrument was constructed by Sonimoraji in Saurastra, Gujarat. It is designed for latitude 22° 30' N. Side A has shadow length and angle scales. Side B functions as a nocturnal.
(Museum of the History of Science, Oxford. O4)

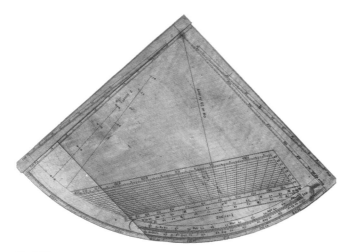

Fig 113
Horary quadrant, English, London, c1672
Boxwood horary quadrant of the type described by William Leybourn in 1672 and called by him 'the Panorganon, or Universal Instrument'.
(Castle Museum, Norwich. N4.31)

Gunter's quadrant

One of a number of distinctive variations of the horary quadrant, developed in England during the 17th century and first described by Edmund Gunter in 1623. It was used for determining time as well as for the solution of astronomical problems, and remained popular until the early 19th century. Frequently the back of Gunter's quadrant is equipped with a rotating disc engraved with an analemma consisting of the orthographic projection of the Roias *astrolabe* marked with the positions of a number of prominent stars, as well as with the outlines of some constellations. The edge of the disc is engraved with a calendar scale over which rotates a graduated cursor. V-shaped projections around the edge of the disc enabled it to be used at night for determining the time, that is it functions as a *planispheric nocturnal*. (fig 115B)

Astronomical quadrant

Renaissance astronomers developed the medieval altitude quadrant in their quest for more accurate measurement. Initially improved precision was related to increased size,

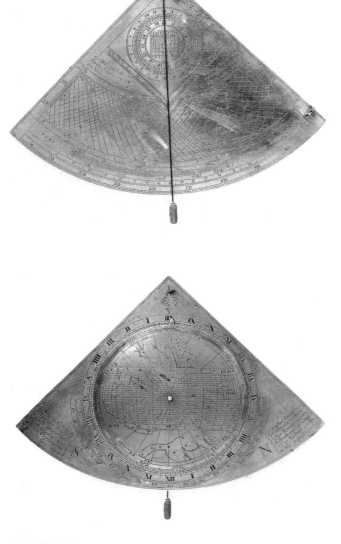

Sinical quadrant

A quadrant engraved with a nomograph of sines and cosines of the angles marked on the arc. It permits a graphical solution of trigonometric problems, and was used in navigation and surveying. The sinical quadrant also occurs as one of the scale on the back of an astrolabe. The sinical quadrant originated in early Islam, and its most frequent occurrence on astrolabes is on those of Islamic origin. It is found on Turkish quadrants of 18th-century date (fig 112B). In Europe a form of sinical quadrant occurs on late Renaissance surveying instruments. It was later used for solving some navigational problems.

Surveyor's quadrant

During the 16th and early 17th centuries quadrants frequently occurred on which the main scale was the shadow square, although an horary quadrant might also be included in a subsidiary position.

Fig 115A & B
Horary quadrant and planispheric nocturnal, by H Sutton, London, 1656
On side A is Gunter's projection for an horary quadrant for use at 51° 30' N, with additionally a perpetual almanac, lunar calendar, and a shadow square. On side B is a planispheric nocturnal.
(Science Museum, London. Lo22.16)

but with the use of telescopic sights and refinements in scale division, relatively small portable quadrants were used for astronomical and related geodetic observation. (figs 116 and 117)

Mariner's.quadrant

An altitude quadrant, as described above, i.e. an instrument in the form of a quarter circle whose arc is graduated from 0 to 90°. It was used for measuring the altitude of a star as a means of determining latitude.

Fig 116
Astronomical quadrant, English, c1700
An equatorially mounted quadrant on an iron and wooden stand. From the collection of the Department of Natural Philosophy, University of Edinburgh. During the 19th century the instrument acquired a false provenance associating it with John Napier of Merchiston, the inventor of logarithms. The design is certainly not of his period, being similar to the quadrant made by John Rowley for Trinity College, Cambridge, c1707.
(Royal Museum of Scotland (Chambers Street), Edinburgh. E8.I.2)

distant landscape or of sunspots) onto the opposite wall of a room. Alternatively a *screw-barrel microscope* could be mounted on the scioptric ball inside the shutter, and the image produced by the microscope thrown on to a screen, allowing the object to be studied by several people at the same time. (See also *solar microscope*.)

Lit: CLAY/COURT (1928) pp214-5; BRADBURY (1967) p153

Fig 118
Scioptric ball, English, 1700-1750
A basic scioptric ball for shutter mounting, turned from ebonised hardwood.
(Science Museum, London. Lo22.12c)

Fig 117
Astronomical quadrant, English, c1820
A three-foot radius quadrant of brass on a heavy braced mahogany base. The design uses a pillar and plate frame so that it is rigid without being excessively heavy (see also Fig 123). This instrument resembles a quadrant by Troughton illustrated in Rees' Cyclopaedia (1819). This particular example belonged to the geologist Henry de la Beche (1796-1855).
(National Museum of Wales, Cardiff. C8.1)

Scioptric ball (fig 118)

A device to provide solar illumination for microscopes and other physical apparatus. It could also be used as a *camera obscura*. It consists of a spherical brass or mahogany ball through which passes a cylindrical aperture into which a long focus lens has been set. The ball is normally mounted in a square or circular wooden framework which can be fixed in an aperture in a shutter. The ball can be turned in any direction within the framework to receive the Sun's rays and direct them into the room. As a camera obscura the scioptric ball would direct the image (of a

Sector (figs 119-120)

An aid to calculation developed in the late 16th and early 17th centuries, the sector continued in use until well into the 19th century. It consists of two flat arms hinged together and engraved with a variety of scales for the solution of problems relating to geometry, surveying, dialling, gunnery, etc. The scales were initially made symmetrical on each arm and include squares, cubes, reciprocals, chords, tangents and metal densities. The solutions are based on the principle of similar triangles, and the instrument is used in conjunction with a pair of dividers.

Mounted on a tripod and equipped with sights, the sector could be adapted to practical surveying. Many late 16th- and early 17th- century surveying instruments were based on similar principles, and incorporate some of the scales which occur on the sector. (fig 121)

There are distinct variations between sectors produced in various European countries – England, France, Germany and Holland. The commonest in British collections are the English and French type of sector.

Lit: CITTERT (1947); WATERS (1958) p356ff; GARVEN (1964); ROSE (1968); SCHNEIDER (1970); DRAKE (1977, 1978)

Fig 119
Sector, by A Sneewins, Delft, mid-17th century
This sector was designed for use by a military engineer; the friction leaf is engraved with tables used to design fortifications.
(Museum and Art Gallery, Leamington Spa. L2.1)

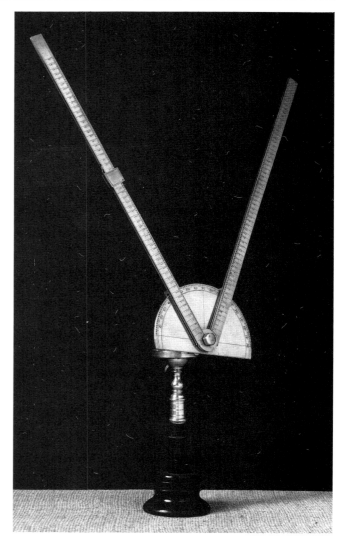

Fig 121
Recipiangle, ? Prague, early 17th century
A gilded brass surveying instrument consisting of a graduated semicircle from whose centre pivot two long arms. The instrument uses the principle of similar triangles, and was one of a number of such triangulation instruments devised during the early 17th century. This particular form of instrument was described by Leonard Zubler in 1607.
(Whipple Museum of the History of Science, University of Cambridge. C6.II.15)

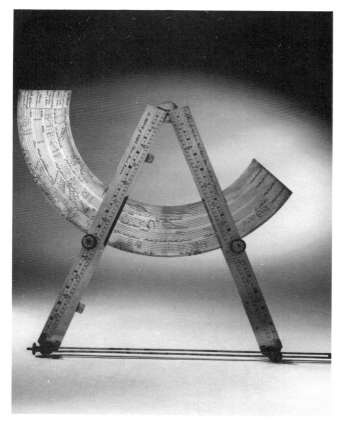

Fig 120
Architectonic sector, by G Adams (elder), London, mid-18th century
Silver, with 12in-radius limbs. The arc, which passes through slits in the limbs of the sector, is engraved with scales for the proportions of the four classical orders of architecture.
(Royal Institute of British Architects (Drawings Collection), London. Lo19.6)

Sextant (figs 122-124)

Evolved from the *octant* in about 1758, the sextant is capable of measuring angles by reflection more accurately than the octant, and in particular permits the measurement of lunar distances. It is normally graduated into 120° or more (fig 122). When the arc is extended to 150° it is known as a quintant (fig 123). In order to achieve greater accuracy, the instrument was normally constructed of brass. Until late in the 19th century the sextant was used by navigators for taking lunar distances as a means of finding longitude and for measuring meridian altitudes, while the less accurate octant was used solely for the latter purpose. See also *circle*.

Box sextant

A small pocket form of the sextant invented in the third quarter of the 18th century and primarily used for surveying. It was normally 3 to 4in in diameter (fig 124). Miniature versions of the sextant were also popular during the late 18th and early 19th century.

Surveying sextant

On occasions the sextant was adapted for surveying, in which case it was often mounted on a stand such as that in fig 20.

  Lit: COTTER (1968) p87ff; LONDON (1971) section 28; MAY (1973) p147; COTTER (1983) pp132-187

Fig 124
Box sextant, by J Allan, London, c1813
The instrument is fitted with a telescopic sight; some merely have a peep sight.
(Royal Engineers Museum, Chatham. C10.2)

Fig 122
Sextant by L Casella, London, mid-19th century
Sextant with ebony frame and ivory scale. There is an assortment of eyepiece telescopes.
(Science Museum, London. Lo22.10)

Slide rule (fig 125)

Although the origins of the slide rule may be traced back as far as the early 17th century (Gunter's logarithmic scale (1623), Oughtred's linear (1633) and circular slide rules (1632) and Delamain's slide rule (1631)), it was only during the latter part of the 17th century that the slide rule began to achieve acceptance. At that stage, however, it was an instrument primarily used by timbermen, gaugers and navigators rather than by engineers, architects or mathematicians. Only during the 19th century did the slide rule finally come into common use; by the middle of that century, with the development of the Mannheim rule (1851), it finally succeeded in displacing the sector.

  Lit: CAJORI (1909, 1920); GARVAN (1964); BAXANDALL (1975)

Fig 125
Excise slide rule, by E Roberts, London, c1775
A 12in boxwood slide rule, for gauging. There are two sliders. The design follows that described by Thomas Everard, in his Sterometrie of 1684. The maker's signature: *EDWd ROBERTS Maker Dove Court Old Jewry LONDON*, is on the stock. More frequently, the signature is to be found under one of the sliders.
(Museum and Art Gallery, Birmingham. B11.18)

Fig 123
Quintant, by E Troughton, London, c1800
Pillar and plate frame, the 150° arc divided on a strip of silver let into the brass. The double-frame sextant was patented by Edward Troughton in 1788, and continued in use until about 1830. The design provided a rigid yet light instrument (see also Fig 117).
(Department of Physics, University of St Andrews. S2.14)

Spectroscope (fig 126)

A spectroscope enables the observation of emitted or absorbed spectral lines. The chemist William Hyde Wollaston (1766-1828) and physicist Joseph von Fraunhofer (1787-1826) discovered in 1802 and 1814 respectively that the solar spectrum was crossed with dark lines. The latter adapted a theodolite for his investigations.

Perhaps the first important work, developing spectrum analysis as a useful technique for chemical investigations, was performed between 1859 and 1861 by RW Bunsen (1811-1899) and GR Kirchhoff (1824-1887) at the University of Heidelberg. Shortly afterwards commercial instruments became available, a number being displayed at the International Exhibition in London in 1862. These simple instruments consisted of a slit at the end of a collimating tube, a prism and a telescope.

Developments towards the end of the 19th century were rapid. Instruments with sets of several prisms to increase separation of the lines were used by research workers, especially for astronomical measurements.

Lit: BENNETT (1984 A and B)

Fig 126
Spectroscope, by Spencer, Browning & Co, London, c1863
An early, simple example of the Crooke's type. Presented to W Johnson of the Newcastle upon Tyne Mechanics Institute by his students in 1863. (Science Museum, London. Lo22.12c)

Telescope (figs 127-130)

Although the history of the telescope goes back to the early years of the 17th century, few survive which were made before the last decades of that century; most date from the 18th century or later. The first telescopes of the Galilean, Keplerian and Schyrlean types, were refractors. Early refracting telescopes were normally constructed of pasteboard, with paper-, leather- or vellum-covered tubes, and bone or wooden lens mounts. Since it was observed that the effect of aberration was lessened by reducing the curvature of the object glass, some early telescopes employed as many as five draw-tubes, a device which permitted an instrument of long focal length to be fully portable. (fig 127) Even after the reflecting telescope (first constructed by Newton in 1668), the refractor was preferred until about 1730, when for about thirty years the reflector gave a markedly improved performance. The construction of an achromatic object glass by John Dollond brought the refractor back into use during the second half of the 18th century, since it eliminated one of the great drawbacks of the early refractors – chromatic aberration. In 1758 John Dollond patented his invention of an achromatic doublet object glass, combining a lens of flint glass with a convex lens of crown glass. By the early 19th century the quality and size of object glasses had been greatly improved, and refracting telescopes enjoyed more popularity than reflectors.

The reflector, classified according to the Gregorian, Newtonian and Cassegrain optical systems, continued to be used in physical astronomy because its greater aperture provided larger light gathering power. Observatory measuring instruments however were fitted with refracting telescopes.

Galilean telescope
First constructed by Galileo in 1608, it produces an image by combining a plano-convex object glass with a plano-concave eyepiece.

Keplerian telescope
First designed by Johannes Kepler in 1611, it produces an inverted image by combining a convex objective and a convex eyepiece.

Fig 127
Astronomical telescope, by J Yarwell, London, c1700
This is a 'reverse taper' instrument, with the eye-piece at the larger end. In use with all the draws extended, the telescope is about 5ft long. The drawn tubes are made of pasteboard covered with vellum; the outer body is stained green and covered with gold tooling, including the Royal Arms.
(Science Museum, London. Lo22.12b)

Fig 128
Reflecting telescope, by E Nairne, London, mid-18th century
Gregorian reflecting telescope on simple pillar and tripod stand. The image is focused by adjustment of the secondary mirror via the rod attached to the barrel.
(Science Museum, London. Lo22.12b)

Fig 129
Reflecting telescope, by B Tilling, Dundee, mid-19th century
A 6in aperture Gregorian reflector, with altazimuth mounting. The maker was a lecturer at the Watt Institution in Dundee.
(City Museum and Art Galleries, Dundee. D9.18)

Schyrlean telescope
The eyepiece, invented by Schyrlaeus de Riheita in 1645, consists of three double convex lenses, and produces an erect image, making it suitable for terrestrial observation.

Gregorian telescope
Proposed by James Gregory in 1663, although for technical reasons his proposals were not put into practice until 1726, long after Newton had constructed his first telescope. The Gregorian telescope employs a large concave paraboloid mirror, perforated in the centre to allow the introduction of an eyepiece lens, and a secondary, concave ellipsoid mirror. (figs 128-9)

Newtonian telescope
Employs a primary spherical mirror with a flat secondary mirror reflecting the image to the eyepiece lens, at right angles to the long axis of the telescope tube. (fig 130)

Cassegrain telescope
Similar arrangement of lenses and mirrors to the Gregorian, but employs a convex secondary mirror. Proposed by the Frenchman G Cassegrain, 1672.

Fig 130
Reflecting telescope, by W Herschel, Slough, late 18th century
A typical 7ft-focus Newtonian reflecting telescope constructed by Sir William Herschel. It was with a telescope of this size and design that Herschel discovered the planet Uranus in 1781. Subsequently Herschel made about a hundred of these instruments for wealthy amateur astronomers and for observatories throughout Europe.
(Science Museum, London. Lo22.12b)

Altazimuth telescope mountings

The telescope can be turned in azimuth about a vertical axis and in altitude around a horizontal axis. (fig 129)

Equatorial telescope mountings

In its simplest form the azimuth axis is tilted so that it points to the celestial pole, thus permitting a celestial object to be followed closely with one motion of the telescope, and not the two motions required with an altazimuth mounting.

A more elaborate form of equatorial mount was used for some large observatory instruments, from the late 17th century onwards (fig 116) and for small portable telescopes of the 18th century (fig 53).

Lit: REPSOLD (1908, 1914); KING (1955); RIEKHER (1957); HOWSE (1975); MADDISON (1963) for bibliography of early astronomical instruments; TURNER (1969) for bibliography of optical instruments.

Terrella
see **lodestone**.

Theodolite (figs 131-133)

A surveying instrument used for simultaneous measurement of horizontal and vertical angles. It consists of a horizontal plate mounted on a tripod and graduated in degrees, and about the centre of this plate is pivoted a rule which carries a vertical semicircle. Readings were made from the vertical semicircle in one of two ways: either the sighting device was attached to the diameter of the vertical semicircle, in which case the whole superstructure inclined between two supports, or the semicircle was fixed and the sighting device was attached to a rule pivoted at its centre.

The earliest representation of a theodolite dates from the early 16th century, but the earliest surviving instruments belong to the latter part of the 16th century. Most are of

Fig 132
Theodolite, by E Nairne, London, 1750-75
This is the instrument shown in Fig 23, with the open sights removed and replaced by a vertical semicircle with telescopic sight and bubble level. (Smith Art Gallery and Museum, Stirling. S20.1)

Fig 133
Transit theodolite by Troughton and Simms, London, mid-19th century
In comparison with the altazimuth theodolite, the vertical circle is set high enough to allow the telescope to turn through 360° in the vertical plane. This allows a back-sight to be made without adjustment of the horizontal alignment.
(Science Museum, London. Lo22.15)

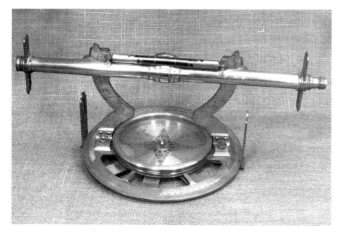

Fig 131
Theodolite, by T Heath, London, c1730
Altazimuth theodolite of the type described as 'Mr Heath's new improved theodolite' in S Switzer, An introduction to a General System of Hydrostatics and Hydraulics (London 1729).
(Science Museum, Newcastle upon Tyne. N2.19)

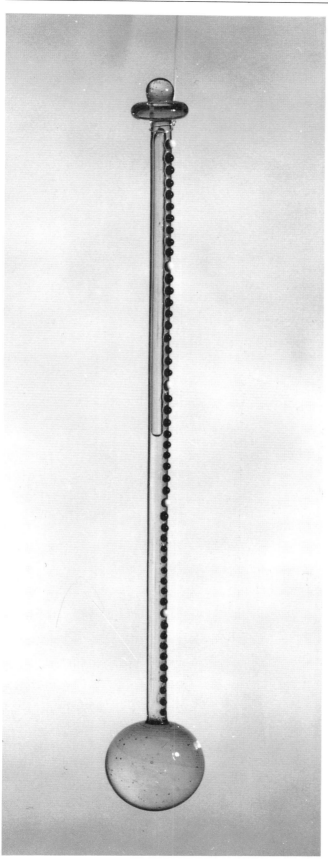

Fig 134
Thermometer, Florence, c1660
A spirits of wine thermometer, formerly part of the equipment of the Academia del Cimento in Florence; probably made by A Mariani, glass-blower to the Grand Duke of Tuscany. A cache of scientific glassware belonging to the Academy was discovered in the 19th century. A few pieces were presented to distinguished foreign savants; this thermometer was given to Charles Babbage. The Royal Museum of Scotland has a similar example presented to JD Forbes.
(Whipple Museum of the History of Science, University of Cambridge. C6.II.17)

18th- or 19th-century date. A 19th-century instrument on similar principles used in mining is normally called a *miner's compass* or *miner's dial*. (fig 91)

The term theodolite was applied during the 16th-18th centuries to instruments capable only of measuring horizontal angles, i.e. *circumferentors*. Some circumferentors were so designed during the second half of the 18th century that a vertical semicircle and a sighting mechanism could be added at will and the instrument thus be used as a theodolite. (fig 132)

A theodolite capable of measuring greater angles of altitude than the usual 45° on either side of the vertical is termed a transit theodolite. In such an instrument the telescope may be revolved through 360°. (fig 133)

Lit: KIELY (1947); RICHESON (1966); BROWN (1982A)

Thermometer (fig 134)

The thermoscope and the thermometer have a history dating from the early 17th century. The earliest surviving instruments are those made for the Accademia del Cimento in Florence (active 1657-1667) (fig 134). However, the vast majority of thermometers found in museum collections are much later in date, associated with domestic barometers, and are not noted separately in this survey.

From the proliferation of thermometer scales used in the 18th century that of Reamur became the most widespread, except in Holland, Scandinavia and the United Kingdom, where the Fahrenheit scale was preferred. The centigrade scale, erroneously attributed to Celsius, became widespread as a result of the growing adoption of metric units by the scientific community as the 19th century progressed.

Lit: MIDDLETON (1966)

Transit instrument (fig 135)

An astronomical instrument designed specifically for determining the moment when a star passes across the meridian of the observer. It consists of a telescope pivoting on a horizontal E-W axis so that it always points to the meridian. The horizontal axis rests, in large observatory instruments, on stout masonry piers. When the transit instrument is correctly aligned to the meridian, the star appears in the centre of the field of view at a sidereal time corresponding to tabulated right ascension, and the accuracy of a clock can be checked. The earliest transit instrument was constructed by Romer. By the early 19th century portable transit instruments were not uncommon.

Lit: REPSOLD (1908) p56ff; HOWSE (1975) p31

Fig 135 (left)
Portable transit instrument, by T Jones, London, 1850-75
Brass instrument set on a cast iron base. Purchased for use on Admiralty surveying expeditions and was, for example, part of the equipment of HMS Sylvia, 1885-89.
(Science Museum, London. Lo22.12b)

Fig 136 (below left)
Traverse board, ? North Germany or Scandinavia, 19th century
Evidence for continued use of the traverse board by British mariners in the 19th century is slight. Surviving examples are often ascribed, as here, to conservative seamen operating from the Baltic seaboard.
(Town Docks Museum, Hull. H6.55)

Traverse board (fig 136)

The traverse board was used for keeping a record of the course of a ship during the watch. It was in existence by the early 16th century, but surviving examples are nearly all of 19th-century date.

In its earliest form the board consisted of a disc marked with the 32 points of the compass, each of the points having eight holes. The course of the ship during each half hour of the watch was marked on the board with a peg.

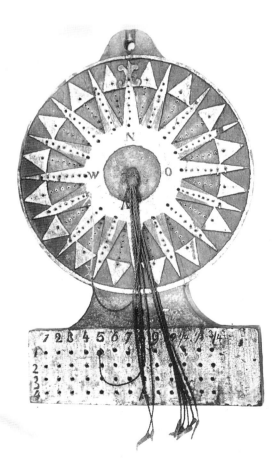

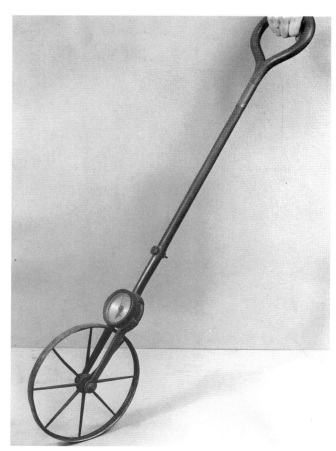

Fig 137
Waywiser, by F Watkins, London, mid-18th century
An unusual small-wheel waywiser. In three revolutions of the wheel, one yard is covered. The silvered-brass dial plate is calibrated in yards, poles, furlongs, and miles.
(Welsh Folk Museum, St Fagans. S4.24)

Later (*c*1600) a series of rows of holes and a further set of pegs were added which permitted the estimated speed of the ship to be recorded. At the end of the watch the mean course steered and the speed would be written up in the ship's log and the board cleared for use in the next watch.

Lit: WATERS (1958) pp36-8

Waywiser (figs 137-8)

Also known as a perambulator. An instrument used for measuring distances, particularly for surveying estates and roads. It consists of a wheel, mounted on a handle, which can be pushed along the ground. Revolutions of the wheel, typically half a pole in circumferences, are counted by a dial attached to the handle or mounted on the wheel itself. The earliest waywisers date from the end of the 16th century. Most are of 18th- or 19th-century date.

Wheel-cutting engine (fig 139)

The clockmaker's wheel-cutting engine seems to have been developed during the second half of the 17th century. The engine consists of a horizontal index plate mounted on a heavy vertical arbor. The index plate is marked with a number of concentric circles divided for cutting wheels with varying numbers of teeth. The wheel is fixed at the centre of the vertical arbor and it, together with the index plate, can be revolved the space of one tooth and arrested by means of an arm with locking pin. The cutter frame is then lowered into position and the tooth is cut. Although there were national and regional variations on the design of the wheel-cutting engines, they changed in principle very little between the late 17th and the second half of the 19th century.

Lit: WOODBURY (1958); GILBERT (1966); CROM (1970)

Fig 138
Waywiser dial plate, by T Wright and W Wyeth, London, c1740
Dial plate from a conventional wooden-wheeled waywiser with a diameter likely to be 38in giving a circumference of about 112in, that is, 15 links of the standard 100-link chain of 22 yards, or four poles.
(Science Museum, London. Lo22.15)

Fig 139
Wheel-cutting engine, clockmaker's, English, late 18th century
This Lancashire-pattern wheel-cutting tool is made to the design associated with John Wyke (1720-87), of Prescot, who had an extensive business supplying watch and clockmaking tools. In the Prescot area of Lancashire many small workshops made parts for clocks and watches which were supplied to 'makers' in London and elsewhere.
(Science Museum, London. Lo22.17)

Inventory of instruments in collections

For information about the content and structure of entries, please refer to the Preface.

A note about dates

A date with no prefix (eg 1786-1852) indicates the dates of birth and death of the individual concerned.

A date preceded by 'fl' (eg fl. 1801-56) indicates that the maker concerned was known through signed instruments, or documentation, to have been active between those dates. In the case of a partnership (eg Troughton and Simms) it refers to the dates between which the partnership is known to have operated.

Where a single date is given (eg I B Paton 1792) this indicates that the individual is known solely from a dated instrument.

Where an approximate date is given (eg *c*1800) this indicates that the maker's period of activity is deduced from extant instruments.

The inventory was prepared using Adept, a word-processing package running on the Prime computer at the Science Museum. This software is not able to handle accents, so these have been omitted in the inventory, and also in the bibliography and indexes.

A

A 1 ABERDEEN
Anthropological Museum

Marischal College, University of Aberdeen, Broad Street, Aberdeen AB9 1AS

0224 273132

In 1871 Dr Robert Wilson bequeathed a collection of Classical and Eastern antiquities to Marischal College, including a few scientific instruments. It was not until 1907 that many specimens, dispersed in various parts of the University buildings, were brought together in the Museum in Marischal College by Professor RW Reid. This collection was augmented by important gifts and bequests of archaeological and ethnographic material and an extensive coin and church token collection.

Three instruments survive which were used by Wilson on his travels through Egypt, Nubia, Palestine, Syria, Arabia, Mesopotamia, Persia and India during the years 1820-22/3 (nos. 6,7,8 and possibly 3). Two electrical machines (nos. 11 and 12) may well have formed part of the collection of instruments assembled at Marischal College during the course of the 18th century for use in the teaching of Natural Philosophy.

1 Gregorian reflecting telescope of brass on a pillar and tripod stand. 3½in aperture, length of tube 22in. Third quarter 18th century. (357⁴)

2 Culpeper compound microscope. Ray-skin outer tube, vellum-covered inner tube. Discontinous brass legs, sunken stage, turned ebony eyepiece. Ht 14in. In pyramidal wooden box containing the trade card of Edmund Culpeper. *c*1730. (357²)

3 Simple microscope. Brass pillar screws into lid of wooden box. Signed: *Banks 440 Strand London*. Possibly used by Wilson in his eastern travels in the early 1820s. (357¹4)

4 Nocturnal. Wood, diameter of circular disc 6¾in, length 11½in. Engraved on the reverse: *MDXI*. Despite the date the nocturnal is probably of fairly recent construction. (357³)

5 Octant. Ebonized wood, brass arm and ivory scale. Radius 13½in. Signed: *Francis Gill 1792*. (357¹2)

6 Pocket sextant. Brass, radius 3½in. The index arm is operated by a toothed quadrant and pinion. Signed: *Dollond London*. Early 19th century. Belonged to Dr Wilson. (357⁹)

7 Box sextant. Brass, silver scale, diameter 2¾in. Signed: *Schmalcalder 82 Strand London*. In round leather box. Used by Dr Wilson 1820-1823. (357¹0)

8 Inclining dial. Brass, silver scale, diameter 4¼in. Signed: *Dollond London*. Used by Dr Wilson 1820-1823. (357¹1)

9 Equinoctial ring dial. Brass, unsigned, *c*1700. (Main Catalogue 357)

10 Ring dial. Brass, diameter 2in. 18th century. (357¹)

11 Small globe electrical machine. White glass globe 3½in diameter with a continuous horizontal axis through the globe. Mounted between high H-shaped supports on a wooden base with a lower intermediate support. Mid-18th century. May possibly have formed part of the philosophical apparatus of Marischal College in the 18th century. (357⁷) (fig 47)

12 Small globe electrical machine with a 4in diameter glass globe mounted on a continuous axis between two turned wooden supports on an 8in wooden base. A cushion fitted to the base of the machine rubs the underside of the globe. Gearing between handle and globe. 18th century. Probable origin as no. 11. (357⁸)

A 2 ABERDEEN
Cromwell Tower Observatory

King's College, University of Aberdeen, High St, Aberdeen AB9 2UB

0224 272507 (Department of Natural Philosophy)

The Observatory, which comes under the care of the Department of Natural Philosophy, houses only one item of original equipment – a mean-time clock. All other items have been transferred to the Museum of the Department of Natural Philosophy, and are listed as nos. 6,7,9,10 and 36 under that heading.

An observatory seems to have existed on the top of the Cromwell Tower as early as 1790 but there is little evidence of the first instruments connected with it. In 1852 a telescope stated by Gill (1913) to be a Dalmayer but actually, if correctly identified, by A Ross, was acquired by David Thomson at the instigation of the clockmaker David Gill, later to become Astronomer Royal at the Cape Observatory. During the 1860s an old transit instrument was resurrected and other telescopes and clocks acquired. Subsequently, the observatory served, until 1947, as a meteorological observatory for the Royal Society's Meteorological Committee. In 1976 it was renovated and is now used as an upper atmosphere observatory.

1 Mean-time clock. Signed on the dial: *D. Gill & Son Aberdeen Watchmakers to the Queen*.

A 3 ABERDEEN
Department of Anatomy, University of Aberdeen

Marischal College, Broad Street, Aberdeen AB9 1AS

0224 40241 ext 243

The Department contains a small museum, which includes two older instruments, as well as two pieces of more recent apparatus of some importance.

1 Culpeper-type microscope. Brass. Signed: *Adams London*. Late 18th century.

2-3 Binocular microscopes, one complete with accessories. Signed: *Baker 244 High Holborn London*.

4 Medicine chest. Signed: *Squire Chemist to the Queen 277 Oxford Street London*. Mid-19th century.

5 Set of trepanning instruments. 18/19th century.

6 Hamilton's freezing microtome made by *Gardner, Edinburgh*. Hamilton was Professor of Pathology at Aberdeen.

7 Original model of McKenzie Davidson's cross-wire localizer for testing foreign bodies within the eye, 1897. Davidson was Honorary Ophthalmologist in Aberdeen Royal Infirmary.

A 4 ABERDEEN
Department of Physics, University of Aberdeen

Fraser Noble Building, University of Aberdeen, Aberdeen AB9 2UE

0224 272507

In common with other former Natural Philosophy Departments in the older Scottish universities, the Aberdeen Physics Department possesses considerable quantities of old equipment. This ranges from demonstration equipment for magnetism, mechanics and electricity of the late 18th and early 19th century, to numerous examples of late-19th century apparatus such as telegraphic and electrical demonstration equipment, optical and acoustic apparatus, and elaborate glassware.

Although natural philosophy was taught earlier in the 18th century (and indeed in the 17th century) at both Marischal and King's College, and a small collection of apparatus existed at both Universities, it was under Patrick Copland, Professor at Marischal College from 1775 to 1822, that a Department of Natural Philosophy firmly established its identity. Old instruments were restored, and new ones acquired. An inventory drawn up in 1823 just after Copland's death permits the identification of some of the surviving instruments as part of the teaching and demonstration equipment at Marischal College in the early years of the 19th century. Copland was also responsible for establishing an observatory on Castle Hill in 1781. The life of the observatory was short (it was demolished in 1795) but the instruments were transferred to the top floor of the tower in Marischal College, and remained there until it too was demolished in 1838. Once again, the 1823 inventory permits the identification of some of the instruments. After Copland's death some instruments from his collection were purchased for Marischal College for the sum of £700.

Until the middle of the 19th century natural philosophy was taught both at King's and at Marischal College. In 1860, however, the two colleges were amalgamated to form Aberdeen University, and David Thomson, already Professor of Natural Philosophy at King's College, became the first Professor of Natural Philosophy for the University. As a result of his interest in astronomy, Thomson had, as early as the 1850s, acquired a telescope for the *Cromwell Tower Observatory* at King's College. Subsequently other instruments were acquired for the observatory. Recently, all save one item from the Cromwell Tower Observatory have been transferred to the Museum of the Physics Department.
Lit: FINDLAY (1935); REID (1982); REID (1983)

The following abbreviations will be used to identify the sources of the instruments: MC, Marischal College; CHO, Castle Hill Observatory, Marischal College; CTO, Cromwell Tower Observatory, King's College.

1 Universal equatorial instrument, with 2in aperture refracting telescope. Unsigned, but closely resembles Ramsden's patented universal equatorial instruments, British Patent 1112 (1775). The only record of an equatorial instrument at Aberdeen is of an instrument by Sisson, divided anew by Ramsden, which was presented by Lord Bute to CHO in 1781. Later that instrument was transferred to MC Observatory (fig 53). (Reid 2)

2 Gregorian reflecting telescope, 3in aperture, 12in focus, altazimuth mounting on pillar and claw stand. Signed: *James Short London 1740* with serial number *5/258 = 12*. Telescope may have been purchased by Professor John Stewart, Professor of Mathematics at MC 1727-1766. It may be identical with the 12in Gregorian reflector later used in CHO. (Reid 6)
Lit: BRYDEN (1968)

3 Refracting telescope, 2⅜in aperture, on pillar and claw stand. Signed: *Barre [?] London Late Ramsden*. Early 19th century.

4 Refracting telescope 3⅜in aperture, with equatorial mounting and clock drive. Signed: *A. Ross London*. Reputedly made by Dallmeyer of London and purchased in 1863 by David Thomson, for the CHO.

5 Refracting telescope 3⅝in aperture, with altazimuth mounting on wooden tripod stand. The telescope is signed: *Ross London 869* and the stand *Invenit et fecit W. Archer 12 Whitcombe Street London*.

6 2¾in refracting telescope, 2¾in aperture, metal tube 3ft 8in long. Equatorial mounting on substantially braced wooden tripod stand. Signed: *Dollond London*. From CTO.

7 Refracting telescope, 3⅝in aperture, wooden tube, length 3ft 9in Altazimuth mounting on tripod stand. Signed: *Dollond London*. From CTO.

8 Transit telescope, 1⅞in aperture, with lamp and level. Used in class experiments in Natural Philosophy Department 1900-1910.

9 Transit telescope, 2½in aperture, signed on level: *Thomas Jones Charing Cross Road, London*. Optics missing. Purchased or resurrected in the 1860s. Was mounted in its original position in the CTO until the early 1970s.

10 Sidereal clock, 1879-1881. Signed: *Sangster & Dunningham Aberdeen*. From CTO.

11 Astronomical clock. Closely resembles the astronomical clock described by James Ferguson in *Select*

Mechanical Exercises, 1775. Probably the clock constructed by Professor Patrick Copland and his assistant J King in the Natural Philosophy Department at MC. Placed in MC Observatory before 1818.

12 Astronomical clock by Marriott, London, with gridiron compensated pendulum. Originally in CHO, later transferred to MC Observatory. (Reid 7)

13 Dip circle invented by John Lorimer. Mounted in octagonal glazed wooden case. The circle is inscribed: *J. Lorimar invt. Sisson fecit*. A brass plaque on the box records: *Instrumentum hocce Navalem ad rem atque Philosophicam perutile Ioan Lorimer M.D. ingenio excogitatum. Academiae Marischallanae Abredonensi grato animo erga Almam Matrem Inventor ipse dono dedit A.D. 1795.* Included in 1823 inventory of MC Observatory. (fig 41)

14 Azimuth compass in gimbals in wooden box. Signed: *McCulloch Compass-maker to his Royal Highness Prince William Henry No. 38 Minories London Patent*. British Patent 1663 (1788). MC, included in 1823 inventory.

15 Azimuth compass by Dudley Adams London. MC, included in 1823 inventory.

16 Compass with paper compass card, octagonal with rectangular extension at one edge, for use with a plane table. Signed: *D. Adams London*.

17 Chinese compass. Presented to King's College, 1833.

18 Set of mechanical pulleys in a mahogany frame. Signed: *W.&S. Jones N. 10 Holborn London*. Early 19th century. (fig 95)

19 Sovereign balance. Affixed to no. 20.

20 Apparatus to illustrate various types of levers. In the form of two shelves supported on turned wooden columns. On the shelves are mounted various levers with weights. Signed: *W. & S. Jones N. 30 Holborn London*. Early 19th century. (fig 96)

21 Anamorphoscopes. Two semicircular mirrors of speculum metal with ebony bases and caps contained in tin boxes. (Used to rectify distorted pictures.) One is signed: *W. & S. Jones, 135 Holborn London*. Late 18th century.

22 Model dead-beat escapement made by J King and Patrick Copland. MC, included in 1823 inventory.

23 Model of Arnold's detached escapement with temperature compensation. Early 19th century. MC.

24 Watch balance wheel and escapement. Signed: *C. Lunan*. Early 19th century. MC, included in 1823 inventory.

25 Model eye of brass, the lenses of glass, with concave and convex lenses mounted on an arm so that each can be swivelled in front of the eye. Signed: *W. & S. Jones 135 Holborn London*. Late 18th century. (fig 97)

26 Model of the cornea and lens of the eye. Unsigned, 19th century or earlier.

27 Optical model to illustrate the optical effect of vision. The path of the light ray from the object to the image on the retina is indicated by threads. Unsigned, early 19th century.

28 Standard Measure of the British Empire, 'taken from the Measure in the Marischal College of Aberdeen by Professor P. Copland by his own hand'.

29 Standard foot. Divided by Jeremiah Sisson and 'fix'd' by Nevil Maskelyne (Astronomer Royal, 1765-1811). CHO. Listed in 1823 inventory.

30 Chinese dotchins (steelyard balances), three examples.

31 Counterpoise balance. Brass, weighing up to 16oz. Signed: *G. Booth Aberdeen 1844*.

32 Pyrometer. Expansion of the metal rod is magnified by a wheel and pulley. Signed: *Claxton & Morton London, Sold by Taylor & Walton, Upper Gower Street*. 19th century.

33 Guyton's gravimeter. Silver and glass. Late 18th/early 19th century. (fig 64)

34 Wet and dry bulb hygrometer. Signed: *W. Duncan Optician Aberdeen*.

35 Portable siphon barometer in wooden mount. Thermometer missing. Probably one of the barometers listed in MC 1823 inventory.

36 Sunshine recorder. Signed: *L. Casella Maker to the Admiralty*. CTO.

37 Proportional compass. Brass with steel points. Signed: *C. Lunan*. MC, included in 1823 inventory.

38 Station pointer. 19th century.

39 Centrifugal railway. Unsigned, 19th century.

40 Apparatus to demonstrate capillarity. The two inclined pieces of glass form a wedge-shaped capillary. 19th century.

41 Apparatus to demonstrate air resistance. Consists of two spindles set at appropriate angles, and set spinning equally by means of a trigger and spring. Unsigned, 19th century. MC, included in 1823 inventory.

42 Perpetuum mobile with Zamboni piles. Probably first half 19th century.

43 Set of specific gravity beads. Unsigned. Probably early 19th century.

44 Octant. Ebony frame, radius 10in. Unsigned but with anchor engraved on ivory arc. Early 19th century.

45 Model Boulton and Watt independently framed steam engine. Early 19th century.

46 Coulomb torsion balance. Purchased 1823 by Professor Knight from a London maker for two guineas.

47 Compound binocular microscope with complete accessories and box. Signed: *Smith, Beck and Beck, 31 Cornhill London, No. 3520*. c1865.

48 French alphabetical railway telegraph. Signed: *Breguet, Brevete, SGDG*. Sender no. 29310, receiver no. 42257. Mid-19th century.

A 6 AIRDRIE
Airdrie Museum (Weavers' Museum)

Wellwynd ML6 0AG

0236 60937

A small observatory is associated with the Museum, which is situated on the upper floor of the Library building. It houses one item of relevance to this study.

1 Micrometer eyepiece said to have been used by the astronomer and telescope maker, William Lassell (1799-1880).

A 7 ALLOWAY
Burns' Cottage Museum

Alloway, Ayr KA7 4PY

0292 41215

The birthplace of Robert Burns (1759-98) containing a Museum with Burns relics. It houses only one item of a marginally scientific nature.

1 Gauging rod (spile rod) measuring 60in and 300 gallons. In six sections contained in a portable wallet. Said to have been used by Robert Burns when a gauger in Dumfries in the 1790s. Signed: *Hughes Maker to the Excise London*.

A 8 ALTON
Curtis Museum

High Street, Alton GU34 1BA

0420 82802. Enquiries to Hampshire County Museum Service, Winchester (q.v.)

This Museum, together with others at Basingstoke and Christchurch, is one of the constituent museums of the Hampshire County Museum Service. It is a regional museum for local geology, botany, zoology, archaeology and history; craft tools, dolls, toys and games. In the Local History Gallery on the first floor there is a small two-case display on the Courages Brewery at Alton which contains:

1 Saccharimeter by J. Baverstock (1741-1815)

A 9 ANSTRUTHER
Scottish Fisheries Museum

St Ayles, Harbourhead, Anstruther KY10 3AB

0333 310628

This small Museum contains exhibits relating to all aspects of fishing and the fishing community. It includes whaling exhibits, model fishing vessels, ship's gear, and material illustrative of the life of the fishing communities

throughout the centuries. In this context a collection of navigation instruments is not unexpected. In addition to the items listed below there are several small hand-held telescopes of no great antiquity, none of which is signed.

1 Octant. Ebony frame, radius 18in. Signed: *I. Urings Fecit London Capt. Willm. Cavin in Leith*.

2 Octant. Ebony frame, radius 12in. Unsigned (ivory label missing) but marked *SBR* on degree scale.

3 Octant. Ebony frame, radius 11in. Sighting telescope. Signed: *G. Bradford 99 Minories London*.

4 Octant. Ebony frame, radius 10in. Signed: *J. Somalvico & Co. London*.

5 Octant. Ebony frame, radius 10in. Unsigned. The case bears the label of : *J. Pyott, 74, West India Road*.

6 Quintant. Radius 7⅜in. T-section lacquered brass frame. Signed: *J. Carstens Hamburg*.

7 Sextant. Brass frame. Radius 7¼in. Signed: *Cail & Sons Newcastle*.

8 Sextant. Brass frame. Radius 7⅜in. Signed: *W.F. Cannon 175 Shadwell High Street London*.

9 Compass in rectangular wooden box. The sights are mounted on slides. Signed: *Alex. Oliphant 20th June 1804*.

10 Three draw brass refracting telescope on wooden tripod stand. Unsigned. Signed: *G.S. Wood late Abraham & Co. Optician Liverpool*. Presentation date 1888.

11 T. Walker's Patent Harpoon Ship Log A4.

12 T. Walker's Patent Harpoon Ship Log A3.

A 10 ARMAGH
The Observatory

College Hill, Armagh BT61 9DG

0861 522928

The Armagh Observatory was founded in 1791 by Richard Robinson, Archbishop of Armagh, and was probably connected with an unfulfilled plan to found a University for Ulster. The first astronomer was the Revd James Hamilton and of the original equipment, one instrument, the Troughton equatorial, and three clocks survive (nos. 1-4). An order for two further instruments from Ramsden was countermanded by Archbishop Robinson's successor.

No further additions were made to the Observatory equipment, other than a locally-made transit which does not survive, until a transit instrument (1827) and a mural circle (1831) were obtained from Thomas Jones. In 1840 the Observatory acquired a number of instruments from the private Observatory of King George III at Kew. These included the Short reflector used by George III to observe the 1769 Transit of Venus and the Shelton mean-time clock. Several instruments from this source, however, no longer survive. In addition to instruments from these three known sources, a number of other items have come to the Observatory over the years, some of them of considerable interest.

In 1964 a Planetarium was established at Armagh, and it is there that most of the scientific instruments – other than those still *in situ* – are displayed or stored.

In the following list the early equipment of the Armagh Observatory appears in order of acquisition (nos. 1-7), followed by the instruments from the King George III collection (nos. 8-14); instruments from other, often uncertain, sources appear as nos. 15 onwards.
Lit: The main accounts of the history of the Observatory are DREYER (1883); MOORE (1967); and BENNET (1990) (All make reference to several of the older instruments.) For the King George III collection instruments see LINDSAY (1969).

1 Troughton equatorial installed 1795; 2½in. aperture, 3ft focal length. Still in position on two stone piers resting on a massive pillar.
Lit: REES (1819) vol. 1 'Astronomical Instruments', plate XVI, and vol. 13 'Equatorial'; DREYER (1883) p5; and MOORE (1967) pp6, 19.

2 Transit clock by Thomas Earnshaw, in use since 1794. Originally with a gridiron pendulum of brass and steel, the clock was modified by TR Robinson in 1830 by the addition of a mercurial pendulum, and in 1832 by a pair of compensating barometers.
Lit: ROBINSON (1833); ROBINSON (1859) p18; DREYER (1883)

3 Sidereal clock by Thomas Earnshaw.

4 Transit instrument by Thomas Jones, 1827, aperture 3¾in, focal length 63in.
Lit: ROBINSON (1829) pp1, 2; ROBINSON (1859) pp11-16

5 Mural circle by Thomas Jones. Installed 1831. Diameter of circle 56in. The original telescope had an aperture of 3¾in. but was later replaced by a refractor made by Grubb.
Lit: ROBINSON (1836), (1859) p23

6 Cup anemometer invented by Thomas Romney Robinson c1846, still in position on the Observatory roof.
Lit: ROBINSON (1850), MIDDLETON (1969) p214

7 Mirror (9in diameter) of 10ft focus from a Newtonian telescope by William Herschel. Accompanied by instructions written by Caroline Herschel. (George III Coll.)

8 Reflecting telescope adaptable for use as a Gregorian, Newtonian or Cassegrain reflector; 6in aperture, 24in focus. Signed: *Thomas Short London 2/1371 = 24*. (George III Coll.)
Lit: BRYDEN (1968) pp33, 34

9 Achromatic telescope (approx 3¾in aperture), lacks object glass and eyepieces. (George III Coll.)

10 Reversing apparatus for a transit instrument by Adams (the instrument itself no longer survives). (George III Coll.)

11 Astronomical quadrant. Brass on mahogany stand composed of eight pillars. Quadrant of 20in radius, signed: *J. Sisson London*. (George III Coll.)

12 Astronomical clock, signed: *Recordon late Emery*. (George III Coll.)

13 Mean-time clock by Shelton, London. (George III Coll.)

14 Achromatic telescope. Signed: *Dollond London*.

15 Reflecting telescope, 4in aperture, on pillar and tripod stand. Signed: *G. Adams No. 60 Fleet St. Instrument Maker to his Majesty*.

16 Reflecting telescope, 4in aperture, on pillar and tripod stand. Brass, with solid turned altitude circle. Probably third quarter 18th century.

17 Polisher for a 15in speculum. *c*1820-30.

18 Small micrometer signed: *Troughton & Simms London*. Second quarter 19th century.

19 Equatorial telescope, 2½in aperture, signed: *Gilkerson & Co*.

20 Miniature equatorial instrument on the lines of Megnie's instrument for Bochard de Saron. The telescope is mounted at one end of the polar axis, the further end of which carries the declination circle. Signed: *Adams London*.

21 Universal equinoctial ring dial. Brass, diameter 8in. Unsigned, 18th century.

22 Orrery on pillar and tripod stand. Brass, signed: *Gilkerson & Co Tower Hill London*. Complete with attachments for use as a lunarium and a tellurium.

23 Orrery, signed: *Philip & Son Fleet Street, London*.

24 Sextant. Brass T-section frame, radius 8in. Signed: *Dollond London*

25 Pyrometer. Mahogany frame, metal rods missing. Signed: *J. Whitehurst Derby*.

26 Barometer. Signed: *J. Newman 122 Regent Street London*.

27 Portable transit theodolite. Signed: *Dollond London*. Used for the Ordnance Survey of Ireland (commenced 1840).

28 Celestial globe, signed: *Cary London*, *c*1800.

29 Celestial globe, signed *W. Newton*.

30 Celestical globe, signed: *Bardin, sold by W. & S. Jones*. Late 18th century.

A 11 AYLESBURY
Buckinghamshire County Museum

Church Street, Aylesbury HP20 2QP

0296 88849

This collection originated through the Buckinghamshire Archaeological Society (founded 1848) which in 1862 hired rooms for a Museum. In 1908 the location in Church Street was changed and the Museum now occupies two adjacent houses.

The collections are primarily local in character, dealing with the history, natural history, industry and local crafts of Buckinghamshire. Archaeological material dates from prehistoric times onwards. The collection contains a small number of scientific instruments, some with local connections.

1 Gregorian reflecting telescope with pillar and tripod stand. Eyepiece missing. Late 18th or early 19th century. From Nether Winchendon House. (373/67)

2 Compound microscope, box mounted. The microscope body is of Jones type, the pillar derived from a Cuff model. Signed: *R. Bancks London*. Early 19th century (188/68)

3 Compound microscope. Signed: *J. & C. Robbins 146 Aldersgate St. London.*. Stamped with serial number: *941* (1/23)

4 Elton's transparent orrery, consisting of a mahogany box on lionpaw feet, open at back. Visible in a window at the front is a paper roll manipulated by a handle at the side of the box, on which are representations of the solar system, etc. The representations of Sun and Earth stand out against the blue ground of the paper when the apparatus is lit by means of a candle behind the box. From Stowe House, *c*1817. A printed label inside the back

records: *To Deane F. Walker Esq. Hiself and Family having by their lectures diffused a general taste for the Sublime Science of Astronomy Elton's Transparent Orrery, is, with his permission, dedicated as a small tribute of respect by his obedient humble servants. The Publishers, Princes Street, Cavendish Square.* (249/13)

5 Magnetic dial. Turned circular box, floating compass card. On label in lid: *S. Porter's Magnetic Sundial*. (31/76)

6 Navigational teaching aid? Bronze disc with compass rose and centrally pivoted needle on one side, and on the other a list of latitudes. French, 18th century. (152/67)

7 Sector, Brass, signed: *Gourdin A Paris 1786*. Last quarter 18th century. (174/19)

8 Waywiser. Single handle, brass dial. Unsigned.

9 Drainage level. Inscribed: *Blundell's drainage level Registered No. 60. 7 June 1847*. Ivory label of: *Bishop & Co. 66 Cheapside London*. (119/61)

10 Fusee cutter. From Olney. Probably late 18th century. (68/63)

11 Clockmaker's wheel-cutting engine. Wheel mounted in rectangular frame. From Olney. Late 18th century. (69/63)

B

B 1 BANFF
Banff Museum

High Street, Banff. Correspondence to: Museums Service, Peterhead Arbuthnot Museum, Peterhead, Aberdeenshire AB4 6QD

0779 77778

This small museum contains a collection of antiquities, local history and natural history, and is of interest from the point of view of this survey, since it houses several items associated with the instrument-maker and lecturer in National Philosophy, James Ferguson (1710-1776), a native of Rothiemay in Banffshire, all of which were purchased by Ferguson's biographer, Henderson, between 1830 and 1861, and were presented by him to the Banff Museum in 1864. Further paper instruments associated with Ferguson were transferred from the Royal Society of Edinburgh to the National Library of Scotland, Edinburgh, in 1990.
Lit: HENDERSON (1867) pp476, 477 (nos. 2, 5, 6, 9 and 14 in Henderson's list were not seen); CALTHORP (1932)

1 Three draw refracting telescope with parchment covered cardboard tubes and horn ferrules. Length closed 12in, fully extended 33in. Henderson states 'Made by Ferguson in the year 1732'. (Henderson 4)

2 Astronomical clock by Nicholas Vallin, *c*1600. Circular table clock with terrestrial globe mounted on an inclined axis and geared to revolve once every 24 hours. The present globe is early 19th century in date, but the wheelwork is apparently original. Work signed: *N Vallin*. The clock is inscribed with the following owners' names: *John T. Des[?]uliers: LLD Lecturer on Nat. and Exp. Phil: London MDCCXXIX Benjamin Franklin Lld; FRS 1757 James Ferguson FRS 1766 Kenneth McCulloch 1774 G.W.* (Henderson 1)
Lit: TURNER (1973) pp36, 45; KING/MILBURN (1978) pp63, 64

3 An assortment of items in a glazed gilded case, listed by Henderson as:

a A boxwood wheel cut by Ferguson.

b A hollow paper axis.

c Boxwood handle of a winch (all three said to have been part of the satellite machine described in Ferguson's *Tables and Tracts*).

d A semicircular graduated card in the autograph of Ferguson.

e A wheel of 45 teeth, made of thick Bristol board, cut by Ferguson.

f A trundle wheel of 15 cogs, mounted on a triple crank axis, which was driven by the above wheel of 45 teeth, both said to have belonged to Ferguson's electrical apparatus.

g The apogee pointer of Ferguson's 'Mechanical Paradox'. (Henderson 15)

4 *A Table of the Equation of Time, showing how much a clock should be faster or slower than the sun, every day of the year – By James Ferguson FRS AD 1776* and inscribed: *Sold by the Author at N... (Bol)t Court Fleet Street....*

5 *The Universal Analemma.* Ink on Bristol board. An orthographic projection of the celestial sphere over which revolves a disc which has been cut away leaving a rim and a diameter which acts as a horizon, and is marked with degrees and wind directions. The rim of the disc is provided with an altitude scale in the upper semicircle while the lower carries declination and calendar scales. (Henderson 10)

6 *An astronomical rotula* showing the times of all the new and full moons and eclipses of the Sun and Moon for 12,000 years. Consists of three movable circles and an index. Ink on Bristol board. (Henderson 8)

7 Horary quadrant. Ink on Bristol board. (Henderson 11)

8 *Astronomical rotula.* Engraved on paper, lacks some of the revolving discs. When complete shows places of Sun and Moon, Moon's nodes and the times of eclipses of the Sun and Moon from 1730 to 1800 (complete example *London, NMM*). (Henderson 7)

9 Manuscript diagram, *The motion of Saturn, Jupiter and Mars in respect of the earth.* Paper mounted on board. Not in Henderson.

10 Volvelle, engraved paper. Consists of three movable circles, the bottom one engraved with calendar and the phases of the moon, the middle one with the North celestial planisphere, calendar and signs of the zodiac; the upper disc is partially cut away to reveal the planisphere beneath, the cut away area being linked along the NS axis by a rule passing through the pole of the planisphere and engraved with an altitude scale. The edge of the upper disc has hour divisions and indications of Moon age, while the interior contains tables of Dominical letters, etc. (Henderson 12)

B 2 BARNARD CASTLE
Bowes Museum

Barnard Castle, Co. Durham DL12 8NP

Barnard Castle 37139

The Museum, with its collections of paintings, tapestries, furniture, glass, porcelain and metalwork, was the gift of John Bowes of Streatlam Castle, Co. Durham, and his wife the Countess of Montalba. The building, designed to house the collections, is by E. Pellechet of Paris. The foundation stone was laid in 1869, and the Museum was opened to the public in 1892. In recent years Durham County Council has taken on responsibility for the Museum. The original Bowes collection contains nothing more scientific than an 18th-century barometer and thermometer in matching carved, painted and gilded cases, but a more recent addition to the collection, listed below, is relevant to this survey.

1 Backstaff. Lignum vitae and boxwood. Inscribed: *John Stephenson 1749*, and signed: *Made by Charles Digby near the Hermitage Stairs London.*

B 3 BARROW-IN-FURNESS
The Furness Museum

Ramsden Square, Barrow-in Furness LA14 1LL

0229 20650

The Museum, opened in 1930, contains collections relating to the Barrow-in-Furness region, including prehistoric archaeological material (especially Bronze Age), Lake District bygones and a collection of ship models.

1 Specific gravity beads in turned wooden (? mahogany) box, diameter 2⅛in. Seventeen beads survive out of nineteen. Signed: *Made and Sold by John Pochaine New-castle.*

2 *Concise chronologer, calculated to answer all the purposes of an almanack from 1752 to the year 1900. By John Atkinson Gateshead.* A perpetual calendar of printed card with movable parts.

3 Nautical telescope. Wood and brass outer tube, length 24in. Signed: *Watkins & Hill Charing Cross London. Night.* Inscribed on brass plate: *Presented by Admiral the Right Honourable Lord Clarence Paget GCB to his friend Llewellyn Turner....* The inscription records the use of the telescope during the capture of a Spanish slaver, *The Vengador,* in 1838.

B 4 BASINGSTOKE
Willis Museum

Old Town Hall, Market Place, Basingstoke RG21 1QD

0256 65902. Enquiries to Hampshire County Museum Service, Winchester

This Museum, together with others at Alton and Christchurch, is one of the constituent museums of the Hampshire County Museum Service. In addition to displays relating to local geology, biology, archaeology, history, and the decorative arts, one room of the Museum is devoted to clocks, watches, watchmaker's tools and a few scientific instruments. The watches include an interesting German watch of *c*1575, unsigned but probably made in Nuremberg, while among the clockmaker's tools there is a watchmaker's wheel-cutting engine, watchmaker's turns, as well as several smaller items.

1 Universal equinoctial ring dial. Brass, diameter 10in. Signed: *Nairne London.* (WOC 2064)

2 Horizontal compass dial. Circular brass box, folding gnomon. Signed: *Stokes fecit 1742* (Possibly Gabriel Stokes (1682-1768), Dublin). (BWM 1958.4)

3 Universal equinoctial sundial. Gilt brass, rectangular.

Signed *LTM* (Ludwig Theodor Muller (1710-70) Augsburg.) (BWM 1964.338)

4-6 Three sandglasses, probably 18th century. One with a wooden frame, and the other two with metal frames, one rectangular and one circular. (ACM 1953.152, WOC 1073, AOC 703)

7 Pair of globes (celestial and terrestrial) by Newton & Son, London, 1857. Table model with stand mounted on four bun feet. (WOC 3038/3039)

8 Celestial globe on table height tripod stand. By Malby & Co., London, 1848. (BWM 1958.44)

9 Waywiser. Mahogany frame and iron bound wheel. 'Y' handle signed: *West, London.* Early 19th century. (WOC 4645)

10 Spirit-testing apparatus used by Hampshire Constabulary *c*1880. (BWM 1955.7)

B 5 BATH
The American Museum in Britain

Claverton Manor, Bath BA2 7BD

0225 60503

The Museum, opened in 1961 in an attractive house designed by Sir Jeffrey Wyatville in 1820, was established by two Americans, Dallas Pratt and John Judkyn, in order to interpret the history and art of the United States. The displays, which cover the decorative arts, folk art, and the history of the United States, include the following scientific instruments.

1 Backstaff. Lignum vitae and pear wood, lacking vanes. Signed: *Made by Will Garner for Thomas White 1717.* Loan from *National Maritime Museum, London.* (NMM Inventory, Section 5, Ref. no. S 161/54-8).

2 Telescope, wood and metal, length of main tube 15in. Signed: *Spencer Browning & Rust London.* Early 19th century.

3 Sextant. Brass with silver scale graduated in 110°. In wooden case. Signed: *J.P.Moller, 118 Minories, London.* With trade card of Graham & Parks, Southside Customhouse, Liverpool, inside lid of case. Mid-19th century.

4 Terrestrial globe on stand. Diameter of globe 4¾in. Signed: *E.C. & A. Murdock, New Wallaston, Mass. c*1820 *Lit:* YONGE (1968) p49

B 6 BATH
Pump Room

Abbey Churchyard, Bath BA1 1L2

0225 28411

The Pump Room, erected 1789-99 by Thomas Baldwin, replacing an earlier Pump Room built in 1705, contains two important items by the clockmaker Thomas Tompion, an equation clock and a sundial. A bronze plaque dated 1764 records the presentations.

1 Long case clock with month equation movement. Presented by Tompion in 1709. (Entry in Bath Chamberlain's accounts, 'Paid to Mr. Tompion's man, by order of the Corporation, £5.7.6') *Lit:* SYMONDS (1951) p49ff, figs. 46, 47, 67, 93, 94

2 Horizontal dial. Bronze. Signed: *Tho. Tompion. London.* Presented with the equation clock in 1709 and used to establish solar time in order to set the clock. This sundial was later removed and mislaid, but was rediscovered in a farm at Corsham, and presented to the Pump Room by Brigadier Meyrick Neilson in July 1971.

B 7 BATH
Royal Photographic Society National Centre of Photography

Royal Photographic Society of Great Britain, The Octagon, Milsom Street, Bath BA1 1DN

0225 62841

The Royal Photographic Society was founded in 1853 as a result of the interest engendered by both the 1851 Great Exhibition and the Society of Arts photographic exhibition held in December 1852. Sir Charles Eastlake was the first President, Roger Fenton was Secretary and Queen Victoria and Prince Albert were patrons. A collection of photographs was initiated but it grew slowly. A major donation of research apparatus was made in 1915, when early photometers and actinometers (and manuscripts) connected with the work of Ferdinand Hurter (1844-1898) and Vero Driffield (1848-1915) were bequeathed to the RPS Museum, then in Russell Square, London. In 1936 the RPS was given a portion of W H Fox Talbot's apparatus (other beneficiaries were the *Science Museum, London* and the *Royal Museum of Scotland (Chambers Street), Edinburgh*). A selection from the collection was loaned to the *Science Museum* in the late 1920s until the RPS moved to its present home in Bath in 1979, where there is a museum of its rich collection of apparatus and photographic prints.

The collection of early cameras is unparalleled. These are not strictly within the scope of this guide but because of their significance and experimental nature they have been listed. The overall collection is now large, comprising 6,000 items, though few are strictly scientific instruments and most material is of the 20th century. There is no published catalogue. FERGUSON (1920) gives a brief listing of the Hurter and Driffield material (and also reprints the papers describing the photometry and actinometry experiments). Some 87 items lent to the Science Museum were published in that Museum's photography catalogue (THOMAS (1969)). The numbering applied in that work is given in parentheses after the entries below.

For a brief history, published on its centenary, see ROYAL PHOTOGRAPHIC SOCIETY (1953); also HOPKINSON (1980).

1 Solar microscope, unsigned. Used by Talbot for photogenic drawing. Early 19th century. (4)

2 Camera lucida support missing, unsigned. Also used by Talbot for drawings. (3)

3,4 Two small camera obscuras. (1)

5-7 Three small cameras, wood. Constructed in 1835 by Joseph Foden of Lacock for Talbot for photogenic drawing. Lenses are those used in microscopes. (5)

8 Daguerre-Giroux camera, French, acquired by Talbot in 1939, together with apparatus to prepare plates, sensitising boxes (1839-1845) and developing boxes (1839-1845).

9 Daguerreotype camera by Lerebours of Paris, used by Talbot for his calotypes, 1841. (19)

10 Three calotype cameras, one with curved glass to hold paper to correct image distortions, another with a ratchet system for tilting camera upwards. Used by Talbot. (6, 7)

11 Camera developed by George S. Cundall, 1844. The first to be made with a focusing scale.

12 Camera designed by Frederick Scott Archer, 1853, for wet collodion process. (23)

13 Folding sliding box camera, by Horne and Thornwaite, London, 1856.

14 Panoramic camera with water-filled lens, developed by Thomas Sutton, constructed by Ross, 1861. (645)

15 Collodion wet plate camera, designed by Charles Piazzi Smyth (Astronomer Royal for Scotland) in 1865. Constructed by Bryson of Edinburgh. (38)

16, 17 Portable dark tents with cameras, for processing photographs int he field. (440, 441)

18 Microscope objective in circular green box labelled *Chevalier Nov 1825 M Talbot.*

19 Camera lens inscribed *LENS USED BY D.O. HILL RSA ..... GIVEN ..... TO J. CRAIG ANNAN CHRISTMAS 1911.* Probably made by Thomas Davidson, (D.O. Hill and Robert Adamson were pioneer photographers in Scotland.) (167)

20 Large glass concave mirror in wooden frame. Formerly the property of Hill and Adamson.

21 Experimental photometers and actinometers developed by Hurter and Driffield.
Lit: FERGUSON (1920)

B 8 BEDFORD
The Bedford Museum

Castle Lane, Bedford MK40 3XD

0234 53323

The Museum of the Bedford Modern School forms the basis of the collection. Set up in 1884 in a building in Harpur Street by Mr Charles Prichard as a memorial to his mother, the collection has expanded greatly since its foundation, and the Museum now contains an important collection of local antiquities, local history and natural history, as well as collections of coins, maps and prints. In addition to clocks, medical and surgical items, the Museum owns a number of scientific instruments.

1 Gunter's quadrant, with planispheric nocturnal on the reverse. Brass. Radius 3¾in. Inscribed on the front: *Tho. Miller* (probably name of owner). Mid-17th century.

2 Small brass astronomical telescope in box. Approximate length 12in. Signed: *Elliott Bros. London.*

3 Stadiometer. *'Invented by G.H. Blakey, R.N. Made by W. Heath, Optician in Devonport'.* Mid-19th century.

4 Backstaff. Lignum vitae, radius 23⅛in. Unsigned, but inscribed with name of owner: *Thady Flaherty 1740.* Lacks vanes.

5 Octant. Mahogany frame, radius 18in. Engraved decoration on brass arm. Unsigned, last quarter 18th century.

6 Octant. Ebony frame, radius 14in. Signed on ivory label: *Spencer Browning & Rust London.* Marked: *SBR* on arc. *c*1790.

7 Sextant. Ebony frame, radius 12in. Signed: *Worthington & Allan London.* Marked *I R* on degree arc. Second quarter 19th century.

8 Cary type microscope. Unsigned. Early 19th century.

9 Withering's portable microscope. Early 19th century. Unsigned.

10 Nuremberg tripod microscope. Boxwood, cardboard tubes, the outer covered with embossed paper. *IM* (Junker of Magdeburg) branded on the base, 19th century.

11 Boxwood slide rule. Signed: *T. Hawkins 1758.*

12 Navigator's rule, Gunter-type. Boxwood. Signed: *Cary London* and inscribed: *Lady of St. Kilda Yacht T.D.A. 1837.* (Other example in *Whipple Museum, Cambridge.*)

13 Slide rule. Boxwood. Marked: *Taston original maker.*

14 Set of drawing instruments (ivory rule and compasses) in black leather case. Marked: *J. & W. Watkins Charing Cross.*

15 Cabinet containing the Bedford Humane Society's apparatus for restoring the apparent drowned, 1807. (See similar example, fig 72)

B 9 BELFAST
Ulster Museum

Botanic Gardens, Belfast BT9 5AB

0232 668251-5

The Museum houses a collection of Irish antiquities, local history, and industrial technology, as well as material illustrative of the natural history of Ireland. There are collections of old master and modern paintings, as well as Irish silver, glass and pottery. A unique collection of material was rescued in 1968 from the wreck of a ship from the Spanish Armada by the diver Robert Stenuit. The galleas *Girona* was returning with the Armada to Spain on 26 October 1588, when it broke up on rocks off Lacada Point, close to the Giant's Causeway, Northern Ireland. From this wreck come the two fragmentary mariner's astrolabes, nos. 2-3 in the ensuing list, as well as several smaller items of nautical equipment (nos. 4-11).

In addition to the items listed below the collection includes Edison phonographs and cylinders, vintage typewriters from 1883 onwards, and some radio equipment from the Belfast transmitting station dated to the 1920s and 1930s.
Lit: STENUIT (1971)

1 Roman bronze and lead steelyard, the weight in the form of the bust of a helmeted warrior. 1st century AD.

2 Mariner's astrolabe, diameter 7½in. Suspension ring and alidade lacking. Scales eroded. From the *Girona*.
Lit: ANDERSON (1972) no. 5; STIMPSON (1988) no.26

3 Mariner's astrolabe, diameter 6⅞in. Frame in nine pieces, suspension ring and alidade lacking. From the *Girona*.
Lit: ANDERSON (1972) no. 6; STIMPSON (1988) no. 27

4-8 Five pairs of brass navigational dividers. From the *Girona*.

9-11 Three sounding leads from the *Girona*.

12 Lead steelyard weight. From the *Girona*.

B 10 BIRMINGHAM
Assay Office

Newhall Street, Birmingham B3 1SB

021 236 6951-2-3

The Birmingham Assay Office was opened on 31 August 1773. In the two centuries since its foundation it has acquired important collections of Birmingham and other silver ware, medals and tokens. There is a fine specialized library in the art and craft of the silversmith and related topics and techniques. The large collection of correspondence of Matthew Boulton now in the possession of the Matthew Boulton Trust is housed at the Birmingham Reference Library.
Lit: BIRMINGHAM (1936)

1 Assay balance in glazed mahogany case. The scales are regulated by a cord and pulley attached to the wooden upright. Steel beam with swan-neck ends shallow brass pans. Movable brass rests for pans. Dimensions of case: 23½ x 18 x 15in. The balance was in use at the Assay office from 1773 until about 1860. (fig 15)

2-6 Five balances signed: *L. Oertling London*. Late 19th century.

B 11 BIRMINGHAM
Birmingham Museum and Art Gallery

Chamberlane Square, Birmingham B3 3DH

021 235 2834

The Museum collections originated in the 1860s but the present building was opened in 1885, and enlarged in 1902 and 1919. The important and extensive collections cover, in addition to the fine and decorative arts, archaeology, local history, ethnography and natural history. In addition to the Central Museum and Art Gallery there are several branch museums throughout the city. The scientific instruments come under the care of the Department of Archaeology, Ethnography and Natural History. The majority formed part of the Pinto Collection of Wooden Bygones, acquired for the museum in recent years. In addition to the items listed below, there are a number of Dutch and English 18th-century jeweller's and coin balances.
Lit: PINTO (1969) illustrates a number of instruments from the Pinto collection.

1 Multiple dial in the form of a casket. Pinewood carcase veneered with marquetry set in oak. Sundials on vertical and horizontal surfaces. Inset compass mounted beneath sliding lid. Unsigned. Possibly German, 17th century. (T 5048)
Lit: PINTO (1969) fig 294 H,J

2 Universal equinoctial ring dial. Brass, partly silvered, diameter 9⅛in. Signed: *G. Adams Fleet Street London*. (130'41)

3 Universal equinoctial dial, gilt brass, octagonal, with silvered compass dial. Blind stamped leather case, velvet lined, with disc inside the lid engraved with a list of latitudes. German Augsburg (?) 18th century. (M2'76)

4 Cube dial. Wooden cube with engraved and coloured paper dials on five faces. Brass gnomons, turned beechwood stand. Signed: *I.G. Kleininger*. Late 18th century. (T. 5090)
Lit: PINTO, fig 294 C

5 Pillar dial. Boxwood with brass gnomon. Unsigned, 18th century. (T 5096)

6 Butterfield dial. Silver. Octagonal signed: *George AParis*. (120'41)

7 Horizontal dial. Oval, brass, signed *George AParis*. (128'41)

8 Horizontal string-style dial in oval lignum vitae case. The inside of the lid is painted dark blue and bears in gold a list of latitudes for 13 cities, 12 of which are French. There are hour rings for latitudes 42, 45 and 48°. French, early 17th century. (156'75)

9 Magnetic dial in turned lignum vitae box with paper dial and compensation tables. Unsigned, early 19th century. (T 5087)

10 Compass in rectangular mahogany box with lid. Engraved paper compass dial. Unsigned, *c*1800. (T 5085)

11 Pocket compass in turned sycamore case with paper dial. Unsigned, 19th century. (T5091)

12 Compass in square mahogany box with lid. Engraved paper dial. Unsigned, *c*1800. (T5084)

13 Compass in turned brass case, silvered dial. Unsigned, early 19th century. (T5317)

14 Culpeper-type microscope, early form. Lignum vitae eyepiece and snout, black ray-skin-covered outer tube. Complete with objectives, slides, fish plate and pyramidal oak box. (T5099)

15 Nuremberg copy of Culpeper-type microscope. Turned fruitwood with cardboard tubes. *I.M.* branded into base (Junker of Magdeburg), 19th century. (T5092)

16 Pantograph. Ebony limbs with ivory scales. Label of *Benjamin Cole Mathematical Instrument Maker at the Orrery* pasted into lid of box. Inscribed on label in ink: *no. 136.* (T.3523 A) (fig 107)
Lit: PINTO (1969) fig 270

17 Napierian rods. Boxwood, in wooden box. *c*1700. (T.5053 31)

18 Excise slide rule, Everard-type. Boxwood. Signed: *Edwd. Roberts Maker Dove Court Old Jewry London.* Owner's name: *Owen Gill. c*1775. (T.5110) (fig 125)
Lit: PINTO (1969) fig 279 B,C

19 Excise slide rule. Signed: *Cook Maker to the Excise Late Wellington Crown Court Soho London 3522.* (T5111)

20 Excise slide rule. Signed: *Dring & Fage Tooley St. London.* (T5113)

21 Publican's stocktaking slide rule, designed by Farmer, 19th century. (T.5108)

22 Excise slide rule signed: *Loftus Maker 146 Oxford Street, London.* (T5112)

23 Excise slide rule signed: *Lewis and Briggs Makers No. 52 Bow Lane Cheapside London.* (T5107)

24 Four-section exciseman's measuring rod signed *Buss Maker 48 Hatton Garden London.* (T339)

25 Six-section exciseman's gauging rod (spile rod) in leather case signed: *Hughes Maker to the Excise London.* (T6117)
Lit: PINTO (1969) fig 297 p278

26 Specific gravity beads. Set of twelve beads in velvet-lined mahogany case. Maker's label in lid: *Joseph Sala & Co. 33 High Street Paisley.* (T.5097)
Lit: PINTO (1969) fig 300D, p278

27 Hydrometer in case, labelled *Sikes' Hydrometer Josh. Somalvico and Co., London.* (T5080)

28 Chondrometer. Signed: *Watkins & Hill.* (T.5102)

29 Dissected cone. Boxwood, 18th or early 19th century. (T. 5055)
Lit: PINTO (1969) fig 294E

30 McFarlane's calculating cylinder. Wood covered with engraved paper. *c*1840. (T.5082)
Lit: PINTO (1969) fig 294A, p273

31 Diamond merchant's balance, silver scale and pans, in walnut box with tinted engraving inside the lid, brass stand. Italian, *C*1650. (T. 5083)
Lit: PINTO, fig 311, p287

32 Diagonal barometer incorporating a thermometer and hygrometer. Signed by Edward Scarlett, London. On display at Aston Hall. (M. 154'70)
Lit: GOODISON (1977) pp216-9

B 12 BIRMINGHAM
Dollond and Aitchison Museum

Dollond and Aitchison Group Ltd, 1323 Coventry Road, Yardley, Birmingham B25 8LP

021 706 6133

This is a company museum, accessible to the public by prior application. Of the several companies which have been amalgamated to form the Dollond and Aitchison Group the earliest component was founded by Peter Dollond (1731-1820), of Huguenot extraction, who in 1750 established a workshop in Spitalfields, London. His father John Dollond (1706-1761) joined him in 1752 and he experimented with combinations of lenses made from different glass to overcome achromatic effects.

Most of the instruments in the Museum are optical and nearly all bear the signature of the Dollond firm. In addition to those items listed below, there are substantial collections of binoculars, field glasses, opera glasses and spy glasses. There is a major collection of spectacles assembled by Gerald Hamblin, d.1952, and donated by his son in 1983. Also in the collection are portraits in oil of Peter Dollond and John Clark (the Scottish maker of microscopes). John Dollond's gold Copley Medal of the Royal Society, awarded in 1758, is preserved, as is a silver spectacle case presented to the Clerk of the Company of Spectacle Makers in 1778 when Peter Dollond was its Master.

Numbers following entries refer to a card catalogue.
Lit: MITCHELL (1975); BARTY-KING (1986) – a company history

1 Microscope, Cuff type, signed: *P.Dollond Strand London*, mid-18th century. (130)

2 Microscope, Dollond double reflecting type, in case with lenses and fish plates. Signed: *Dollond London.* (20)

3 Microscope, Culpeper type, brass barrel, signed: *Dollond London*, with pyramidal wooden box, mid-18th century. (131)

4 Microscope, Wilson screw-barrel type, in fish-skin covered box, *c*1730. (125)

5 Microscope, Cary type, in fitted wooden box, signed: *Dollond London*, 19th century. (132)

6 Microscope, 'Jones improved' model, signed: *Dollond London*, early 19th century. (2)

7 Microscope, 'Jones improved' model, signed: *Dollond London*, with wooden case, early 19th century. (124)

8 Solar microscope, in box, signed: *Dollond London*, date 18th century. (133)

9 Telescope, five draw refractor covered in gold-stamped green vellum, early 18th century. (67)

10 Telescope, hand-held refractor, signed: *King Bristol*. (66)

11 Telescope, hand-held refractor, signed: *Watkins & Hill Charing Cross London*. (65)

12 Telescope, hand-held refractor, signed: *Dollond London* and *MHP* on eyepiece mounting. (61)

13 Telescope, hand-held refractor, signed: *Dollond London* and inscribed with names of four generations of the Lambert family, with one date (1812). Also with the place-name *Lisboa* inscribed on barrel close to maker's signature. (69)

14 Telescope, hand-held refractor, signed: *Dollond London* and inscribed *From James Yeaman Miln Esqure of Murie to Robert Constable Esqr Hill Jany 1st 1852*. (59)

15 Six further telescopes, hand-held refractors, all signed: *Dollond London*, 19th century. (60, 62, 63, 64, 70, 71)

16 Telescope, refractor, mounted on tall brass tripod, signed: *Dollond London*, mid-19th century. (127)

17 Two pocket telescopes, with small brass folding stands, signed: *Dollond London*.

18 Telescope incorporated in a walking stick, signed: *Thos Harris & Son London*. (58)

19 Telescope mount in fish-skin covered box. (117)

20 Two camera lucidas in boxes, signed: *Dollond London*, early 19th century. (124)

21 Equinoctial dial, silvered finish, signed: *Dollond London*. (118)

22 Equinoctial dial, brass, signed: *Dollond London*. (9)

23 Compass in silver case, hallmarked 1817, signed: *Dollond London*.

24 Kater's hygrometer, signed: *Dollond London*. (10)

25 Barometer, Kew Standard of *c*1855 design, signed: *Dollond London*. (56)

26 Waywiser, iron-bound wooden wheel, signed: *Dollond London*.

27 Sykes hydrometer in fitted box with two wooden slide rules, signed: *Dollond London*, mid-19th century.

28 Pocket aneroid barometer, signed: *Dollond London*, *c*1890.

29 Brass sextant, signed: *Dollond London*, early 19th century.

B 13 BIRMINGHAM
Birmingham Museum of Science and Industry

Newhall Street, Birmingham B3 1RZ

021 236 1022

Established in 1950, the Museum assembled an important technological collection. It contains steam engines, machine tools, small arms, early motor cars, aircraft, cycles, motor cycles and industrial machinery. There is an interesting range of mechanical musical instruments, and a comprehensive collection of pens and writing equipment; from the range of lathes in the Museum, some of the more important are listed below. There is a small but growing collection of scientific instruments. Among the instruments listed below are a number of items from the Boulton & Watt Collection which were passed to the Birmingham Museum after the closure of the Soho Foundry, and were transferred in 1953 to the Museum of Science and Industry (nos. 9-11, 22-4, 28).

1 Universal equinoctial dial. Brass, partly silvered. Bubble levels and adjusting screws. Signed: *Docteur Arthur Chevalier Opticien 158 Palais-Royal Paris*. (54.19)

2 Theodolite. Brass, silver scales. 5in base. Signed: *Troughton & Simms*. Engraved on objective cap: *G.A. Bell 1841*. Mid-19th century. (53.329)

3 Surveyor's level. Brass. 15⅜in telescope with bubble level below, mounted on base with screw for setting telescope in the horizontal plane. Signed: *PHill Edinr* (P and H in monogram). (53.325)

4 Octant. Ebony frame, radius 10in. Vernier, tangent and fixing screw, three shades. *SBR* engraved on degree arc. Signed: *Spencer Browning & Rust* on ivory label. (56.623)

5 Octant. Ebony frame, radius 10in. Tangent and fixing screw. Vernier, three and two shades. Signed on ivory label: *Julius F. Schierbeck Copenhagen*. (52.23)

6 Circular protractor with folding arms with attached points. Brass, silver scales, diameter 6in. Signed: *Troughton & Simms*. In wooden box. Early 19th century. (53.327)

7 Circular protractor with vernier. Brass. Signed: *J. Gargory Birmingham*, *c*1850. (53.3287)

8 Folding square. Brass, hinged at corner. Cut out for plumb bob. Signed: *Bernier Paris*. (52.202)

9 Standard yard. Brass, inscribed: *Imperial Standard Yard 1835 I Blackburn Maker Minories London. Boulton Watt & Co.* and stamped WR. IIII. In wooden box with brass plate inscribed: *Imperial standard Yard 1835. Boulton Watt & Co.* (51.88/58)

10 Pantograph. Mahogany, brass and ivory. Length of arms 21¾in. (51/88/5)

11 Watt's roller copying press in mahogany box 21⅞ x 13¼ x 6¾in. Label on box (in contemporary hand): *Watt's Patented Roller Copying Press.* (51.88/1).
Lit: ANDREWS (1981-2) pp2-5

12 Gregorian reflecting telescope, 4in focus. Dumpy telescope on altazimuth mounting, pillar and tripod stand. Signed: *John Cuthbert London 1840.* (56.684)

13 Culpeper-type microscope, square base with drawer, black ray-skin covered outer tube, green vellum covered inner tube. Lignum vitae eyepiece, 18th century. (70.2465/3)

14 Cuff type microscope on mahogany box base with drawer. Unsigned. (72.2554/4)

15 Solar microscope. Brass, unsigned. *c*1800. (72.2554)

16 Compound microscope. Signed: *Powell & Lealand Makers London 1846.* Lacks eyepiece, objective and accessories. (56.730)

17 Compound microscope, signed *Pillischer London*, serial no. *1773 c*1866-60. (56.732)

18 Small compound microscope. Baker's 'Students' type. *c*1864. (58.977)

19 Compound microscope, signed: *Ross London.* Incomplete, late 19th century. (D216)

20 Binocular microscope with revolving stage for petrological work. Signed: *J. Swift 43 University St. London WC.* 1880s. (57.769)

21 Binocular microscope with petrological stage. Lacks objectives. Signed: *R. & J. Beck 31 Cornhill London.* (56.731)

22 Steam indicator designed by James Watt. (51.88/15)

23 Steam indicator dated 1785. Belonged to Boulton & Watt agent in Manchester. Boulton & Watt Collection. (51.88/15)

24 Steam indicator. Late 18th century. Boulton & Watt collection. (51.88/15)

25 Balance in vacuum case. Signed: *L. Oertling London.* Late 19th century. From Board of Trade. (56.727/3)

26 Plate electrical machine, Cuthbertson-type. Late 18th century. (58.928)

27 Electrical machine, Wimshurst-type. Late 19th century.

28 Matthew Boulton's lathe. From Boulton & Watt collection.

29 Rose turning engine. Early 18th century. On loan from Museum of the History of Science, Oxford. (53.50)

30 Rose turning engine, once in the workshop of Louis XVI. French, perhaps by Lecroix, mid-18th century.

31 Rose turning lathe by Fieldhouse, 1846. (55.593)

32 Holtzapffel lathe, *c*1870. Loan from Museum of the History of Science, Oxford. (58.893)

33 Fuller's spiral slide rule. Signed: *Stanley Maker London Entered Stationer's Hall No. 574.* (65.1929)

34 Excise slide rule. Boxwood, two brass slides. Signed: *Joseph Long.* (62.2569)

35 Sikes' hydrometer. Brass in wooden box. Thermometer signed: *Oertling.* (56.727)

36 Specific gravity beads in turned wooden box with label of: *Forbes Altrea Glassblower and Optician 162 High Street Edinburgh.* (56.727/4)

B 14 BLAIR ATHOLL
Blair Castle

Blair Atholl, Pitlochry PH18 5TL (The Duke of Atholl)

079681 207

The castle, dating from 1269, is in part open to the public and houses notable collections, including paintings, ceramics and arms and armour. It contains two items relevant to this survey.

1 Waywiser by Dollond, London.

2 Angle barometer set in a frame containing a thermometer, oatbeard hygrometer and a 'Perpetual regulation of Time' (an elaborate form of perpetual calendar). The metal plate on which the thermometer is mounted is signed: *Watkins fecit.*

B 15 BLANTYRE
The David Livingstone Centre

Livingstone Memorial, Blantyre, Glasgow G72 9BT

0698 823140

The Museum, which is housed in the birthplace of David Livingstone, contains items associated with the explorer and missionary. The scientific instruments, therefore, while not of great significance in themselves, are of considerable interest since they were used by Livingstone while in Africa. In addition to the items listed there are two cases of surgical instruments, thermometers, parallel rulers, etc.

1 Dipleidoscope. Inscribed on lid: *F. Dent's patent meridian instrument 61 Strand London.*

2 Artificial horizon, signed: *I.D. Potter (successor to R.B. Bate) London.*

3 Compass with sights. Unsigned, late 18th or early 19th century.

4 Pocket compass in wooden case. Signed: *Cary London.*

5 Compass with sights in brass case, signed: *Troughton London.*

6 Compass in brass case signed: *Elliott Bros. London.*

7 Sextant, 8in radius. Signed: *Troughton & Simms London* Serial number 2310. *c*1830.

8 Equinoctial dial. Signed: *Dollond London.* Early 19th century.

B 16 BOURNEMOUTH
Natural Science Society

39 Christchurch Road, Bournemouth BH1 3NS

0202 23525

This is the private Museum of the Bournemouth Natural Science Society, established about 1900. It contains a natural history collection including fossils, British birds and lepidoptera, as well as a collection of Egyptian and other archaeological antiquities. It is open by appointment.

1   Orrery. Brass mechanism mounted on a stand in the form of a four-legged circular table overlaid with engraved and coloured paper printed with a zodiac calendar. Signed: *Designed for the new portable orreries by W. Jones and made and sold by W. & S. Jones, 30 Holborn, London.* The instrument can be assembled as a planetarium or as a tellurium.

B 17 BOURNEMOUTH
Russell-Cotes Museum

East Cliff, Bournemouth BH1 3AA

0202 21009

The Russell-Cotes Museum, administered by the local authority, contains early Italian paintings and pottery, porcelain, furniture, bygones, ethnography, arms and armour and maritime material. A few scientific instruments are to be found, forming part of the maritime collection. In addition to the two items listed below there are three unsigned 19th-century ship's telescopes.

1   Octant. Ebony frame. Signed: *Spencer, Browning & Rust London.* Early 19th century.

2   Canal maker's level. First half 19th century.

B 18 BRACKNELL
The Meteorological Office

London Road, Bracknell RG12 2SZ

0344 420242

The Meteorological Department of the Board of Trade was formed in 1854. After being housed in a number of London buildings (including, early in the 20th century, the building which now forms part of the Science Museum), the Headquarters moved to Bracknell in 1961.

A small museum (open to the public by appointment only) was set up in 1979. Many of the meteorological instruments preserved date from the 20th century, and hence are outside the scope of this inventory. On the other hand, a few of the earlier instruments can be traced back to having been used at the Kew Observatory, and indeed, one or two may have originally formed part of the King George III Collection (the bulk of which is now at the *Science Museum, London*) A number of items were presented by the Royal Meteorological Society in 1978; they are marked 'RMS'.
Lit: JAY (1980)

ANEMOMETRY

1   Snow Harris's wind gauge, signed: *J. Lilley & Son,* *c*1858. (A1)

2   Hagemann's wind gauge, *c*1879. (A2)

3   Robinson's improved anemometer, possibly by L. Casella. Designed by TR Robinson of Armagh Observatory. *c*1870. (A4)

4   Air meter of type introduced *c*1870. Used by Lieutenant-Colonel E Gold, Royal Engineers, when meteorological adviser to General Haig during the First World War. (A7)

5   Lowne's patent recording anemometer, signed: *Negretti & Zambra.* Type introduced *c*1875. (A10)

6   Beckley-Robinson anemograph, from the Radcliffe Observatory, Oxford. 1893. (A11)

7   Model of Beckley-Robinson anemograph, *c*1856. (A12)

BAROMETRY

8   Portable barometer signed: *Daniel Quare, c*1700. (B1)

9   Marine barometer, signed: *William and Samuel Jones,* *c*1825. (B3)

10   Mountain barometer, signed: *J. Newman, c*1833. Three examples. (B7-9)

11   Portable barometer, signed: *J. Newman, c*1835. (B13)

12   Portable barometer, in bamboo case, *c*1850. Used by Captains JP Basevi (in Tibet) and WJ Heaviside (in India, Aden and Egypt). (B14)

13   Marine barometer, signed: *Ernst, Paris, c*1855. (B15)

14   Syphon barometer, signed: *Adie.* (B16)

15   Syphon barometer, signed: *Thomas Jones.* (B18)

16   Fitzroy's 'marine gun' barometer, signed: *Negretti & Zambra, c*1860. This design can withstand vibration caused by the firing of guns. (B17)

17   Fishery barometer, signed: *T. Salleron, Paris.* Presented to the Meteorological Office by the French Government in August 1861. Fishery barometers were provided by the Board of Trade for display at fishing ports between 1858 and 1910. (B20)

18   Fishery barometer, signed: *Negretti & Zambra,* No. BT99. Originally installed at North Shields in October 1866. (B21)

19   Eight fishery barometers, five by Negretti and Zambra, one by J Hicks and one by pAdie. (B 19,23-29)

20   Portable standard barometer, signed: *Tonnelot, Paris.* Presented to the Meteorological Office by C StC Deville in September 1874. (B30)

21   Station barometer, signed: *Holland,* Utrecht. (B31)

22 Admiral Fitzroy barometer, signed: *S Maw & Son*, c1880. (B32)

23 Sympiesometer, signed: *Cox, Devonport*, c1825. (B60)

24 Storm glass. (B50)

25 Aneroid barometer, signed: *Negretti & Zambra*. Probably used by James Glaisher on his balloon ascents in 1862-66. RMS. (B100)

26 Goldschmid marine aneroid barometer, signed: *Hottinger & Cie, Zurich*. Improved version of 1858 design by J. Goldschmid of Zurich. c1880. (B101)

27 Goldschmid's barograph, c1880. (BG 1)

HYGROMETRY

28 Kater hygrometer, signed: *Thomas Jones*, c1812. RMS (H1)

29 Daniell hygrometer. To 1819 design of JF Daniell. c1860. RMS (H2). Another example signed by J Newman. (H12)

30 Hair hygrometer, signed: *Bate, London*. c1830. RMS. (H3)

31 Dines dew-point hygrometer, signed: *L. Casella*. Devised by G Dines in 1871. (H5)

32 Mason hygrometer, signed: *Tozer, Torquay*, c1870. Dr Abraham Mason of Pentonville devised this wet-and-dry-bulb psychrometer in 1836. RMS (H8)

33 Mason hygrometer, by Henry Barrow, with calibration notes by James Glaisher. c1860. (H9)

34 Regnault hygrometer, possibly constructed by J Hicks, c1875. The instrument was devised by the Frenchman Henry Regnault in 1845. (H10)

35 Balloon thermometer (wet-bulb type), signed *L. Casella*, c1880. (H11)

RAINFALL

36 Beckley recording rain gauge, signed: *J. Hicks*, c1880. Designed by Robert Beckley, mechanical assistant at Kew Observatory, in 1869. This instrument installed at *Armagh Observatory* from 1880 to 1972. (R1)

37 Portable rain gauge, signed: *L. Casella*. Originally designed for use by David Livingstone on his Zambezi expedition of 1858. (R2)

38 Indicating rain gauge, signed: *Negretti & Zambra*, c1890. (R6)

39 Model of Dines's tilting-syphon rain recorder, c1915. (R7)

SUNSHINE AND SOLAR RADIATION

40 Campbell sunshine recorders (three examples), signed: *Negretti & Zambra*, c1880. (SS1)

41 Campbell-Stokes sunshine recorder, possibly by J Hicks. This instrument was in use at the Radcliffe Observatory, Oxford from 1880 to 1975. (SS2)

42 Callender sunshine receiver, signed: *Cambridge Scientific Instrument Company Ltd*, 1913. Of type devised by HL Callender in 1900. (SS10)

43 Solar maximum thermometer, signed: *Negretti & Zambra*, c1885. Of type devised by the firm in 1864. (SR1)

44 Herschel actinometer, signed: *Robinson and Barrow*. (SR6)

45 Angstrom pyrheliometer. Operates by means of thermocouples, originally devised by Knut Angstrom in 1886. (SR11)

THERMOMETRY

46 Thermometer, mercury, signed: *G. Adams*, c1780. Probably King George III Collection (T1)

47 Maximum and minimum thermometer, alcohol, signed: *Thomas Jones*. c1820. (T2)

48 Thermometer, mercury, signed: *Thomas Jones*, c1820. (T3)

49 Thermometer, mercury, signed: *J. Newman*, 1830. (T4)

50 Thermometer, alcohol, signed: *J. Newman*, c1854. (T5)

51 Thermometer, mercury, signed: *J. Newman*, c1854. (T6)

52 Sea thermometer, mercury, signed: *Ernst, Paris*, c1860. (T7)

53 Maximum and minimum thermometer, bimetallic coil, signed: *Hermann & Pfister, Bern*, 1867. (T8)

54 Thermometer, mercury, signed: *L. Casella*, 1854. The first thermometer ever supplied by Casella to the Meteorological Department of the Board of Trade. (T9)

55 Maximum thermometer, mercury with capillary constriction, signed: *Negretti & Zambra*, 1873. (T10)

56 Maximum thermometer, mercury with graphite index, signed: *Newcombe & Co*. (T11)

57 Minimum thermometer, alcohol with glass index, signed: *Negretti & Zambra*, 1891. (T12)

58 Absolute maximum thermometer, mercury with capillary constriction, signed: *Negretti & Zambra*, 1909. (T13)

59 Recording thermometer, Negretti & Zambra's patent, c1898. (T14)

60 Fluctuation thermometer, mercury, signed: *L. Casella*, c1856. An experimental instrument designed by Balfour Stewart at Kew Observatory to measure the sum of fluctuation in temperature. (T15)

61 Cylinder jacket minimum thermometer, alcohol, designed by J. Hicks in 1874 to have a very quick response to temperature changes. (T16)

62 Thermometer, alcohol, signed: *J. Hicks*, made for Captain Scott's Antarctic expedition of 1901. (T18)

63 Rutherford minimum thermometer, signed: *Baudin, Paris*, 1877. (T20)

64 Recording thermometer, signed: *E. Richter & Wiese, Berlin.* Type patented by Negretti & Zambra in 1897. (T21)

65 Kew Observatory standard thermometer (20 examples), mercury, range of dates 1851-1889. (T17,22-37,41-43)

66 Thermometer with two expansion chambers, by Adie. (T49)

67 Thermometer, alcohol, by Adie. (T53)

68 Minimum thermometer, alcohol, bifurcated bulb, signed: *Aschoff, Breslau.* (T55)

69 Thermometers (nine examples) by Adie and Newman, from St Helena, 1849. (T56)

70 Thermometer, alcohol, signed: *Greiner, Berlin,* dated 'Mai 1845', the property of Edward Sabine. (T68)

71 Calibration thermometers (two examples), signed: *Salleron, Paris.* (T74)

72 Earth thermometers, Six's maximum and minimum (two examples). (T81,82)

73 Casella patent minimum thermometer, signed: *L. Casella.* Type patented by Louis Marino Casella in 1861. RMS (T85)

74 Thermograph, bourdon tube type, signed: *Richard Freres Paris.* This instrument was supplied by the Hydrological Office for the Antarctic Expedition of 1901. (T86)

75 Recording photothermographs (two examples), used at Kew and Eskdalemuir, *c*1850-1950.

MISCELLANEOUS

76 Balloon meteorograph, designed by WH Dines in 1908 to record changes in temperature and pressure through the atmosphere. (UA3)

77 Standard metre, signed: *Deleuil a Paris.* (M1)

78 Specific gravity bottle with label 'Presented by R.H. Scott 15/11/1884'. (M11)

79 Chinese sundial. (M12)

80 Meteorologist's rule, ivory, various scales, by Stanley. (M15)

81 Plate electric machine (plate 6in diam.), *c*1800. Probably King George III Collection. (EL32)

82 Pantograph, brass, signed *Adie, 15 Pall Mall, London,* mid-19th century.

83 Apparatus for collecting gases, glass bulb with brass tag, early 19th century.

84 Electrometers, signed *Thomas Jones, 62 Charing Cross, London,* early 19th century.

85 Magnetometer, signed: *Elliott Bros, London.*

B 19 BRADFORD
Bolling Hall

Bolling Old Road, Bradford BD4 7LP

0274 723057

Bolling Hall is largely a 17th-century house incorporating a late medieval peel tower, with a wing remodelled in the late 18th century by John Carr of York. The Hall was opened as a period house and museum of local history in 1915. It is now primarily filled with furnishings contemporary with the house, amongst which are some scientific instruments. Though not numerous, several (nos. 1 to 8 below) are particularly important since they were constructed by the mathematician and astronomer Abraham Sharp who acted as assistant to John Flamsteed, first Astronomer Royal, at the Royal Observatory, Greenwich, over the period 1684 to 1690.

Sharp was born at Little Horton, near Bradford, in 1653 and died there in 1742. Few of his instruments have survived. As well as those cited here, a universal equatorial instrument presented to the Yorkshire Philosophical Society in 1835 has been lent to the *National Maritime Museum, London.* An astronomical and trigonometrical calculator is in the *Science Museum, London.*

In addition to the material below, Bolling Hall possesses two diagonal barometers (one single, the other double) by Charles Howorth of Halifax, and a Newton's New and Improved Terrestrial Globe of 1824 in an ornate stand.

Lit: BRADFORD-BOLLING HALL MUSEUM (1963); references to entries in this catalogue of an exhibition dealing with Abraham Sharp are indicated in brackets below.

1 Brass rete to 'Mathematical Jewel'. (47)

2 Brass horary quadrant of Sutton type. Engraved on one side only for use at the latitude of 53°51' (Sharp's home at Horton). (48)

3 Instrument in the form of a trapezium for use as a quadrant. It is engraved along one edge with the names of the first six months of the year, and along the opposite edge with the names of the last six months. The two remaining sides are engraved with divisions from VI to XII and from XII to VI. These divisions are linked by a network of lines. Presumably the instrument was used with a plumb line as a kind of horary quadrant. (51)

4 Horary quadrant. The main part of the instrument consists of a quadrant of an orthographic projection of the celestial sphere on to the colura of the solstices. The radii of the instrument have been extended and are engraved with degree and other scales. The reverse of the quadrant is engraved with diagrams of astronomical co-ordinates and other data. (50)

5 A composite instrument consisting of a calibrated wooden rule to which is attached a brass segment of an arc engraved with a section of an orthographic projection of the celestial sphere on the colura of the solstices. To one end of this arc is pivoted one end of a folding brass rule. A rule in the form of an arc or a circle is mounted on the brass segment. The instrument is clearly experimental. (49)

6 Eccentric chuck. Believed to have been constructed by Sharp and used with his lathe for making scientific instruments. (45)

7 Collection of wooden polyhedra made by Sharp and used to illustrate his *Geometry Improv'd* (Part 2), published under the pseudonym AS Philomath. (53)

8 Horizontal sundial. Bronze, octagonal. Inscription (partly legible): *Mr. Sharp Delinie. Latit. 53°13′ 1722*.

9 Sextant. Ebonized wood. Signed: *Parnell London*. Label of Parnell inside box. Early 19th century.

10 Waywiser. Iron-bound wooden wheel. Signed: *Wm. Harris & Co. 50 Holborn London*. Early 19th century.

11 Lodestone. Brass mounted, the mounts with scalloped edges. Suspension ring. Possibly 17th century.

12 Surveyor's compass in rectangular mahogany box with lid. Folding sights on N-S axis. In the base of the compass box, an engraved paper disc with 16-part windrose and stereographic projection of the world. Signed on the paper disc: *J. Jones London*. Third quarter 18th century.

13 Personal weighing scale and height measurer, signed: *Merlin Princes Street Hanover Square*.
Lit: LONDON (1985) pp68-70 for similar examples

B 20 BRADFORD
National Museum of Photography, Film and Television

Prince's View, Bradford BD5 0TR

0274 727488

The Museum was established as an outstation of the *Science Museum, London,* in cooperation with the City of Bradford Metropolitan Council in 1983. Galleries so far opened show the role of photography through working models, equipment, reconstruction and pictures themselves. Most material lies outside the scope of this survey.

In February 1984 Kodak Limited announced the company's intention to donate the Kodak Museum, Wealdstone, Middlesex, to the National Museum of Photography. This includes about 15,000 items of historical equipment. Again, although most of this material falls outside the scope of the inventory, the few items which can be included from this source are marked: (Kodak).

1 Camera lucida, in box stamped: *Newman, 24 Soho Square*, *c*1820. Ex Wellcome Collection (A645039). (ScM 1981-268)

2 Camera obscura, folds into mahogany box. Lens 25in focal length. *c*1800. On loan from *Museum of the History of Science, Oxford*. (NMP 1982-83)

3 Camera obscura, mahogany. Lens 8in focal length. *c*1830. (NMP 1982-6)

4 Camera obscura, replica of WH Fox Talbot's example in the *Science Museum, London*, original *c*1830.

5 Actinograph, Hurter and Driffield type, for calculating exposure times. *c*1892. (ScM 1907-100)

6 Metric radius gauge, for measuring curvature of spherical surfaces. Made by WG Pye and Co, Cambridge. (1990-5036/3392) (Kodak)

7 Imperial radius gauge, for measuring curvature of spherical surfaces. Made by Pillischer, London. (1990-5036/3393) (Kodak)

8 Camera obscura, rosewood. Box construction with mirror reflector and horizontal ground glass. Simple lens in square section sliding tube. Hinged hood/cover with supporting strut. *c*1800. (1990-5036/6982) (Kodak)

9 Camera obscura, in mahogany box with metal hood. Ground glass screen 2⅓ x 2⅔in. Lens 3⅛in in sliding mount 2¾ x 2⅔ x 3in. *c*1800? (1990-5036/0641) (Kodak)

10 Camera lucida, brass, 'patent' in shagreen covered case. G-clamp fitting, telescopic tube. Two pivoting correction lenses. (1990-5036/4967) (Kodak)

11 Camera lucida, nickel plated, in fitted case. G-clamp fitting, telescopic shaft. Ten correction lenses and filters. Made by Stanley, London. (1990-5036/4966) (Kodak)

B 21 BRECON
Breconshire War Memorial Hospital

Cerrigcochion Road, Brecon LD3 7NS

0874 2443

The Hospital houses two early stethoscopes, both of them with local connections, and one of considerable interest since it is a rare and early type.
Lit: KYLE (1970)

1 Stethoscope, Laennec type. Cylindrical in shape, and composed of three sections fitting into one another. Hard wood, perhaps mahogany, 12in in length, and about 2in in diameter, channel down the centre of the cylinder of approx ¼in. Hole in the chest piece lined with brass tubing which projects slightly and fits into the wooden channel of the next section. A small silver label on the side of the stethoscope is inscribed: *This stethoscope was presented by Laennec to Dr. Prestwood Lucas*. Thomas Prestwood Lucas, physician in the Brecknock Infirmary from 1840 until 1871, studied in Paris under Laennec in 1825 and 1826, after having qualified in Edinburgh, and must have been presented with the stethoscope during his Paris sojourn. It is of the type invented by Laennec in 1819.

2 Stethoscope, *c*1860. Belonged to Dr J Balfourd Jones, who practised in Brecon from 1865 to 1887.

B 22 BRECON
The Brecknock Museum

Captain's Walk, Brecon, Powys LD3 7DW

0874 4121

The Museum was established in 1928 by the Brecknock Society. In 1950 it was leased to the Brecon County Council and was administered by the Library and Museum Committee of the County Council from 1966. The collection includes natural history, archaeology, folk life, coins, costumes, ceramics, prints and paintings. It houses three items of relevance to this survey.

1   Octant. Mahogany frame, brass arm, ivory scale. *I.W.P.* engraved on arm, *c*1760.

2   Nautical telescope, octagonal wooden tube, *c*1750.

3   Waywiser. Iron-bound wooden wheel. Signed: *Heath & Wing London*. mid-18th century.

B 23 BRIDGWATER
Admiral Blake Museum

Blake Street, Bridgwater, Somerset

0278 456127

This small museum contains relics associated with Admiral Robert Blake (1599-1657) and the battle of Sedgmoor. The house is reputedly the Admiral's birthplace. The Museum also contains collections illustrative of the archaeology and local history of the area.

1   Surveyor's compass. Circular brass compass contained in an octagonal pinewood box. Semicircular ears project on an axis of the compass and are provided with screw holes. Two wing screws, of a size to fit these holes, fit into recesses in the wooden box. Presumably two arms with sights could be screwed onto the compass. The compass is reputedly associated with Admiral Blake but a handwritten sheet pasted inside the lid commences 'Hollewell page the 19th Sturmy page IV Seaman's Kalender page the 1st and so onwards'; these are references to three late 17th-century works on navigation and surveying published after Blake's death.

B 24 BRISTOL
City of Bristol Museum and Art Gallery

Queens Road, Bristol BS8 1RL

0272 299771

The Museum, which shares a building with the City Art Gallery, traces its origins to the collection of the Bristol Institution for the Advancement of Science and Art, founded in 1820. Its collection became the property of the Museum in 1894.

The collections include zoology, geology, botany, archaeology and anthropology, local history, technology and transport. A number of branch museums, as well as the main museum building, come under the care of the Museum and Art Gallery.

From the 17th century some instrument-making was carried on in Bristol, as might be expected of a major port. The collection of scientific instruments, mainly housed in the Department of Technology, is small but growing. It includes navigation instruments, and some items by Bristol makers. In addition to those items described is a group of instruments from the local Customs and Excise Department (including callipers, slide rules and other miscellaneous devices, some signed by Dring and Fage of London).

1   Screw barrel microscope. Brass with ivory handle, four objectives and a set of ivory slides all contained in a black fish-skin box. Unsigned, 18th century. (J 1540)

2   Culpeper microscope, form reputedly constructed by Matthew Loft. (J 1539)

3   Solar microsope. Brass, mounted on later wooden stand. Signed: *Jones & Son Holborn*, *c*1785.

4   Lucernal microscope. Mahogany with brass fitting. Stand with gilt lion's-paw feet. Signed: *S. Washbourn London*. (J902)

5   Compound microscope on folding tripod foot. Triangular pillar inclinable by means of a compass joint at the centre of the tripod. A development of Benjamin Martin's microscope. Probably *c*1780. (J 1554)

6   Binocular microscope. Lister limb mounted in trunnions. Signed: *Smith & Beck 6 Coleman St. London*, Serial no. 1313. Inscribed with presentation date, 1857. (J 1570)

7   Binocular microscope in mahogany box with accessories. Stand signed: *Thomas King Bristol serial no.223* and inscribed with date 1857. Around binocular tube: *Wenham's Binocular by Ross London*. Presumably the instrument was modernized in the 1870s. (J 445)

8   Compound microscope. In mahogany with two cases of accessories. Signed: *King & Coombs Opticians Bristol*. Serial no. *171*. (J 739)

9   Gregorian reflecting telescope. Brass on pillar and tripod stand. Length of barrel 30in. Signed: *Nairne London*. Third quarter 18th century. (J 744)

10   Refracting telescope. Brass, on wooden tripod stand with wooden box. Outer tube leather-covered, single draw tube. Closed length 52in. Signed: *Dollond London*. Mid-19th century. (J 691)

11   Refracting telescope. Single-draw tube, brass, outer tube sheathed with wood. Length (closed) 25in. Signed: *Berge late Ramsden*, inscribed: *Belonging to Lord Viscount Nelson 21st October 1805*. (600 M. Loan).

12   Spyglass/telescope. Single-draw tube, brass, with leather-covered outer tube and shade tube. Closed length 5½in, open 6½in. Signed: *Carpenter & Westley 24 Regent St. London*, inscribed: *Genl Breton United Services Club*. *c*1840. (J 1552)

13 Refracting telescope. Brass with leather-covered outer tube and leather carrying case. Extended length 27½in. Signed: *J.H. Steward 406, 66, 457 The Strand and 54 Cornhill London*. Late 19th century. (J 743)

14 Refracting telescope. Brass, with ivory-covered outer tube, and leather carrying case. Four draw tubes. Closed length 6½in, fully extended 20½in. Signed: *Husbands 8 St. Augustine's Parade Bristol*. (J 1567)

15 Horizontal plate dial. Brass. Signed: *T.W. 1793*. From Chippenham, Wiltshire. (J 1536)

16 Horizontal dial for use in one latitute only. Square mahogany box with lid, 3 x 3in. Includes engraved paper scale of equation of time. Signed: *W. Watkins Optician St. Augustine's Back Bristol*. Early 19th century. (J 1537)

17 Universal equinoctial ring dial. Brass, diameter 6in. Unsigned, 18th century. (J 1765)

18 Universal equinoctial ring dial. Brass, diameter 9in. Signed: *Heath & Wing London*. Third quarter 18th century. (J 1596)

19 Universal mechanical equinoctial dial. Gilt brass 3½in octagonal base plate with pewter hour ring and glazed minute dial. In blind-stamped leather case. Unsigned. German, second half 18th century. (J 1538) (fig 32)

20 Nocturnal. Boxwood, length 10in. Inscribed: *Both Bears*, English, *c*1700. (J 1113)

21 Backstaff. Pearwood and lignum vitae. Signed: *H. Gregory near ye India House London*. Lacks two vanes. Third quarter 18th century.

22 Octant. Ebony frame, radius 10in. Fixing and tangent screw. Ivory arc marked: *SBR*. Signed on ivory label: *Braham Bristol*. Second quarter 19th century.

23 Sextant. Brass with silver degree arc. Radius 7in. Signed: *Elliott Bros. London [1060]*. Inscribed: *L.H.K. Hamilton R.N.* Box with label of George Lee & Son Portsmouth. (590 M. Loan)

24 Reflecting circle, diameter 11in, with counterpoised brass stand. Signed: *W. & S. Jones 30 Holborn London*. (J 746)

25 Mariner's compass. Brass, in gimbals in wooden box, 10¾ x 10¾ x 6in. Diameter of compass 7¼in. Unsigned, early 19th century. (J 1565)

26 Mariner's compass. Brass bowl in gimbals in wooden box, 7⅝ x 7⅝ x 5in. Label on underside of sliding lid: *The Property of James Hoskin 48 Cabin Gt. Western July 1840*. (J 410)

27 Tide gauge designed by JG Blunt. (J 1764)
Lit: WHEWELL (1838)

28 Orrery with tellurium and lunarium. Brass, on pillar and tripod stand. Calendar/plate 8¾in diameter. Signed: *W. & S. Jones Holborn London*. Globe signed: *J. & W. Cary Strand April 1791*. Late 18th century. (J 1082)

29 Dividers, two pairs, brass and steel. (a) Length 5⅛in. Brass, steel points no longer survive. (b) Length 4¾in. Lacks points. Both dredged up in Bristol harbour. (J 1594-5)

30 Dividers. Brass, length of arms 12in. Signed: *Julianus Venturin fecit Romae 1785*. (33/1966)

31 Drawing instruments in green ray-skin case with silver mounts, length 6⅞in. Parallel rule, square protractor and scale, and sector are of ivory, other instruments are of brass. The protractor and sector are signed: *E. Halse & Son London*. (J 426)

32 Farey's ellipsograph. Brass and steel in mahogany box, 10½in square. Probably mid-19th century. (J 1566)

33 Pantograph. Brass, length 36in, in wooden box. Signed: *W. & S. Jones*. First half 19th century. (J 1555)

34 Pantograph. Unsigned, mid-19th century. Length of arms 35in. Brass, in wooden box. (J 1556)

35 Camera lucida, Wollaston-type. Brass, in black fish-skin case, velvet lined. Instrument marked: *Patent*. (British Patent 2993, 1806); Serial No.1330 engraved on prism. (J 1576)

36 Compass. Diameter 10¼in. With folding hair and slit sights. Signed: *Langford Chronometer & Instrument Maker. Broad Quay Bristol*. Late 19th century. (J 261)

37 Surveyor's compass, incomplete. Diameter 10in. Signed: *W. Langford & Son Bristol*. (J 1561)

38 Brass compass. Circular lid is equipped with a socket for use on a tripod. Unsigned, 19th century. (J 1549)

39 Theodolite. Brass, silver scale. Signed: *Troughton & Simms*, *c*1830-40. (J 1533)

40 Theodolite. Brass, silver scale, height 13½in, diameter of horizontal circle 6in. Unsigned, early 19th century. (J 1548)

41 Surveyor's level on tripod. Box with label of Adie London. Mid-19th century. (J 692)

42 Miner's dial. Compass mounted in gimbals, folding open sights, clinometer. Signed: *Ash and Son Makers 4 Bull Street, Birmingham*. On lid of compass: *James Roe & Co. Manufacturers Birmingham*. In wooden box, *c*1850-60. (J 1577)

43 Miner's dial. Overall length 8¼in, diameter of compass 4½in. Signed: *Elliott Brothers London*. (J 715)

44 Miner's dial. Signed: *J. Archbutt & Sons 202 Westminster Bridge Rd*. In box with label of J Davis & Son. (J 1597)

45 Tate's arithmometer. Made by C & E Layton London. (J 1758)

46 Clarke's hydrometer. Brass, with copper sphere, length 6½in. In cylindrical tin-plate case, containing a booklet describing the hydrometer. The frontispiece is dated 1746. Hydrometer signed: *Clarke London*. (J 750)

47 Sikes' hydrometer. In mahogany box inscribed: *Sikes' hydrometer Loftus 6 Beaufoy Terrace London*. Serial no. on hydrometer 17638. (J 1572)

48 Hydrometer. Glass, paper scale in crude box. Signed: *Josh Long 43 East Cheap London*. (J 1575)

B 25 BRISTOL
University of Bristol Library, Special Collections

Tyndall Avenue, Bristol BS8 1TJ

0272 303030

The Special Collections house a group of mid-19th century drawing instruments associated with the engineers Marc Isambard Brunel (1769-1849), his son Isambard Kingdom Brunel (1806-1859) and grandson HM Brunel. The instruments were presented with Brunel papers in 1950 by the Dowager Lady Noble.

1   Set of 17 drawing instruments, in mahogany box, signed: *Elliott, London.* Each instrument inscribed: *I.K.Brunel.*

2   Set of 30 drawing instruments in mahogany box. Two instruments inscribed: *I.K.B.,* five inscribed: *Henry Brunel* and two inscribed: *H.B.*

3   Parallel rule, inscribed: *I.K. Brunel 1846.*

4   Steel rule, inscribed: *I.K. Brunel 1846.*

5   Two parallel rules, ebony, signed: *Elliott* and *J.W. Hammond.*

6   Set of 17 drawing rules in mahogany box, signed: *Elliott* and *J. Tree.* Inscribed: *H.M. Brunel.*

7   Parallel rule, ivory, in case, inscribed: *I.K. Brunel.* The rule inscribed: *H.B. from A.B.H. Sept. 1858.*

8   Set of six drawing rules in wooden box, two signed: *Elliott* and four signed: *Stanley.* Inscribed: *H.M. Brunel.*

9   Slide rule, wood, signed *W. & S. Jones London.*

10   Set of eight chain scales in case, signed: *Holtzapffel & Co. London.* One scale inscribed: *Plan Office 18 Duke Street.*

11   Set of 23 drawing scales, ivory, signed: *W. Elliott* and *Elliott* and *Watkins & Hill.* Nineteen scales inscribed: *I.K. Brunel.*

12   Chain scales (25) in case, signed: *Holtzapffel & Co., London.*

13   Drawing scales (34) in case, signed: *Holtzapffel & Co., London.*

14   Callipers, steel, inscribed: *H.M. Brunel.*

15   Double dividers, steel.

16   Protractor, brass, signed: *Gardiner, Bristol.*

17   Binoculars, in leather case inscribed: *H.M. Brunel.*

18   Pocket barometer, signed: *Adie, London.*

B 26 BROADHEATH
The Elgar Birthplace

Crown East Lane, Lower Broadheath, Worcester WR2 6RH

090566 224

A Museum containing personal relics, manuscripts, letters, etc., of the composer Edward Elgar (1857-1934), brought together in his birthplace. It is governed by the Elgar Birthplace Trust and was opened in 1938.

1   Compound microscope. Signed: *A. Ross London No. 233.*

2   Barometer: Signed: *Martin Baskett & Co. Cheltenham.* Mid-19th century

3   Incomplete sulphuretted hydrogen apparatus designed and said to have been patented by Elgar.

B 27 BROADWAY
Snowshill Manor

Snowshill, Broadway WR12 7JU

0386 852410

A small manor house, of 18th-century and earlier date, which was presented to the National Trust in 1951 by the late Charles P Wade (1883-1956), together with the vast and varied collection of items he had assembled. To mention only major categories, the manor contains furniture, musical instruments, costumes, toys, oriental material including Samurai armour, bicycles and other transport items, agricultural implements, material relating to industries such as lace-making and weaving, as well as clocks, watches and scientific instruments.

In addition to the instruments listed below there are a small number of medical and surgical instruments (including cupping sets and scarificators), and a number of jeweller's and coin balances.

No catalogue has been published of the instruments, but many of them are listed in a manuscript catalogue prepared by Wade and preserved at Snowshill Manor. More recently, in 1974, the greater part of the scientific instruments have been photographed and catalogued by Eric Voice, and his numbering has been retained in nos. 1-59 of the ensuing list. Items not listed by Voice appear as nos. 60-77. The numbers in Wade's catalogue appear in parentheses after each entry.
Lit: VOICE (1974)

1   Armillary sphere. Beechwood and engraved paper on pasteboard. Ebonized stand. Height 20in, diameter 12⅝in. French, early 19th century.

2   Copernican armillary sphere of similar dimensions and with ebonized stand as 1. Pewter horizon and meridian rings. Crude gearing for Earth, Sun and Moon. French, unsigned, early 19th century. (Wade 140)

3   Orrery with planetarium and tellurium. Brass mechanism, wooden base, diameter 17¼in covered with engraved and varnished paper. English, mid-19th century. (Wade 153)

4 Orrery, planetarium only. Wooden base diameter 9½in covered with engraved and varnished paper, brass mechanism. Engish, mid-19th century. (Wade 150)

5 Tellurium, pair to 4. Wooden base diameter 9½in covered with engraved and lacquered paper. Brass mechanism. English, mid-19th century. (Wade 149)

6 Pocket globe in fish-skin case, diameter 3¼in, lined with engraved paper celestial sphere. Signed: *Newton's Improved Pocket Celestial Globe*. (Wade 26)

7 Pocket globe. Spherical fish-skin case, diameter 3in. A celestial sphere lines the case which contains a terrestrial globe. Signed: *Newton's New & Improved Pocket Celestial Globe 66 Chancery Lane London*. (Wade ii)

8 Pocket globe in spherical case covered with red leather, diameter 3in. Terrestrial globe housed inside the case which is lined with a celestial sphere. Signed: *Dudley Adams*, late 18th century. (Wade iii)

9 Pocket globe. A spherical black fish-skin case, diameter 3in, contains a terrestrial globe with brass meridian circle so that the globe can be revolved inside its case. A celestial sphere lines the case, and is signed: *Newton's Improved Pocket Celestial Globe*. Second quarter 19th century.

10 Pocket globe. Terrestrial globe in black fish-skin-covered spherical case, diameter 2¼in. The case is lined with engravings representing the 12 lunar phases, the seasons and the zodiac. Globe inscribed: *Newton's New Terrestrial Globe 1818*. (Wade i). For larger globes see nos. 71-4.

11 Backstaff. Lignum vitae and boxwood, length 25in. Signed: [fleur de lys] over *I.F.* and *Thos Dutch Jany 20 1746*. (Wade 26)

12 Octant. Mahogany frame, ivory scale, brass and mahogany arm, radius 16in. Signed: *J. & W. Watkins Charing Cross London*. Late 18th century. (Wade 59)

13 Octant. Ebony frame, brass arm, ivory arc engraved with anchor, radius 12in. Signed: *W. & T. Gilbert London*. Second quarter 19th century. (Wade 60)

14 Octant. Ebonized mahogany frame, radius 10in. Brass arm, ivory arc. Signed: *Hughes London*. Second quarter 19th century. (Wade 13)

15 Octant. Ebonized mahogany frame, brass arm, ivory arc engraved with anchor. Radius 10in. Box. Unsigned. Very similar to 14. (Wade 19)

16 Sextant. Brass frame, radius 8in. With magnifier for vernier, and telescope. Signed: *Cary London*. Mid-19th century. (Wade 215)

17 Pocket sextant. Diameter 2¾in. Signed: *Smith Royal Exchange London*. c1820. (Wade 166)

18 Bloud (magnetic azimuth) dial, ivory, 3⅛ x 2¾ x ⅝in. Signed: *Fait & Inven. par Charles Bloud A Dieppe*. (Wade 248)

19 Universal equinoctial ring dial. Brass, diameter 9⅛in. Signed: *I. Coggs fecit*. Early 18th century. (Wade 139)

20 Horizontal garden sundial. Brass, 12 x 12in. Signed: *J. Fowler London*. Second quarter 18th century. (Wade 16)

21 Circumferentor. Brass, diameter 12in. Sights on alidade and on one diameter. Signed: *F. Combes Fecit*. Early 18th century. (Wade 148)

22 Miner's dial. 5½in compass with lid, folding slit and thread sights. Overall length 8½in. Signed: *R. Field & Son Birmn*. Third quarter 19th century. (Wade 595)

23 Gunner's quadrant. 7in quadrant with 23in rule attached to one radius. Brass, signed: *Adams Charing Cross*. Late 18th century. (Wade 20)

24 Circular protractor. Brass, diameter 9in. Signed: *J. Simons London*. c1780-90. (Wade 40)

25 Altazimuth theodolite. Brass, height 9¼in, in wooden box. Signed: *W. & S. Jones Holborn London*. c1800. (Wade 622)

26 Pelorus course corrector. Maker: Bain & Ainsley, Patent No. 3299. Brass dial, on gimbals with pillar and sighting tube. Originally invented by MC Friend, and patented 1854 (British Patent 2652); improved version of which this is an example patented by R Bain in 1877. (Wade 89)

27 Artificial horizon. Mercury trough, hood, and mercury jar in box 8⅜in x 6in. 19th century. (Wade 155)

28 Azimuth compass. Brass compass bowl, card diameter 8in, in mahogany box 14½in square. Signed on compass card: *McCulloch Compass maker to his Royal Highness Prince William Henry, No. 38 Minories London. Patent*. Trade card of Kenneth McCulloch in lid of box. English, c1790, British Patent 1663 (1788). (Wade 82)

29 Ship's compass. Brass bowl, diameter 4¼in, mounted in gimbals in a mahogany box 6 x ⅜in square. Signed: *H. Hughes 59 Fenchurch St. London*. Third quarter 19th century. (Wade 273)

30 Compass in turned wooden case, diameter 5½in, paper compass card. Unsigned, late 18th/early 19th century. From a terrestrial globe.

31 Compass in mahogany box with lid, 6in square. Copper compass dial. Unsigned, early 19th century. (Wade 37)

32 Compass in circular turned wooden box, diameter 6 in. Paper compass card, signed: *D. Adams London*. Probably from a globe stand. (Wade 35)

33 Surveyor's cross. Brass, octagonal. Compass with silvered dial mounted on top of cross, height 5⅝in. In box with labels of John Archbutt & Sons, Westminster Bridge Road, Lambeth, and of C Baker. Instrument probably French in origin. Third quarter 19th century. (Wade 101)

34 Telescope, hand-held, two draw tubes, diameter 2in, extended length 32in. Red ray-skin outer tube, green vellum inner tubes, brass fittings. Signed: *Mann & Ayscough London*. (Wade 83)

35 Telescope, hand-held, four draw tubes, diameter 1¾in, extended length 45in. Black ray-skin outer tube, green vellum inner tubes. Lacks objective. Signed: *I. Mann fecit*. Second quarter 18th century. (Wade 84)

36 Telescope, hand-held, diameter 1½in, extended length 27½in. Green ray-skin outer tube, green vellum inner tubes. Signed: *Dollond London*. Last quarter 18th century. (Wade 85)

37 Telescope, hand-held, diameter 2in, extended length 31in. Brass, with mahogany covered outer tube. Signed: *G. Wilson London*. Early 19th century. (Wade 87)

38 Culpeper-type microscope – Matthew Loft form with continuous curving tripod legs. Red ray-skin outer tube, green vellum inner tube. Lignum vitae eyepiece and snout. Circular wooden base mounted on square box foot with drawer containing lenses and accessories, revolving object holder. Height 16½in. In pyramidal case. (Wade 137)

39 Compound microscope on circular brass base and the three objectives can be used separately or combined. Stage focusing. Ht. 8½in. The tube screws into an arm at the top of the pillar. In cylindrical red leather case. Early 19th century. (Wade 138)

40-44 Sandglasses, 18th-19th century. (Wade 90, 91, 648)

45 'Double' or multiple tube barometer in glazed mahogany case (Amontons pattern). 5in wide x 21.6in high. Signed: *Dom. Sala London*. (Wade 697)

46 Marine barometer, brass and mahogany, length 37½in with rotating cover for barometer tube. Unsigned, early 19th century. (Wade 48)

47 Telescope in walking stick. Point and top with inset compass can be unscrewed to reveal telescope. Single draw tube, length 34in. Signed: *Abraham Optician Bath*. Early 19th century. (Wade 46)

48 Excise officer's malt sampler. Incomplete. English, late 18th or early 19th century. (Wade 145)

49 Dividers. Brass and fruitwood in wooden (oak) box. Steel points. A brass arc which can be screwed to one arm, passes through a slit in the other and is engraved with a scale which permits the distance between the points to be read off between 6 and 48in. Length 37½in. Probably used by surveyor or architect. Unsigned; second half 18th century. (Wade 21)

50 Slide rule and folding 2ft rule. Boxwood, with brass hinge, and slide. Signed: *Keep Brothers*. Mid-19th century. (Wade 609)

51 Slide rule and folding 2ft rule. Boxwood, brass hinge and slide. Unsigned. Mid-19th century. (Wade 610)

52 Rule 18in, hinged to fold in four. Ivory with brass hinges and mounts. Unsigned. Probably late 19th century. (Wade 113)

53 Sector, 6in. Ivory with brass hinge and caps. Unsigned. English, early 19th century. (Wade 116)

54 Sector, 6in. Brass with boxwood hinge and caps. Unsigned. English, early 19th century. (Wade 115)

55 Gunner's callipers. Silvered brass, steel points, length 7in. Signed: *Critchton London* and monogram of The East India Company (*EIC* within a heart) and numbered 4. (Wade 24)

56 Cattle gauge (for estimating the weight of stock from measured dimensions). Boxwood, ivory sliding scale, brass cursor and mounts. Length 8in. Inscribed: *Ewart's cattle gauge adapted for any market*. Signed: *T.B. Winter Newcastle on Tyne*. 19th century. (Wade 665)

57 Excise rule, square section with slides on four sides, length 12in. Signed: *Rix Maker in Shrewsbury Court in White Cross Street near Cripple Gate London*.

58 As 57, save in spelling of *Criple Gate*.

59 Beer and wine gauge. Boxwood and brass. Early 19th century. Unsigned. (Wade 144)

INSTRUMENTS NOT LISTED IN VOICE'S CATALOGUE

60 Set of drawing instruments in fish-skin case. Incomplete, signed: *Nairne & Blunt*. Include ivory sector and rule.

61 Magnifying glass in circular mount with handle. Wooden case. Diameter of glass 6in. 18th century

62 Miniature balance in glazed brass case contained in mahogany box. Dimensions of case 8½ x 3¼ x 2¼in. Unsigned. Probably latter part of the 19th century.

63 Hydrometers and a saccharometer in wooden case, signed: *Joseph Cappo 24 Portland St. Belfast; William Twaddel Glasgow; and Buss Hatton Garden* (saccharometer); one unsigned Twaddel hydrometer of German origin.

64 Waywiser, signed: *C.W. Dixey*.

65 Waywiser, signed: *W. & S. Jones*.

66 Waywiser, 16in wheel, signed: *Moore Ipswich*. Separate dials for rods and furlongs.

67 Waywiser. Triple dial mounted at centre of wheel.

68-70 Three waywisers, one lacking the counting mechanism.

71 Terrestrial globe, on stand with four turned black lacquered legs. Diameter of globe 15in. Signed: *Newton & Son 66 Chancery Lane London*.

72 Celestial globe, diameter 12in. Wooden stand with four turned legs. Signed: *Car. Price London fecit*. (Wade 154)

73 Celestial globe, diameter 18in, in wooden pillar and tripod stand. Overall height 42in. Globe signed: *W. and J.M. Bardin, 16 Salisbury Square, Fleet St. London*.

74 Terrestrial globe, 12in diameter, signed: *S.S. Edkins son in law and successor to the late J. Bardin*, dated 1836.

75 Clockmaker's wheel-cutting engine. Brass plate approx 12¾ in diameter, iron frame. Probably late 18th century.

76-77 Watchmaker's lathes, one signed: *T. Jones Prescott*.

B 28 BUCKLER'S HARD
Maritime Museum

Buckler's Hard, Beaulieu, Hants SO42 7XB

0590 63 203

Ships were built at Buckler's Hard as early as the late 17th century. In the second half of the 18th century the second Duke of Montagu decided that Buckler's Hard should be made into a port, and a design exists of the proposed layout on formal lines. However, only a small part, consisting of two terraces of cottages leading down to the slipways, was erected. The plan did not succeed, but the Hard was a centre of shipbuilding from 1749 and many warships of Nelson's time were built there.

The museum contains ship models, designs, and other material relating chiefly to the history of Buckler's Hard, the ships built there, and personages associated with it. There are a small number of nautical instruments.

1 Gunter's quadrant, in wood, radius 7in.

2 Octant, ebony frame, 12in radius. Signed: *Cary London*. Early 19th century.

3 Sextant, ebony frame. Signed: *Cox Devonport*. Late 19th century.

4 Day and night telescope. Signed: *Dollond London Day or Night*. Three draw tubes, outer tube wood/brass, inner tubes brass. Early 19th century.

5 Nautical telescope, single draw, outer tube string bound, inner tube of brass. Reputed to have belonged to Admiral Rodney before 1782.

B 29 BURTON CONSTABLE
Burton Constable Hall

Burton Constable Hall, Sproatly, Nr Hull, HU11 4NL (Mr John Chichester-Constable)

0964 562316

An Elizabethan mansion, modified internally in the second half of the 18th century by Adam, Lightoler, Wyatt, etc., for William Constable (1721-91) who inherited the estate in 1746/7. As well as being an energetic patron of artists, architects and craftsmen, William Constable took considerable interest in scientific matters, and possessed what must have been in his day a very important collection of scientific instruments, some of which still survive, while fragments of others and extant receipts indicate the original extent of the collection. William Constable was elected as Fellow of the Royal Society, on the proposal of Sir Joseph Banks (PRS 1778) whose botanical interests he shared.
Lit: HALL (1970). In this catalogue of an exhibition at Ferens Art Gallery, Hull, Dr Hall has succeeded in identifying some of the surviving instruments as items listed in the receipts.

1 Brass pantograph. Signed: *J. Bennett, London*. Purchased 1768 at a cost of £4.14.0. (Hall 131 and 154)

2 Refracting telescope on pillar and folding tripod stand. All brass, altazimuth mounting. Signed on tube: *Ramsden London*. (Hall 137a)

3 Double-barrel air-pump. Table model. Vertical brass barrels, elevated pump plate. Signed: *Cole fecit*. Purchased 1757 with accessories. (Hall 128)

4 Pressure vessel. Glass jar clamped between brass yoke and oval mahogany base, with brass columns., Signed: *Cole Maker Fleet Street London*.

5 Terrestrial globe after Leonard Cushee and sold by G Adams.

6 Pyrometer in finely carved mahogany case. Silvered dial. Length 36¾in. Signed: *Made by Joseph Finney in Liverpool*. (Hall 136)

7 Waywiser. Straight metal handle. Small iron wheel. Dial mounted on handle. Signed: *Heath fecit*.

8 Waywiser. Cane handle, small iron wheel, with silvered dial (approx 4in) mounted on axis. Signed: *Made by G. Adams Fleet Street London*. (Hall 142a)

9 Set of regular solids. Purchased 1761 from Ann Moor. (Hall 130 and 192)

10 Set of plane figures. Signed by Nairne and Blunt. (Hall 130)

11 Double cone for use with an inclined plane.

12 Mahogany model of a pile-driving machine. Height 20½in. Unsigned, but a model of the pile-driving machine used for constructing Westminster Bridge was purchased from Benjamin Cole in 1757. (Hall 133)

13 Large cylinder electrical machine of the type described by Benjamin Wilson in *Treatise on Electricity* (1750). Cylinder missing.

14 Small plate electrical machine. Plate of smoke grey glass. The prime conductor, originally separate, is now missing. 1770s.

15 Large twin plate electrical machine, Cuthbertson type, *c*1770.
Lit: HACKMANN (1973)

16 Nairne electrical machine (globe type for clamping to edge of table). Incomplete. Signed: *Nairne & Blunt London*.

17 Leyden jars. Two 'electrical jars', purchased from Benjamin Cole in 1759. (Hall 132)

18 Small camera obscura. Unsigned.

19 Burning mirror, diameter 24in, on wooden tripod stand.

20 Nooth's apparatus (incomplete), with engraved number: *2073*.

21 Considerable quantities of glass vessels of various types, mainly chemical.

B 30 BURY ST EDMUNDS
The Gershom Parkington Collection

Angel Corner, Bury St Edmunds, Suffolk, IP33 1UZ

0284 763233 ext 7071

The collection, which was bequeathed to Bury St Edmunds corporation by Frederic Gershom Parkington in memory of his son John, killed in the Second World War, is exhibited in a Queen Anne House belonging to the National Trust. The Borough's own collection contains locally made clocks and watches, and about 30 sundials and other instruments. The earliest clock in the collection, a drum-shaped table clock, is dated 1541, and there is a mid-16th-century clock watch. All the instruments are of a high quality and fine examples of their type. In addition to the items listed below there is a small collection of watch and clock-making tools, including a work bench said to have been used by the Vulliamy family, 1775-1854.

Lit: BEEVERS (1948, reprinted and augmented 1971); MEYRICK (1979). Reference to the last publication is indicated as follows: Meyrick, followed by catalogue number.

1 Pillar dial. Boxwood with brass gnomon. 19th century. (Meyrick 7SD)

2 Horizontal table dial. Gilt brass and silver. Screw feet. A plummet hangs over a crescent scale incorporated in the gnomon, and the screw feet permit some slight adjustment for latitude. Signed: *Johannes May Amsterdam.* Late 17th century.

3 Horizontal string-gnomon dial. Octagonal, gilt brass and silver, inlaid with black and red lacquer. Signed: *Nicholaus Rugendass in Augsp.* (Meyrick 10SD) .
Lit: BOBINGER (1966) p293.

4 Butterfield dial. Silver, octagonal. Signed: *Butterfield A Paris.* (Meyrick 12SD)

5 Butterfield dial, brass, octagonal, in black fish-skin covered case. Signed: *Macquart A Paris.* (Meyrick 22SD)

6 Butterfield dial. Silver, octagonal, black leather covered case. Signed: *P.le Maire A Paris.* (Meyrick 26SD)

7 Including dial. Brass, octagonal, in black fish-skin covered case. Folding gnomon and latitude quadrant. Signed: *Menant A Paris.* (Meyrick 27SD)

8 Diptych dial, ivory with gilt spandrels. Signed: *Paulus Reinman Norimbergae faciebat*, dated 1598. (Meyrick 31SD)
Lit: ZINNER (1956) p486-7

9 Diptych dial, ivory with coloured inlay. Signed: *Liehart Milner 1600* (Liehard Miller). (Meyrick 16SD)
Lit: ZINNER (1956) p447 (where date is given as 1612)

10 Diptych dial, ivory with coloured inlay. Signed: *Mechior Karner* (Nuremberg). Late 17th century. (Meyrick 13SD)

11 Diptych dial, ivory with coloured inlay. Signed: *Liehart Miller* (Lienhard Miller) and dated 1622. (Meyrick 19SD)

12 Diptych dial, ivory, octagonal with coloured inlay. Unsigned but two thrushes engraved on the face of the dial (Hans Troschel, Nuremberg). Early 17th century. (Meyrick 11SD)

13 Diptych dial. Boxwood with magnetic azimuth dial of engraved paper pasted in base of compass. The lid is engraved with the constellations of the northern hemisphere. Unsigned, English, *c*1700. (Meyrick 8SD)

14 Universal equinoctial dial. Gilt brass with silver mounts. Signed: *Johann Willebrand in Augsburg 48.* In black fish-skin case with silver plaque engraved with list of latitudes. (Meyrick 14SD)
Lit: ZINNER (1956) 440; BOBINGER (1966) p272

15 Universal equinoctial dial. Octagonal, brass, gilt and silvered. Black leather covered case. Printed instructions for use and long list of latitudes. Signed: *And. Vogl.* (Andreas Vogler, Augsburg). (Meyrick 15SD)
Lit: BOBINGER (1966) p367

16 Universal equinoctial dial (Crescent dial). Gilt brass and silver, black fish-skin covered case. Signed: *Johann Martin in Augsburg 48.* (Meyrick 29SD)
Lit: BOBINGER (1966) p387

17 Universal equinoctial dial. Octagonal, brass, gilt and silvered, black leather covered case. Signed: *L.T. Muller* (Ausburg). (Meyrick 20SD)
Lit: BOBINGER (1966) p284

18 Universal equinoctial sundial in round watch-like case. The latitude is set on a small dial marked *Gradus Polus* which mechanically sets the equinoctial ring. Signed: *Johann Martin in Augsburg 48.* (Meyrick 25SD)
Lit: ZINNER (1956) p440; BOBINGER (1966) p270

19 Ring dial. Brass, diameter 2½in. Inscribed: *G.G.Z. 1715.* Separate pin-holes for winter and summer. (Meyrick 9SD)

20 Ring dial. Brass, diameter 2¹⁄₁₆in. Pinhole sight on a sliding collar. Inscribed: *The Master IAM 1795.* (Meyrick 24SD)

21 Universal equinoctial ring dial, silver. Signed: *H. Bedford in Fleet Street* (London). Second half 17th century. (Meyrick 28SD)

22 'Bloud-type' (magnetic azimuth) dial. Ivory and silvered brass. Signed: *Fait et Invent. par Charles Bloud a Dieppe, c*1670. (Meyrick 21SD)

23 Magnetic dial. Circular box decorated in gold on green paste. Compass dial of white enamel. Perpetual calendar on underside of box. Signed: *Martin Rue St. Martin no.244.* Second half 18th century. (Meyrick 30SD)

24 Gunter's quadrant. Boxwood, radius 7in. The plummet is housed in an aperture in the side of the quadrant. Constructed for latitude 52°, *c*1700. (Meyrick 6SD)

25 Nocturnal. Boxwood, length 9⅝in. For use with both Bears. Unsigned. *c*1700. (Meyrick 3SD)

26 Nocturnal. Boxwood, length 10⅜in. Inscribed: *For Both Bears. c*1700. (Meyrick 4SD)

27 Nocturnal. Boxwood, length 10⅛in. Inscribed: *Charles Turner 1703*. For use with pointers of Great Bear only. (Meyrick 5SD)

28 Perpetual calendar on box of 2¾in diameter. Brass, silvered and gilt. Signed: *D. Spronk fecit A Cleve*. Engraved with a coat of arms. First half of 18th century. (Meyrick 18M)

29 Perpetual calendar engraved on box of 2⅝in diameter. Silver. Inscribed: *Almanach Universal et Perpetual*. French, first half 18th century. (Meyrick 19M)

30 Pedometer in gilt metal and shagreen case, 1¹¹⁄₁₆in in diameter. Signed: *Spencer & Perkins London*. (Meyrick

16M)

31 Horizontal dial, wooden, found under the floor of the North aisle of the chancel of St Mary's Church, Bury St Edmunds, on the site of the Chapel of St Nicholas. The gnomon is part of a Nuremberg token. (Meyrick 23SD)

32 Chinese sundial, 19th century. (Meyrick 18SD)

33 Ashkadar stick (Tibetan 'shepherd's' timestick), octagonal wooden stick with pointed iron ferrule. Height 4 feet 11ins. 19th century? (Meyrick 32SD)

34-36 Three sand-glasses, heights 6½, 6½ and 7½ins. 19th century. (Meyrick 2-4M)

C

C 1 CAMBRIDGE
Cavendish Laboratory

Department of Physics, University of Cambridge, Madingley Road, Cambridge CB3 0HA

0223 337472

While all early instruments not specifically related to work carried out in the Cavendish Laboratory have been transferred to the *Whipple Museum*, the laboratory still retains many instruments and apparatus connected with research undertaken there. There is apparatus connected with James Clerk Maxwell (Cavendish Professor from 1871-1879), GFC Searle and Joseph John Thomson, as well as CTR Wilson, Lord Rutherford, WH and WL Bragg and FW Aston. In the entry below, which lists only material connected with Rayleigh and Maxwell, the association of each item is indicated in parentheses. Details of material connected with later work may be found in the *Outline Guide to Exhibits*. Not separately noted in this entry are quantities of electrical and other instruments from the latter part of the 19th century, including apparatus associated with the standardization of the ohm, described by the British Association for the Advancement of Science, 1880-81.
Lit: ANON (1954), CAMBRIDGE (1966), GUNTHER (1937A), FALCONER (1980)

1 Apparatus for measuring the viscosity of gases, made by Elliott Brothers London, 1865. (Maxwell)

2 Maxwell's improved zoetrope. Pre-1869.

3 Model to show the perturbation of a ring of satellites. Designed by Maxwell, *c*1858

4 Real image stereoscope, made by Elliott Brothers, *c*1856. (Maxwell)

5 Thermodynamic surface for water made by Maxwell to illustrate the theories of J Willard Gibbs.

6 Colour top and prismatic colour box used by Maxwell for his work on the quantitative analysis of colour vision and colour blindness.

7 Dynamical tops to demonstrate the motion of a riding body rotating about its centre of gravity, 1855-6. (Maxwell)

8 Spinning coil. (Maxwell)

9 Balance arm for the comparison of the electrostatic and electromagnetic units of electricity, *c*1854. (Maxwell)

10 Pseudosphere and deformable cap, 1854. (Maxwell)

11 Wire meter designed by Maxwell and made by Elliott London.

12 Prismatic colour box. Used by Rayleigh in experiments on colour blindness. *c*1880

C 2 CAMBRIDGE
Gonville and Caius College

Trinity Street, Cambridge CB2 1TA

0223 332400

The College was founded in 1348 by Edward Gonville and refounded by Dr Caius in 1557. Of the two astrolabes it retains, one is believed to have belonged to Dr Caius (1510-73).

1 English astrolabe, *c*1350. Believed to have belonged to Dr Caius. Diameter 3½in. Rete for 19 stars. No separate plates but meter engraved for 52° (i.e. London).
Lit: GUNTHER (1932) no. 301, pl. 132, GUNTHER (1937A) p185, PRICE (1955) no. 301

2 European astrolabe, *c*1450. Inscribed: *Hooft*. Diameter 5¼in. Rete for 23 stars. One latitude plate for 49° and 50°, Two paper plates are printed from engraved plates for the latitude of Paris (48°50') and Jerusalem (32°). Six other incomplete brass plates. Leather case stamped with Tudor portcullis and the word 'Gauges'.
Lit: GUNTHER (1932) no. 133, GUNTHER (1937A) p186, PRICE (1955)

C 3 CAMBRIDGE
Department of Earth Sciences

University of Cambridge, Downing Place, Cambridge CB2 3EW

0223 333400

The first Professor of Mineralogy was appointed in 1808. The Department contains several items associated with William Hallowes Miller (Professor in 1829) as well as a total reflectometer designed by WH Wollaston in 1802 which belonged to Sir GG Stokes (appointed Professor of Mathematics in 1849). In addition to the items listed below there are several late 19th century microscopes by Swift, as well as a universal optical instrument by the same maker. Other material from the Department has been transferred to the *Whipple Museum*.
Lit: GUNTHER (1937A) p440

1 Two circle goniometer constructed by Professor Miller in 1874 from a horizontal circle by Troughton and Simms and a vertical circle, signed: *Cary London*.
Lit: GUNTHER (1937A), no.318.

2 Hand goniometer invented by Carangeot. With two sets of removable radial arms. Red leather case inscribed: *W.H. Miller*.
Lit: GUNTHER (1937A), no.319.

3 Apparatus used by Prof. Miller for the approximate measurement of crystals. *c*1840. Constructed from L-shaped wires and corks.
Lit: GUNTHER (1937A), no.320.

4 Total reflectometer designed by WH Wollaston in 1802. Belonged to Sir GG Stokes.
Lit: GUNTHER (1937A), no.317, fig 75.

C 4 CAMBRIDGE
Queen's College

Cambridge CB3 9ET

0223 335511

The College was founded in 1448 by Queen Margaret, wife of Henry VI. It houses only two scientific instruments, although another item, the cabinet of *Materia Medica*, is of considerable interest and therefore included here.

1 Air-pump. Floor-standing model. A double-barrel pump mounted on an oblong stool, the vacuum plate elevated on a vertical column in the centre (Smeaton's type modified by Nairne). Signed: *R. Saunders Salisbury Ct London*. Late 18th century. (Housed in the Library)
Lit: GUNTHER (1937A) p94

2 The Vigani cabinet of *Materia Medica*, 1704 with later additions. Consists of 600 specimens in an oak cabinet. (In the President's Lodge)
Lit: GUNTHER (1937A) pp330, 473

3 Celestial globe of wood with gold applied work, diameter 48 in. Made for Sir Thomas Smith *c*1570. (In the College Library)

C 5 CAMBRIDGE
The Library, Trinity College

The Library, Trinity College, Cambridge CB2 1TQ

0223 338400

Of the numerous scientific instruments which belonged to the College (see GUNTHER (1937A), *passim*), the majority have been transferred to the *Whipple Museum, Cambridge*, together with others subsequently discovered by DJ de Solla Price, and described by him (see PRICE 1952). Only four items remain in the Library (designed by Christopher Wren), while a fifth, a horizontal garden sundial, stands in the Great Court.

1 Parallel ruler, drawing pen and pencil holder in a case. Said to have belonged to Sir Isaac Newton.
Lit: PRICE (1952) p6

2 Equinoctial ring dial. Silver, diameter 6⅜in. Unsigned. Discovered when the observatory at the Great Gateway of Trinity College was demolished in 1797.
Lit: PRICE (1952) p6; GUNTHER (1927) p184, no. 172

3 Horizontal garden sundial. Octagonal. Signed: *Colleg. Trinit. Cantab. Made by John England Charing Cross London 1703*. Originally in the Great Court at Trinity College, but removed in 1795 to be replaced by a sundial by Troughton (no.5). Now in the Library.

4 Prism. Presented to Trinity College in 1947.

5 Horizontal garden sundial in Great Court of Trinity College. Diameter 16in. Signed: *Troughton London*. Installed 1795.
Lit: GUNTHER (1937A) p182, no. 150; PRICE (1952) p9

C 6 CAMBRIDGE
Whipple Museum of the History of Science

University of Cambridge, Free School Lane, Cambridge CB2 3RH

0223 334540

The Whipple Museum is part of the University Department of the History and Philosophy of Science. It is housed in a building which includes the 1624 Free School with its 19th-century wing which was originally the home of the Fitzwiliam Museum. In 1976 major building work was undertaken to provide more adequate housing for the Museum's important collection of instruments and books.
    The basis of the collection is formed by more than 1000 scientific instruments and a similar number of antiquarian science books assembled by Robert Stuart Whipple in the years following 1913 and presented to the University of Cambridge in 1944, together with a small purchase fund which was subsequently increased by an endowment from his estate. Robert Whipple's association

with Cambridge began in 1893 when he became personal assistant to Sir Horace Darwin, the co-founder of the Cambridge Scientific Instrument Company. Ultimately he rose to become Chairman and Managing Director of the Company. The presentation of the collection, intended to form the basis of a museum of the history of science, was marked by the staging of an exhibition in 1944. The collection was exhibited in a series of locations before being moved to the present building in Free School Lane. Over the years it has been considerably enhanced by instruments from Colleges or Departments of the University, and many of the items which appear in RT Gunther's *Early Science in Cambridge* have been transferred or are on permanent loan to the Whipple Museum.

Lit: CAMBRIDGE (1936), GUNTHER (1937A), CAMBRIDGE (1944) for catalogue of material presented by RS Whipple; CAMBRIDGE (1949), CLAY (1951), HALL (1951), BRYDEN (1975 and 1978). Many anatomical, navigational and surveying instruments are described and illustrated in BENNETT (1987).

I   SOURCES OF THE COLLECTION

1   *Whipple Collection* (see introduction)

2   *Cavendish Laboratory.* While the laboratory still retains instruments and equipment specially related to important experimental work carried out there (see *Cavendish Laboratory, Cambridge*) it has transferred to the Whipple Museum a considerable quantity of historic material unconnected with the past research of the department. This material includes early 19th-century apparatus from the collection of WH Wollaston presented to the University by HW Elphinstone in 1876, a small collection of slide rules and other instruments presented to the University in 1939 by Professor AR Hutchinson, and other items such as a 50° thermometer from the Accademia del Cimento (given to Charles Babbage in 1834 and presented to the University by Babbage's son in 1872), a cylinder patent electrical machine by Nairne which reputedly belonged to Henry Cavendish (1731-1810) and a prototype camera lucida by Wollaston.

3   *Fitzwilliam Museum.* The collection of more than 100 sundials, together with three astrolabes and a number of other instruments from the collection of C Holden-White was placed on permanent loan to the Whipple Museum in August 1973.

4   *Heywood Collection.* An important collection of instruments consisting mainly of microscopes, but including also telescopes, optical accessories, orreries, sundials, slide rules and drawing instruments. The collection, which was assembled by the late Professor H Heywood of Loughborough University, was purchased by the Whipple Museum with the assistance of grant-in-aid administered by the *Science Museum, London* exemption of estate duty benefits. The collection includes a fine Marshall-type microscope and several early Culpeper microscopes.

5   *Pembroke College.* Whilst retaining the Castlemaine 'English' globe by Moxon (see *Cambridge, Pembroke College*), the scientific instruments found at the College,

which include a micrometer eyepiece for a telescope by John Rowley, an eyepiece for strapping to a long telescope tube, and a 2¾in lens signed and dated: *Christopher Cock London 1668*, have been placed on loan to the Whipple Museum. These instruments appear to have formed part of the equipment of Trinity College observatory.

Lit: PRICE (1952)

6   *St John's College.* The Whipple Museum has on loan instruments from the College Observatory, erected in 1765, whose equipment, dating from the second half of the 18th century, included refracting and reflecting telescopes, a transit instrument, an equatorial, quadrants, an octant and a sextant, a repeating and a reflecting circle and an altazimuth instrument. The Revd W Ludlam observed there in the late 1760s, and Thomas Catton (1760-1828) from 1791 until 1832. The Observatory was closed in 1859.

7   *Trinity College.* The College retains only five items (see *Cambridge, Trinity College*), all other items having been lent to the Whipple Museum. These include slide rules, sectors, quadrants, protractors, a spirit level, and a Newtonian telescope by George Hearne. Many of these instruments, as well as those on loan from Pembroke College, can be identified from an inventory as part of the original equipment of the first official astronomical observatory built for the use of the Plumian Professor, Roger Cotes, in the first years of the 18th century. Other instruments, nearly all bearing the inscription: *Ex dono Tho: Scattergood Arm*, are to be dated to around 1660 and were presented to the College in May 1677.

Lit: PRICE (1952)

8   *Other University Sources.* These include the Department of Earth Sciences (which, however, still retains a few items), the University Observatory, the Archaeological Museum, the Sedgwick Museum and the Department of Geography.

9   *Cambridge Scientific Instrument Company Ltd.* The contents of the company Museum, presented to the Whipple Museum in 1974, include examples of products dating from the late 19th century, and associated instruction manuals, catalogues and engineering drawings.

10   *WG Pye Ltd.* The Whipple Museum holds some original drawings of scientific instruments manufactured by this Cambridge firm. Access to recent material in this collection is restricted.

II   SCOPE OF THE COLLECTION

The majority of instruments are of 17th-19th-century date, and instruments by English makers form the greater part of the collection, but there are predictably many sundials by continental, and particularly by German, makers. Less expected is an interesting series of microscopes of Continental manufacture. 19th-century material includes microscopes, surveying instruments, spectroscopes, galvanometers and other apparatus associated with current electricity. The library includes many early works on the construction and use of instruments. In addition to the more recent printed catalogues and instruction manuals associated with the Cambridge Scientific Instrument Company collections, there is an extensive collection of material for the period 1910-1940 formed by Professor Pollard of Imperial College, and presented to the Museum by the Scientific Instrument Manufacturers Association in 1972. There is also a small collection of scientific prints, engraved portraits and printed ephemera produced by instrument makers. Since 1982 a series of sectional catalogues has been published.

1 *Astrolabes.* Including three oriental astrobales from the Holden White collection (on loan from the Fitzwilliam Museum), and the King's College astrolabe, the collection now contains eleven oriental and four European astrolabes.
Lit: BRYDEN (1986)

2 *Astronomical instruments (excluding portable telescopes).* These include several instruments from St John's College Observatory (Section 1/6), a 10ft reflector constructed by William Herschel and used at Blenheim Palace Observatory (a gift from King George III to the Duke of Marlborough), a Newtonian reflector by George Hearne, an 18in quadrant by John Bird and a portable equatorial by Nairne and Blunt.
Lit: BENNETT (1983B)

3 *Astronomical models.* Globes include a 17th-century Indo-Persian globe (*Lit:* SAVAGE-SMITH (1985) no.68), a small 16th-century Italian terrestrial globe of silver made after Paolo de Furlani, and a Castlemaine globe on loan from Trinity College. There is a 15th-century armillary sphere and a pair, Ptolemaic and Copernican, by Richard Glynne. A 'grand orrery' by George Adams, of *c*1750 was transferred by the Sedgwick Museum, Cambridge. An important oriental item is a mid-18th-century astronomical screen from the Royal Palace at Seoul, Korea.
Lit: NEEDHAM/LU (1966); BROWN (1983)

4 *Computing instruments.* The collection includes sectors, an example of Galileo's sector by Mazzoleni, slide rules and several sets of Napier's bones. There are two important examples of early circular slide rules – Oughtred's 'circles of proportion', *c*1640, and an example of Thomas Brown's spiral slide rule of 1632, dated 1650. There are examples of Gunner's callipers, as well as the apparently only surviving set of Nicholas Goldmann's 'stylometers' of *c*1660 – six triangular rods engraved with scales giving the proportions of the parts of different orders of columns. There is also a fragment of Charles Babbage's difference engine.

5 *Circular dividing engine.* Ramsden type, late 18th or early 19th century.

6 *Drawing instruments.* Single instruments and sets, including an extensive set of instruments by the Italian maker D Lusuerg, Rome.

7 *Drug jars and mortars*

8 *Electrical instruments.* These include a Nairne medical electrical machine used by Cavendish, a number of other electrostatic generators, an early voltaic pile, electrometers, galvanometers, etc. mainly from the Cavendish Laboratory. Later material includes numerous products of the Cambridge Scientific Instrument Company Ltd. Induction coils include one that belonged to Crookes. There are various gas tubes – Gassiot, Plucker and Crookes tubes, as well as early X-ray tubes.

9 *Magnetism.* Lodestones, bar magnets and compasses, including a very rare amplitude compass signed by the Lisbon maker, J de Costa Miranda and dated 1711.

10 *Microscopes.* The microscope collection has been greatly extended by the acquisition of the Haywood collection, and now includes most main types of English microscopes, including two Marshall type microscopes. Later types include Amici and Cuthbert reflecting microscopes. There are foreign microscopes by J Balthasar, GV Schleenstein, Dellebarre, Amici, Chevalier, etc.
Lit: BROWN (1986)

11 *Navigation instruments.* Included are three backstaves, octants, sextants, circles, navigator's plotting scales, artificial horizons, nautical compasses etc.
Lit: BENNETT (1983)

12 *Optical instruments and toys (excluding microscopes and telescopes).* Includes a prototype Wollaston camera lucida, an example of Varley's patent graphic telescope, a zoetrope, magic lantern, kaleidoscopes, stereoscopes, polariscopes, spectroscopes – an important collection (*Lit:* BENNETT (1984A and B)), 3 Rowland gratings, a photographic replica of a Rowland grating made by Lord Rayleigh, and an interesting group of early microscope micrometers from Cuff to Nobert.

13 *Physics and mechanics.* Hydrometers, apparatus for hydrostatic experiments, stands with pullies, and a group of air pumps of 18th- and 19th-century date.

14 *Sundials.* The collection includes a very wide range of sundials and other time measurement instruments such as horary quadrants and nocturnals (fig 100). In total there are more than 300 sundials, including material from the Holden White collection, and nearly all major makers of sundials are represented.
Lit: BRYDEN (1988)

15 *Surveying instruments.* Several early instruments including a graphometer made by Vernier, 1607, a Zubler recipiangle (an angle measuring instrument used mainly for gunnery) of *c*1600 (fig 121), reputed to have come from the collection of Emperor Rudolph II, and gunner's levels with plumb bobs. Later instruments include circumferentors, waywisers, spirit levels and theodolites of 18th-, 19th- and early 20th-century date.
Lit: BROWN (1982A)

16 *Telescopes, portable.* As well as the instruments listed in Section II/2, there are refracting and reflecting portable telescopes.

a Refracting. Mainly hand-held instruments ranging from a six-draw telescope signed by John Yarwell and bearing the Royal coat of arms (*c*1680) and a small four-draw refractor which bears the signature of Giuseppe Campani on the objective; a telescope objective is signed Christopher Cock and dated 1668.

b Reflecting telescopes include instruments by makers such as Nairne and Blunt, James Short, and Passement of Paris. The 6in aperture reflector by James Short was constructed in 1741 for the Earl of Macclesfield, and was used by Sir William Huggins in the 19th century for solar photography.
Lit: BENNETT (1983B)

17 *Thermometers and hygrometers.* These include a 50° thermometer from the Accademia del Cimento (fig 134), as well as thermometers used by WH Miller while remaking the standard pound and early examples certified at Kew Observatory. A whale-bone hygrometer bears the signature of Hurter, London (late 18th century) and there are a pair of wet- and dry-bulb thermometers by Barrow, calibrated by JWF Glaisher in 1843.

18 *Weights and measures, balances.* In addition to a series of 19th-century short-beam chemical balances, made by TC Robinson, balances used by WM Wollaston and by WH Miller in remaking the standard pound, there is a hydrostatic balance by Adams, a precision balance of 1790 by the Paris maker Nicolas Fortin, and another, *c*1818 by Fortin & Hermann. Weights and measures include WH Miller's standard pound and standard kilogram, nested weights, and money changers' and apothecaries' scales and weights.
Lit: BROWN (1982B)

C 7 CANTERBURY
Dean and Chapter of Canterbury Cathedral

Chapter Office, The Precincts, Canterbury, Kent CT1 2EH

0227 762862 (Chapter Agent)

The Dean and Chapter of Canterbury Cathedral own one very important item of scientific instrument – a Saxon sundial found in 1938 in the soil of the Cloister Garth. A replica is in the Science Museum, London

1 Saxon sundial, 10th century AD. The dial consists of a silver tablet with gold cap and chain. A gold pin surmounted by an animal head with jewelled eyes serves as a gnomon. The pin can be inserted in a hole in the bottom of the dial when not in use.
Lit: WARD (1955) p13

C 8 CARDIFF
National Museum of Wales

Cathays Park, Cardiff CF1 3NP

0222 397951

The primary purpose of the National Museum of Wales is 'to tell the world about Wales and the Welsh people about their own fatherland'. It was established by parliamentary resolution in 1903 and the grant of a Royal Charter in 1907. Building was commenced in 1911 but only in 1927 was the Museum formally opened by the King. The Museum contains wide-ranging collections covering geology, botany, zoology, archaeology, industry and the fine and decorative arts. The scientific instruments are to be found chiefly in the Department of Industry, although the Department of Geology contains one important item (no. 1).

Many instruments, in addition to those listed below, relate to mining and navigation. They include anemometers, mine rescue apparatus and miner's lamps, as well as a number of late 19th-century sextants, artificial horizons, and a deep-sea thermometer. Surveying instruments of the latter part of the 19th, and the early part of this century, include such items as dumpy levels and theodolites.

1 Quadrant. On heavy wooden tripod base. The quadrant, 3 feet in radius, consists of two frames connected by tubular cross pieces. There are two telescopes, one for each arc. The instrument closely resembles a quadrant constructed by Troughton for Bilbao (*Lit:* REES (1819), article 'Quadrant'). The quadrant is associated with the geologist Henry de la Beche, some of whose manuscripts are preserved in the Department of Geology. (fig 117)

2 Indo-Persian astrolabe, dated AH 1062 (AD 1651-2), signed by Diya' al-din Muhammad, Royal Astrolabist of Lahore. Brass, 5¾in diameter. Plates for eight latitudes. (39.573-2)

3 Indo-Persian astrolabe. Brass, diameter 3¾in. Five plates for nine latitudes and a table of horizons. (40.342-6)

4 Celestial globe. Indo-Persian, dated AH 1068 (AD 1657/8). Brass. Inscribed as the work of Diya' al-din Muhammad. (39.573-1)
Lit: SAVAGE-SMITH (1985), no. 23

5 Quadrant. Lacquered wood, operative radius 2½in. On one side a Prophatius quadrant, on the other a sinical quadrant. Turkish, 18th century. (39.573-7) (figs 112A and B)

6 Quadrant. Similar to 5. Operative radius 5½in. (39.573-6)

7 Quadrant. Similar to 5. Operative radius 5¼in, in leather case. (39.573-5)

8 Pocket quadrant. Brass, possibly late 16th century. Initials *C.S.* in one corner. Engraved with shadow square marked 'Right shaddow Contrary shaddow'. Crudely executed. (39.373-4)

9 Equinoctial ring dial. Lacquered brass, signed: *Dollond London.* (40.342-5)

10 Equinoctial ring dial. Brass, English early 18th century (old style calendar). (40.342-4)

11 Equinoctial ring dial. Brass, unsigned but probably constructed in Augsburg, early 18th century. (39.573-9)

12 Bloud (magnetic azimuth) dial. Ivory and silvered brass, incomplete. Signed: *Fait et Inv. par Charles Bloud A Dieppe.* (39.573-9)

13 Octant. Ebony frame, brass arm radius 14in. Elaborate engraved rococo decoration on the arm. Probably Dutch, *c*1770-1780. (14.214)

14 Octant. Ebony frame, brass arm, radius 12in. Nameplate engraved: *D. Heron Glasgow; SBR* engraved on ivory arc.

15 Octant. Ebony frame, brass ribbed arm, radius 10in. Signed: *D.J. Harri Amsterdam.* Telescopic sights. (41.117.2)

16 Octant, ebony frame, radius 10in. Brass arm. Signed: *Spencer Browning & Co. London SBR* engraved on ivory degree arc. (56.204-2)

17 Octant. Ebony frame, brass arm, radius 10in. Signed: *J. Somalvico & Co. London.* (11.118-8)

18 Borda type reflecting circle. Brass. Signed: *No. 308 Jecker A Paris.* In box with label of John Parkes and Sons Liverpool. (35.115)

19 Miner's dial. Signed: *J. Casartelli Manchester.*

20 Miner's dial. Signed: *Troughton & Simms.* (46.328-1)

21 Miner's dials (2): Signed: *J. Gargory fecit 5 Bull Street Birmingham.* (66.202-2)

22 Miner's dials (2): Signed: *Davis & Son Derby.*

23 Miner's dial. Signed: *A. Abraham Liverpool.* (34.349-1)

24 Miner's dial. Signed: *Negretti & Zambra.*

25 Miner's dial. Signed: *Potter Poultry.* (43.349-2)

26 Circular protractor. Signed: *Troughton & Simms.* (11.118-5)

27 Circular protractor. Signed: *Cary London.* In box labelled: *Cary & Co. (Henry Porter) Pall Mall London S.W.* (11.118-6)

28 Station pointer. Signed: *Cary London.* (11.118-12)

C 9 CARLISLE
Tullie House Museum and Art Gallery

Tullie House, Castle Street, Carlisle CA3 8TP

0228 34781

The Museum is housed in a Jacobean mansion with a later extension. There are collections of prehistoric and Roman material, Lakeland birds, mammals and geology, paintings and ceramics. In addition to the scientific instruments listed below there are a number of items relating to pharmacy and surgery, including a bleeding set incorporating two cups, scarificator and syringe, probably of late 18th-century date, as well as balances, weights and measures.

1 Gregorian reflecting telescope on pillar and claw stand. Aperture 2½in. Length of tube and eyepiece 16in. 18th century. (18-1938)

2 Octant. Mahogany frame and arm, the lower part of the index arm encased with brass. Ivory scale and vernier. Fixing screw. Two coloured shades. Radius 17¾in. Inscribed on ivory label: *In° Robinson 1779.* An earlier inscription has been erased. (36-1976.3)

3 Octant. Ebony frame, brass arm, radius 11⅞in. Vernier and fixing screw. Three coloured shades. Engraved on ivory arc *SBR.* Signed: *Spencer Browning & Rust London.* (13.1930)

4 Drainage level. Length 6¾in. Consisting of bubble level, pinhole and thread sights, also small clinometer. (27-1946.3)

5 Specific gravity beads in turned wooden box. Incomplete. Signed: *Baldy Bombaly Whitehaven.* (32-1944.2)

6 Set of drawing instruments in fish-skin case, including an ivory scale. Unsigned. Late 18th century. (47-1964.2)

7 Sikes' hydrometer. Signed: *Bate Poultry London.* Serial No. 2406. (107-1968.2)

8 Bates' saccharometer. Signed: *P.Stevenson 9 Forrest Rd. Edinburgh.* (107-1968.1)

9 Two-draw hand-held telescope. Brass with wooden outer tube. Signed: *J. Strachan London Day or Night.* (36-1976.2)

10 Three-draw hand-held telescope. (36-1976.1)

11 Four-draw hand-held telescope. Brass with wooden outer tube. Unsigned. (177-1975.4)

12 Four-draw hand-held telescope. Brass with wooden outer tube. Unsigned. (21-1974.2)

13 Dissecting microscope with three lenses and rack focusing. Signed: *Baker High Holborn London.* (36-1976.4)

14 Artificial horizon. (108-1968)

C 10 CHATHAM
Royal Engineers Museum

Brompton Barracks, Chatham ME4 4UG

0634 844555 ext 2312

The Museum was set up by the Institution of Royal Engineers in 1885 to provide a visual history of the Royal Engineers and the development of military engineering throughout the ages. Its displays cover the history of the Royal Engineers from the formation of the Royal Sappers and Miners in the early 1770s to the present day. They also record some earlier events which concern the development of the Royal Engineers and military engineering. The activities of engineers in the various campaigns of the British Army are shown and the work of the Royal Engineers in their many Branches such as Survey, Postal and Telegraph, Submarine Mining, Transportation, Tunnelling, Chemical Warfare, Camouflage, and the construction of buildings, ports, airfields and railways are also dealt with. The Museum contains a number of instruments, listed below. Most have an association with individuals of the Royal Engineers or Royal Sappers and Miners. In addition to these instruments there is an important chronometer watch by John Roger Arnold. A theodolite used by Lieut. HH Kitchener (Earl Kitchener of Khartoum), belonging

to the Museum, is at present on loan to the *National Army Museum*, London.
Lit: HARRISON (1912)

1 Miniature theodolite. Brass with silver scale, 4in base plate. Telescope with bubble level. Signed: *Ramsden London*. Late 18th century.

2 Box sextant. Brass with copper scale, diameter 2¾in. Signed: *Allan London*. In mahogany box inscribed 'A present from the Duke of Gordon 1813', i.e. recording presentation by Alexander, 4th Duke of Gordon, to his son Alexander, later Major-General Alexander Gordon, Royal Engineers. (fig 124)

3 Box sextant. Brass, diameter 3in, signed: *W. Gilbert London*. In leather case with inscription recording presentation to Cadet J Auchterlony by the Court of Directors of the East India Company in June 1832. (I. 134)

4 Box sextant. Brass, silver scale, signed: *Gilbert London*. An inscription records that it was presented to Gentleman Cadet Robert Maclagan in 1839. (I. 124)

5 Box sextant, brass with silver scale. In leather case. Signed: *Troughton & Simms London*. Presented to Gentleman Cadet Cowell, 1850 (inscription on lid). (L.61)

6 Kater's improved prismatic compass in leather case. Signed: *Bate London*. (I. 16)

7 Prismatic compass. Brass, circular signed on lid: *Elliott & Sons 56 Strand London*. (Ins. 14)

8 Artificial horizon with mercury bottle. Used by Lieut. Kitchener, 1875-7. (M.237)

9 Circular protractor. Signed: *W.H. Harling London made for J. Pardy Durban*.

10 Pantograph. Brass. Length of arms 30½in. Belonged to CH Brookes, Royal Engineers, (d. 1894). Second half 19th century.

11 Gunner's 2ft rule. Boxwood, folding in four. Unsigned, ? early 19th century.

12 Set of drawing instruments in brass bound wooden box. Some instruments signed *Stanley London*. An inscription on the box records that it was presented to Corp. George Keyte for having designed and executed a monument to the members of the Corp of Royal Sappers and Miners who fell at Sebastopol. Mid-19th century.

13 Abney level. Belonged to Lieut. CH Brookes (d. 1894). (Ins. 15)

14 Watkin's Clinometer, Patent 217. Signed: *J. Hicks Maker 8 Hatton Gdn. London No. 2836*.

15 Prismatic compass. Signed: *J. Hicks*.

16 Chinese ivory dotchin.

17 Geomantic compass. Chinese, 19th century.

18 Hand telescope. Signed: *Watkins Charing Cross London*. With inscription dated 1815.

19 Fuller's spiral slide rule and instruction book, dated 1897.

C 11 CHATSWORTH
Devonshire Collection

Chatsworth, Bakewell, Derbyshire DE4 1PD (Trustees of the Chatworth Settlement)

0246 582204

Chatsworth was built for the first Duke of Devonshire between 1687 and 1707. Substantial additions were made for the sixth Duke, 1820-30. The house contains an outstanding collection of paintings, drawings, books, furniture, etc. In addition there are a number of scientific instruments, some of which are known to have belonged to the chemist and natural philosopher Henry Cavendish (1731-1810), a grandson of the 2nd Duke of Devonshire (nos. 6, 9-13, probably 8, and 5). Others, of earlier date, are remarkable for their material – silver – and the outstanding quality of construction and decoration. A number of instruments were presented to the *Science Museum, London* in 1930.

1 Mechanical universal equinoctial sundial with gear mechanism and minute dial. Square base on three screw feet, inside which revolves by means of gearing a circular disc, permitting the instrument to be adjusted for magnetic variation. Brass, with silvered minute and compass dial. Spirit level in base. Signed: *Made by Tho. Wright Instrument Maker to His Majesty*.

2 Inclining dial. Silver. Circular on three adjustable screw feet, two bubble levels. Outer ring of base with calendar and 'Watch slow, watch faster'. The inner ring is adjustable against the outer by means of a small key with an ivory handle. The equinoctial ring can be set between 40 and 60°. Folding gnomon engraved with the arms of the Cavendish family. Signed: *R. Glynne Londini fecit*.

3 Mechanical universal equinoctial dial. Silver. An amalgamation of the mechanical equinoctial and a variation of the universal ring dial. Consists of circular base with compass and bubble levels adjustable for magnetic variation. Two pillars support the meridian ring (fitted at the top with a suspension ring) inside which is mounted the equinoctial ring. A third ring, engraved and cut away, is equipped with an alidade with sights incorporating lenses moving over a declination scale. A minute dial is mounted on one end of the alidade, the hour dial is on a plane with, but below, the equinoctial ring. The instrument is engraved with the Cavendish arms, and signed: *R. Glynne London Fecit*.

4 Butterfield dial, silver, oval. Signed by P le Maire.

5 Terella in silver mount with semi-circular end pieces. Fine acanthus ornament around suspension ring. Steel pole-pieces. Signed: *Edwᵈ Amory Fecit*. Said to have belonged to Henry Cavendish. (fig 70)

6 Small reflecting telescope on pillar and tripod stand. Brass, tube length 25in. Signed: *Berge London late Ramsden*. Steadying rod from bottom of pillar to eyepiece end of tube with rack and screw for adjusting the elevation of the telescope.

7 Refracting telescope with wooden tube, length 4 feet, aperture about 5in. Signed: *Dollond London.*

8 Insulating stand with four white-glass pillars.

9 Battery of Leyden jars without lids. Originally three rows of six jars in a wooden box, but the central row is now missing.

10 Brass pantograph in case. Unsigned.

11 Set of bar magnets in case

12 Collection of bell jars

13 Henry Cavendish's worktable. Contains set of drawing instruments.

    a Brass set square. Unmarked.

    b Set of boxwood regular solids

    c Boxwood rule

    d Brass rule

    e Brass sector, 11½in. Signed: *J. Sisson London.*

    f Set of plane figures – oval, semicircle, etc.

    g Scale, 8in. Signed: *Ramsden London.*

    h 6in plotting scale. Signed: *Jonathan Sisson London.*

    i 10in protractor. Brass. Inch scale along diameter. On the back a plotting scale (diagonals) and subdivisions of inches into sevenths.

    j Triangular protractor with chamfered edges. Degrees indicated on inner chamfered edge of triangle.

    k Lodestone, steel-mounted, with keeper.

    l Set square and ivory foot rule. Signed: *Fraser London.*

    m Brass parallel ruler, 12in. Unsigned.

    n Brass plotting scale, 13in. Wood backed. Signed: *J. Morgan London.*

    o Four ivory rules, unsigned, with divisions into sixteenths, twentieths, ninths, etc.

14 Surveying quadrant, brass, 4½in radius, with bubble level. Signed: *Heath and Wing, London.* Second half 18th century. Thought to have belonged to Henry Cavendish.

15 Travelling stick barometer, suspended in gimbals within a mahogany tripod stand. Height 43½in (closed). Unsigned.

C 12 CHELTENHAM
Cheltenham College Laboratories

Bath Road, Cheltenham, Gloucestershire GL53 7LD

0242 513540

Cheltenham College was founded in 1841 as a preparatory school. In addition to a collection of geological material, the laboratories house several

instruments of 18th-century and 19th-century date. Cheltenham College provided science teaching from a relatively early date (*c* 1857), initially in association with military studies.

1 Air pump, table model. Receiver missing. Signed: *W. & S. Jones No.30 Holborn London.* First half 19th century.

2 Gregorian reflecting telescope on pillar and tripod stand. Signed: *Jn Bennett Londini Fecit.* Mid-18th century.

3 Gregorian reflecting telescope. Pillar and claw stand. Signed: *Harris and Co. 50 Holborn London.* First half 19th century.

4 Theodolite. Signed: *Troughton and Simms.*

5 Induction coil, marked: *Apps Patent Coil* and signed: *Newton and Co. Fleet St*

C 13 CHELTENHAM
Cheltenham Art Gallery and Museum

40 Clarence Street, Cheltenham Gloucestershire GL50 3JT

0242 237431

The museum contains paintings, watercolours and prints, a large collection of English pottery and porcelain, as well as Chinese ceramics, geological, natural history and archaeological material, period rooms, and Cotswold furniture. It owns a number of relatively common microscopes of 19th-century date in addition to a small quantity of late 19th-century photographic material.

1 Cary-type microscope signed: *Hawes London.* Rectangular mahogany box. The microscope support screws into the short side of the open box. Accompanied by a copy of C Gould, *The Companion to the Compound Oxy-Hydrogen and Solar Microscopes...and a description of C. Gould's improved pocket compound microscope,* London 1837.

2 Cary-type microscope, signed: *Cary London.* In mahogany box. Incomplete, the microscope lacks the horizontal arm into which the tube screws.

3 Cary-type microscope. Signed *Cary London.* Mounted on a (?later) circular pressed metal base of high Victorian design.

4 Brass drum microscope in mahogany box. Unsigned. Mid-19th century. Label in box: *From C. Baker's Optical and Mathematical Warehouse. 244 High Holborn London.*

5 Compound monocular microscope. Signed: *Pillischer London no. 1111. c*1850-60.

C 14 CHESTER
Grosvenor Museum

27 Grosvenor Street, Chester CH1 2DD

0244-21616/313858

The collections are primarily archaeological, relating in particular to the Roman period, but include natural history and bygones. In addition to the items below there are a few 19th-century dental and surgical instruments, weights and measures and a late 19th-century microscope.

1   Ivory sector. Signed: *A. Abraham Liverpool*. First half 19th century.

2   Glass alembics (2) and cucurbits (3), found in 1930 at St John Street, Chester. Early 17th century.
Lit: NEWSTEAD (1931)

C 15 CHIPPING CAMPDEN
Woolstapler's Hall Museum

High Street, Chipping Campden, Gloucestershire GL55 6HB

0386 840289

Housed in a 14th-century building is a varied collection which includes militaria, apothecary's material, early typewriters and vacuum cleaners, photographic equipment, material relating to ballooning and parachuting, as well as considerable quantities of brass, ceramics and silver. The collection also contains a number of scientific instruments, mostly fairly recent in date. These include microscopes, telescopes, sextants and hydrometers. The more important items are listed below.

1   Portable transit instrument on iron frame on three screws. Refracting telescopes of 1½in aperature. Signed *W. & S. Jones 30 Holborn London*. Early 19th century.

2   Refracting telescope on pillar and claw stand. All brass. Inscribed: *Bernard Ryde I.W.* 19th century.

3   Astronomical refractor. Probably second half 19th century. Signed: *P. Trambouze Opticien. Rue de Rennes 92. Paris.*

4   Brass single-draw refracting telescope with leather outer tube. Signed: *W. Gray 10 Crooked Lane London.*

5   Petrological microscope with accessories. Inscribed: *Charles Neale.*

6   Drum microscope. Brass. Unsigned. Mid-19th century.

C 16 COLCHESTER
Colchester and Essex Museum

The Castle, Colchester, Essex CO1 ITJ

0206 712490

The Colchester and Essex Museum covers natural history, archaeology, and social history. Broadly, natural history is housed in All Saints Church, archaeology in the Castle (a Norman keep) and social history in Holy Trinity Church and in The Holly Trees, a house built in 1718 close to the Castle. The latter includes such items as clocks, scientific instruments, toys, costumes, etc. Not generally on show is a considerable collection of material of the latter part of the 19th century which belonged to

the late Charles Benham of Colchester. This includes a quantity of photographic and optical items such as stereoscopes and cameras, as well as electrical instruments.

1   Brass Culpeper microscope. Signed: *T. Harris & Son. 52, Gt. Russell Street, Bloomsbury, London.*

2-3   Drum microscopes (2). Both unsigned. First half 19th century.

4   Folding botanical microscope. Brass with ivory handle. Early 19th century.

5   Spyglass. Single draw. Ray-skin outer tube, leather inner tube with gilt tooling. Signed: *Dollond London.*

6   Spyglass. Single draw. As no. 4 but signed: *G. Adams London.*

7   Cruciform sundial. Painted wood. On modern base. ?18th century.

8   Chinese diptych dial. Lacquered wood. Probably 19th century.

9   Floating magnetic dial. Signed on dial: *Porter fecit April 7 1823* and on label: *Saml. Porter Norfolk Place. Shacklewell London.*

10   Compass in rectangular mahogany box with lid. Engraved paper compass card. Unsigned. Late 18th century.

11   Theodolite. Base plate with central compass. Vertical semicircle with toothed edge across which moves an alidade equipped with telescopic sight and bubble level. Signed: *J. Gilbert Tower Hill London.*

12   Level. Telescope with bubble level fixed to the side and open sights mounted above. The level may be corrected by means of a screw which works against it. Unsigned. 18th century.

13   Waywiser. All wood construction. Signed: *J. Pritty*. Mid-19th century.

14   Octant. Ebony frame. Radius 12¼in. Ivory degree scale and label missing. Vernier and tangent screw. Late 18th century.

15   Octant. Much damaged, radius 13¾in. 18th century.

16   Octant. Ebony frame. Radius 13¾in. Vernier and tangent screw. Signed: *Heather London.* Late 18th century.

17   Box sextant. Signed: *Troughton & Simms.*

18   Sector. Brass. Signed: *Charles Chevalier Ingenieur Opticien brevete Palais Royal 163, Paris.*

19   Square protractor. Brass. Signed: *J. Cuff London.*

20   Set of drawing instruments in leather-covered and lined box. Including set square, compasses, dividers and protractor. Dated: *1740.*

21   Set of mathematical instruments in ray-skin case. Boxwood sector and protractor signed: *W. Harris & Son 50 High Street Holborn London.*

22   Pantograph. Brass and ivory. Signed: *S. Whitford London.*

23 Camera lucida. Unsigned. Early 19th century.

24 Specific gravity beads. In circular plywood box. Unsigned, *c*1800.

25 Chondrometer. Unsigned, early 19th century.

26 Sikes' hydrometer. Signed: *Dring & Fage 20 Tooley St. London.*

C 17 COVENTRY
Herbert Art Gallery and Museum

Jordan Well, Coventry CV1 5RW

0203 832381

The Museum contains primarily collections relating to the locality, with an emphasis on local natural history and industry, especially transport. There is a particularly extensive collection of early cars and motor cycles. A branch museum covers archaeology and local history. Although no deliberate attempt has been made to collect scientific instruments, there are a few relevant items, notably a very fine level by Benjamin Cole, and an early slide rule and an octant.

1 Octant. Ebony frame, brass arm, ivory scale, radius 15 in. Vernier with zero to right. On ivory label: *Brown Maker London Robert Carson.* An anchor is engraved at the centre of the ivory arc. *c*1770-80.

2 Surveyor's level. Brass, length of telescope tube 19¾in. A compass is mounted at the centre of a horizontal bar with a tripod attachment beneath. A telescope surmounted by a bubble level is attached by

means of a compass joint to the horizontal bar, and can be levelled by means of a screw working against the telescopic tube. The compass dial is signed: *Cole fecit.* A label in the box reads: *Made by Benjamin Cole Mathematical Instrument Maker, at the Orrery. Next to the Globe Tavern in Fleet Street, London. c*1780. (fig 68)

3 Engineer's level, made by WF Stanley, Great Turnstile, Holborn, London, serial No. 8448. Length 17¼in. (H/193/65)

4 Circumferentor. Brass plate, diameter 4½in, graduated at edge. Primitive alidade with open sights. Unusual device – a pointer – for reading from degree scale. 18th century.

5 Boxwood slide rule, 12in. Signed: *I. Cooke fecit 1706.*

6 Ivory slide rule: Signed: *J. Long 20 Little Tower Street London.*

7 Cylinder electrical machine. Cylinder, approx 13½in long mounted in wooden box 31in long and 25in wide. A woven belt connects the axis of the cylinder with an 8-spoked wheel whose axis, passing through the side of the box, was originally connected to a handle (now missing). Accessories include an insulating stool, two Leyden jars, two medical directors, one straight and one curved, and two Lane electrometers. *c*1780.

8 Compound microscope by Joseph Gutteridge, Coventry, with accessories in corner display cabinet, and box of slides. Exhibited at Coventry & Midland Manufacturers' Art and Industrial Exhibition, 1867.

9 Compound microscope by Joseph Gutteridge, Coventry. Exhibited at opening of Coventry Market Hall, 1867.

D

D 1 DARLINGTON
Darlington Museum

Tubwell Row, Darlington, County Durham DL1 1PD

0325 463795

The Museum contains collections of local and natural history as well as an important section dealing with railways. It possesses a few scientific instruments of an unusual degree of interest.

1 Scarlett form of Culpeper microscope (continuous straight legs). Height (closed) approx 16in. Black ray-skin outer tube, green vellum inner tube with gilt tooling. Lignum vitae eyepiece and snout, five objectives in stained ivory mounts. Circular object carrier. Complete with pyramidal box.

2 Gregorian reflecting telescope on pillar and tripod

stand with ball joint. The leg unscrews to allow the instrument to fit into its mahogany box. Black fish-skin covered tube. Speculum 2⁹⁄₁₆in in diameter. Length of tube 13in. Signed around eyepiece: *Matthew Loft fecit.* The number '8' occurs on the tripod in two places. Inside the lid of the box is pasted a slip of paper on which is written in ink in an 18th-century hand: *The Metals of the Telescope were new ground by Mr. John Bird in Octobr 1761.*

3 Glass mirror for a reflecting telescope. Diameter 8⁹⁄₁₀in. Traces of jewellers' rouge round the edge. In a black circular cardboard box. Perhaps this mirror may be identical with that mentioned in a notebook, also in the Museum, of observations made by the local astronomer John Readshaw (1826-1912) with his 9in telescope. He refers to an unsilvered lens which he used in this telescope for observing the Sun.

4 Universal equinoctial dial. Octagonal (2¹⁄₁₆in). Silvered and gilt brass. Signed: *Johann Martin in Augsburg 48.*

5 Equinoctial ring dial. Diameter 9¼in. Brass. Signed: *E. Culpeper fecit.* The suspension ring is attached to a sliding wire ring: at the opposite point of it is attached a hook for affixing a lead weight.

6 Ivory diptych dial incorporating a horizontal dial and a lunar volvelle. Unsigned: French, late 17th century. The lunar volvelle which fitted in the upper leaf of the diptych within the circular hour scale is missing.

7 Octant. Ebonised mahogany frame. Brass arm. Radius 15in. Adjusting screw and vernier. The initials IP (or TP) surrounding an anchor are engraved on the ivory arc. The ivory label is uninscribed. *c*1800.

D 2 DARTFORD
Dartford Borough Museum

Market Street, Dartford, Kent DA1 1EU

0322 343555

The Museum contains local geological, archaeological, industrial and natural history collections. In addition to the two instruments listed below the collection contains two more recent microscopes, one of which is signed by Baker.

1 Octant. Ebony frame, radius 10in. Signed: *Spencer Browning & Co. London*. The box bears the label: *S. Thaxter & Son importers of and dealers in scientific instruments, 125 State Street Boston*. Recorded as having been in use on the Peruvian warship *Huascar*, 1879-1883.

2 Brass Culpeper microscope with rack focusing. Height 14½in. Part of the base is missing, also mirror and box. Signed on stage: *Smith Royal Exchange*, 1820-30.

D 3 DERBY
Derby Museum and Art Gallery

The Strand, Derby DE1 1BS

0332 293111 ext 782

The Derby Town and County Museum and Natural History Society was inaugurated in 1836. Subsequently it absorbed the Derby Philosophical Society, before it was taken over by the town in 1871. As might be expected from its history, it has important collections, chiefly locally orientated: works by local artists, Derby porcelain and other local ceramics, local history collection, minerals, natural history and industrial collections. The Museum owns a small number of clocks and watches and a few scientific instruments. There are several clocks and instruments from the famous workshop founded by John Whitehurst (1736) in Derby. Among the paintings by Joseph Wright of Derby is the famous 'A philosopher giving a lecture on the Orrery'. In addition to the items listed below there are several later anemometers, miner's dials, and steam engine indicators, made or sold by the local firm, Davis of Derby. There are several clocks in the collection, and also weights, scales and medical and surgical dental instruments, some by Derby makers. The collection of clockmaker's tools includes a watchmaker's lathe, a threading machine, clockmaker's wheel-cutting engines and a watch-maker's turn.

1 Jones' 'improved' microscope, signed: *W. & S. Jones Holborn London*. The instrument belonged to Dr Erasmus Darwin (1731-1802).
Lit: GUNTHER (1937) p398; BIRMINGHAM (1966) p39; DERBY (n.d.)

2 Drum microscope in box. Made by John Davis, Derby.
Lit: BOOTH (1964)

3 Compass in mahogany box with lid. Signed: *Whitehurst Derby*.

4 Miner's dial with sights. Signed: *W. & S. Jones, Holborn London*.

5 Miner's dial. Signed: *Davis & Son London & Derby*. Mid-19th century.

6 Miner's dial in wooden box. Folding sights. Signed: *Cary London*.

7 Miner's dial. Signed: *W. & S. Jones Holborn London*.

8 Wollaston-type camera lucida. Early 19th century.

9 Indoor dial of a wind vane by John Whitehurst.

10 Clock-maker's wheel-cutting engine by John Whitehurst.

11 Cylinder electrical machine in a contemporary box with accessories, which include: electric pistol, flicker tube, medical Leyden jar with two other Leyden jars, spark gap, electrometer, electric whirl, discharging tongs and voltaic cannon. This latter is signed: *Adams. c*1790.

12 Battery of eight Leyden jars – six large, and two small.

13 Wimshurst machine.

14 Anemometer. With fabric sails. Made by Davis, Derby, *c*1830.

15 Dickinson's pressure plate anemometer by Casartelli, Manchester, *c*1820.

16 Equinoctial ring dial. Brass. Signed: *W. & S. Jones*.

D 4 DEVIZES
Museum of the Wiltshire Archaeological and Natural History Society

41 Long Street, Devizes, Wiltshire SN10 INS

0380 77369

The Museum owns important archaeological and geological collections relating to Wiltshire, including the collection of prehistoric material assembled by Sir Richard Colt Hoare of Stourhead. It also houses a number of weights and scales, a set of bird-stuffing instruments which belonged to the ornithologist and taxidermist Colonel George Montagu (1755-1815), a pair of simple ring dials of uncertain date and two early watches (one *c*1600). A rare item is the 13th-century steelyard found at Huish (*Wiltshire Archaeological and Natural History Magazine* 63 (1968) 66-71).

1 Waywiser. Ironbound wooden wheel, dial mounted below handle. Signed: *Watkins & Smith London. c*1770.

D 5 DORCHESTER
Dorset County Museum

High West Street, Dorchester DT1 1XA

0305 62735

The Dorset Natural History and Archaeological Society, to whom the collections belong, set up the Dorset County Museum in 1846. In 1883 the Museum found a home in the present building which was erected for the purpose by public subscription. The Museum is basically of a regional nature, covering Dorset geology, natural history, archaeology and bygones, with a section devoted to Thomas Hardy and 'Dorset Worthies'. There are collections of works by local artists and photographers.

In addition to the items below there are weights and measures, including the 1601 bushel of the Borough of Dorchester (1884.4) and a set of 1835 (1935.11), many scales and guinea balances, the Bere Regis parish church turret clock of 1719, hour glasses and some medical items.

1 Horary quadrant, with sights, brass, dated 1398. Engraved with the rebus of King Richard II, as is the example in the *British Museum, London* (1973.7).
Lit: TURNER (1985)

2 Traverse board. Rectangular with arched top carved with Gothic tracery. Inscribed: *Systrane Stockholm.* The ship Systrane was shipwrecked off the English coast in 1809.

3 Octant, in box. Ebony frame, brass arm, ivory scale. Signed on ivory label: *Made by Ripley & Son, Hermitage London for George Gilgour Jany 26th 1807.* (1940.20.1)

4 Waywiser. Mahogany, iron-bound wheel. Signed on brass dial: *Made by Tho. Wright Instrument Maker to his Majesty.*

5 Gauging slide rule, four-sided, square section, ink on boxwood. Signed: *Isaac Carver fecit 1697.* (1941.11.2)

6 Ross radial binocular microscope to the design of Francis Wenham. Signed: *Ross London.* Serial no. *1770.*

D 6 DOUGLAS
Manx Museum

Douglas, Isle of Man

0624 75522

The Museum was established by Act of Tynwald in 1882 and established in its present building in 1922. The Museum is devoted to assembling and displaying material relating to the Isle of Man. Its collections include archaeology, natural history, folk life, works by Manx artists and topographical maps, prints and photographs. In addition to a number of clocks and watches, including

longcase clocks by local makers and several sundials, mainly of stone, the Museum houses four scientific instruments, one of which is on loan (no.2).

1 Equinoctial ring dial. Brass, probably late 17th century (old style calendar). (2505)

2 Octant. Lignum vitae frame. Signed: *George Christian Liverpool.* (loan 21563)

3 Quintant. Brass frame arc graduated 0-145. Signed: *Harris & Co. London.* Belonged to Capt. Wm Kelly, RN (1771-1823). (3289)

4 Microscope. Presented in 1854 to the Rev. Theophilus Talbot, a locally noted botanist, whose collection is housed in the Museum.

D 7 DOVER
Dover Museum

Ladywell, Dover, Kent CT16 1DQ

0304 201066

The Museum is housed under the Connaught Hall which adjoins the Maison Dieu. It houses displays of local history, Roman ceramics, lepidoptera, zoology, geology and Victoriana. In addition to clocks and a collection of watches, there are the following instruments in the collection.

1 Octant. Ebony frame, ivory scale, brass arm, radius 18in. Inscribed: *Capt. Stanclif.* Possibly *c*1780.

2 Cuff microscope, height 14½in, in mahogany pyramidal box. Unsigned; *c*1765-70.

3 Tobacco box and log. Dutch, 18/19th century.

D 8 DUMFRIES
Dumfries Museum

The Observatory, Church Street, Dumfries DG2 7SW

0387 53374

The Museum, which contains collections of natural history, archaeological and folk-life material, is centred around the old Observatory building which was opened in 1836 and housed the astronomical instruments of the Dumfries and Maxwelltown Astronomical Society – notably a camera obscura and a 10in reflecting telescope. The Museum still possesses these instruments (nos. 1-6), as well as a few others acquired more recently.
Lit: TRUCKWELL (n.d.)

1 Camera obscura supplied by Morton of Kilmarnock in 1836. Still in use.

2 10in Gregorian reflecting telescope of brass on elaborate carved mahogany tripod stand. Made by Morton of Kilmarnock and supplied in 1836 at a cost of £73.

3-4 3½in refracting telescopes (2).

5 Anemometer by Baird and Son, Glasgow. Purchased in 1836.

6 Brass microscope, unsigned, presented by Grierson of Dalgonar in 1836.

7 Compound microscope in case. Signed: *Gilbert & Co. London.*

8 Octant. Ebony frame, radius 13in. Signed: *Spencer Browning & Rust London.*

9 Octant. Ebony frame, 11in radius. Signed: *Spencer Browning & Rust* and *SBR* on the ivory degree arc.

10 Octant. Ebony frame. Radius 11in. Early 19th century.

11 Octant. Ebony frame, Radius 10in. Unsigned, but late 18th or early 19th century.

12-13 A field camera and a studio camera, both known to have been used by a Mr Crombie, a Dumfries architect, in 1846.

14 Equinoctial sundial of silurian schist, 24 x 20in. Signed by *John Bonar 1623.*

D 9 DUNDEE
Dundee Art Galleries and Museums

Albert Square, Dundee DD1 1DA

0382 23141

The City's collections are housed in the main Museum building as well as in several branch museums. They cover local history, natural history, archaeology, botany, geology and painting. The Barrack Street Museum contains shipping and industrial material; another branch museum relates to ecclesiastical and local history. There is a golf museum and a whaling gallery.

The Museum contains a number of scientific instruments, relating chiefly, but not entirely, to navigation. The most important items came from the Watt Institution, founded in 1824, whose collections were presented to the City Museum when it opened in 1874. (nos. 1-3)

1 Mariner's astrolabe. Cast brass, dated *1555* and inscribed: *Andrew Smyton 1688.* Probably Spanish or Portuguese, it is attributed by Destombes (see below) to Lopo Homem (fig 67). A shipmaster and owner, Andrew Smyton is recorded in the shipping records of the Port of Dundee in the 1650s and 1660s.
Lit: PRICE (1956 A and B); WATERS (1966); DESTOMBES (1969); ANDERSON (1972) no. 1; STIMPSON (1988) no.2

2 Backstaff. Signed: *Benjamin Macy.* Inscribed: *George Sheppert* and dated: *1731.*

3 Backstaff. Dutch, inscribed: *Iohannes Eekstrom 1754 op-de-Adelaar.*

4 Backstaff. Signed: *Jno. Gilbert, London*, 1763. Radius 25in. (1974-568)

5 Octant. Mahogany frame, radius 10in. Signed: *J.P. Cutts and Sons Opticians to Her Majesty Sheffield.* (70-306-2)

6 Octant. Ebony frame, radius 15⅞in. Inscribed on ivory plate: *William Allen 1792.*

7 Octant. Ebony frame, radius 11in. Signed: *Bon Dundee.* Early 19th century.

8 Octant. Ebony frame, signed on ivory label: *C. Wilson late Norie & Son, London.* Mid-19th century.

9 Octant. Signed: *D.W. Laird Leith.* Radius 11in. (1974-409)

10 Octant. Signed: *J.P. Cutts and Sons Opticians to Her Majesty Sheffield*, radius 10in, *c*1840. (1973-1003)

11 Sextant. Signed *P.A. Feathers Dundee.* Radius 9⅝in. (AB-1975-1883)

12 Sextant. Brass frame, radius 7¾in. Signed: *Robt. Anderson Aberdeen.* 1870-80.

13 Sextant. Brass frame, radius 8in. Signed: *Feathers Dundee.*

14 Miniature quintant. Brass frame, radius 5⁹⁄₁₆in. Unsigned. Early 19th century.

15 Nautical telescope. Brass with leather covered tube. Signed: *J.P. Cutts Optician to Her Majesty Sheffield.*

16 Telescope. Brass, diameter 1⅜in. Signed by Ramsden, London. (1974-304)

17 Telescope. Brass, maximum diameter 2⁹⁄₁₆in. Signed by Lowden Dundee. (974-560)

18 6in Gregorian reflecting telescope constructed by Bowman Tilling (1799-1862). The maker was a lecturer at the Watt Institute Dundee. (fig 129)

19 Sounding machine. Signed: *Edwd Massey Patentee London 2782.*

20 Bain and Ainslie Course Corrector. By Lawrence and Mayo, London. (70-306-4)

21 Marine mercury barometer salvaged from the ship *Forfarshire* in 1838. Signed: *Steele & Son No.9 Duke's Place Wapping Liverpool.*

22 Compound monocular microscope. Signed: *G. Lowden Optician Dundee.*

D 10 DURHAM
Dean and Chapter Library

The College, Durham DH1 3EH

091 386 2489

The Estate Office houses a number of early to mid-19th-century mine-surveying instruments, which were used to survey the Dean and Chapter Estates, and their colliery at Ferryhill.

1 Miner's dial. Signed: *Cail Newcastle upon Tyne.* In a box with the label of John Cail where the printed address, No.2 Bridge Street, has been amended in ink to 61 Pilgrim Street, Newcastle. *c*1840-50. The dial belonged to the Alnwick Construction and Engineering Company.

2 Miner's dial. Signed on instrument: *J. Cail Newcastle upon Tyne*. In box with label of TB Winter, 21 Grey Street, Newcastle.

3 Lean's miner's dial. Signed: *G. Davis Leeds*. Label of TB Winter in box.

4 Dumpy level. Four screw. Signed: *Troughton & Simms London*.

5 Drainage level. Signed: *Gardner & Co. 21 Buchanan St. Glasgow Registered no. 2602 Dec. 28th 1850*. Label of John Cail, Newcastle upon Tyne in box. Believed to have been used in the construction of the North Sunderland – Chathill Railway Line.

E

E 1 EDINBURGH
Astronomical Society of Edinburgh

The Observatory, Calton Hill, Edinburgh, EH7 5AA

031 556 4365

The foundation stone of an observatory was laid on Calton Hill in 1776. The site had been provided by the City, who also promised some financial help. Thomas Short, the instrument maker, provided £400 placed at his disposal by the Trustees of a fund established by the Earl of Morton and Professor Colin Maclaurin in the 1740s. The observatory was to house the 11ft focus telescope constructed by James Short. Observatory and instruments were to be the property of the City but Thomas Short and his heirs were to be permitted to make a charge for the use of the instruments. However, money ran out and Thomas Short died in 1788. In 1792 the city completed the half-built observatory, which was taken over in 1812 by the newly founded Astronomical Institution. A Transit House was erected shortly afterwards, and the Troughton transit set up, together with a mean-time regulator (nos. 1 and 2). The construction of a new observatory on the Calton Hill was begun in 1818. However, funds were exhausted on the building itself and only several years later could instruments be acquired (nos. 3-6).

In 1832 financial considerations forced the Astronomical Institution to offer the Royal Observatory, as it was now known (this title had been granted in 1822), to the State. The offer was accepted and the post of Astronomer Royal for Scotland was created in 1834. In 1894 the Royal Observatory was transferred to its present premises on Blackford Hill. The Calton Hill Observatory is now the property of Edinburgh Corporation and is the base of the Astronomical Society of Edinburgh.
Lit: BRYDEN (1968) pp32-3; BRUCK (1972); WOOD (1972); BRYDEN (1990)

1 Mean-time regulator. Signed on back plate: *Reid & Auld Edinburgh Invt & Fecit 1813*.
Lit: WOOD (1972) p57

2 Transit telescope by Troughton, *c*1813. This was originally housed in a specially erected transit house, together with the Reid and Auld regulator.

3 Mural circle, six feet in diameter, by Troughton and Simms, dated 1833 (now on loan to *Royal Museum of Scotland (Chambers Street), Edinburgh*).

4 Altazimuth telescope by Troughton & Simms, 1830. 3½in object glass, divided circles. Purchased for £460.

5 Transit telescope. Ordered from Fraunhofer & Utzschneider, but Fraunhofer died when only the object glass had been completed. The body of the telescope was constructed by Repsold & Son, Hamburg. The instrument cost £500, and was set up in 1833.
Lit: WOOD (1972) p55, fig 5

6 Sidereal clock by Robert Bryson, Edinburgh. Originally fixed to a stone pillar near the transit instrument. Signed: *Robt. Bryson, Edinburgh*. Installed 1831 and in use as the principal clock of the observatory until 1855 when it was replaced by a regulator by Dent (the latter now in *Royal Museum of Scotland (Chambers Street), Edinburgh*).
Lit: WOOD (1972) pp58-60

7 Object glass, perhaps by W Wray, from a 21in equatorial of 28½ft focus built by George Buckingham. It was exhibited at the 1862 exhibition, then installed in an observatory on Walworth Common. Later it was transferred to Calton Hill, and was finally dismantled in 1928.
Lit: KING, *History of the Telescope*, p254

8 Small altazimuth instrument. Unsigned, second half 19th century.

9 Deep-earth thermometer, 7ft long, by Adie & Son, Edinburgh, 1877 (now on loan to *Royal Museum of Scotland (Chambers Street), Edinburgh*).

E 2 EDINBURGH
Department of Chemistry

University of Edinburgh, Edinburgh EH8 9YL

031 667 1011

While the Robinson balance, dating from the 1820s or early 1830s is now on loan to *Royal Museum of Scotland (Chambers Street), Edinburgh* (*Lit:* STOCK/BRYDEN (1972)), the Chemistry Department Museum retains three relevant items, in addition to tubes with gas samples filled by their discoverer Sir William Ramsay (He, Ne, Ar, Kr, Xe), powder for razor strops, prepared by Joseph Black MD (Professor of Chemistry 1766-99), and specimens in glass flasks with decorated stands of strontia and baryta prepared by TC Hope (Professor of Chemistry 1795-1843). There are also three items associated with CTR Wilson.

1 Direct vision spectroscope, designed by CD Liveing and J Dewer, and built by Adam Hilger. The spectroscope was ordered by Professor Crum Brown for the Chemistry Department in 1879 at a cost of around £35.
Lit: For the design see LIVEING/DEWER (1879)

2 Box of brass weights. Belonged to Thomas Andrews (1813-1885).

3 Model of a solid prepared by A Crum Brown, Professor of Chemistry 1869-1908.

4 Bronze mortar, diameter 12in. Ornamented with hunting scenes and the maker's inscription,: *Garit Schimmel me facit Daventriae Anno 1664* (on loan to *Royal Museum of Scotland (Chambers Street), Edinburgh*).

5 Glass eudiometer; probably early 19th century.

E 3 EDINBURGH
Heriot Watt University

Riccarton, Edinburgh EH14 4AS

031 449 5111

The origins of the University go back to 1821 when the Edinburgh School of Arts was founded 'for the Education of Mechanics in such Branches of physical Science as are of practical Application in their several Trades'. In 1851 the name was changed to Watt Institution and School of Arts, and was changed again in 1885 when the endowments of the Watt Institution were amalgamated with those of George Heriot's Hospital. Henceforth, until the University was established by Charter in 1966, the College was known as the Heriot-Watt College.

The University has assembled a series of photographs, books, medals, and original documents relating to the history of the University, and James Watt's letter-copying machine, patented in 1780. A collection of nine surveying instruments, formerly in the Department of Mining Engineering, has been acquired by the *Royal Museum of Scotland (Chambers Street), Edinburgh*.

E 4 EDINBURGH
Huntly House

142 Canongate, Edinburgh EH8 8DD

031 225 2424 during office hours; 031 225 1131 after 5pm and at weekends

Huntly House is administered by the Library and Museum Department of Edinburgh District Council, Central Library, George IV Bridge, Edinburgh, EH1 1EH. A timbered dwelling of 1517 houses the principal museum of local history, which contains important collections of Edinburgh glass and Scottish pottery. The collection contains only four scientific instruments.

1 Set of specific gravity beads in circular wooden case. Signed: *Made by P. Massino glass-blower St. Mary's Wynd Edinburgh.*

2 Balance in glazed case. Iron beam, brass pillar. Made by: *Browne & Smith Edinr.* The balance belonged to the Burgh of Leith.

3 Fish-skin and silver case which once contained a set of draughtsman's instruments. Of these, only one remains: a combination of parallel ruler with a square protractor, diagonal scale and a series of units with subdivisions. Signed: *T. Heath London.* The case of instruments belonged to the architect, James Craig, planner of the New Town of Edinburgh.

4 Balance beam, 5ft long with Dutch end suspensions, dated 1642, formerly in use at the Tron, or municipal weighhouse.

E 5 EDINBURGH
Lauriston Castle

Cramond Road South, Edinburgh EH4 5QD

031 336 2060

The Castle is under the care of the Library and Museum Department of City of Edinburgh District Council. Lauriston Castle is a 16th-century house with 19th-century additions. It contains fine furniture, Flemish tapestries, ceramics, Blue John and *objets d'art*, among which are included two or three scientific instruments. There is a fine garden sundial as well as a mid-18th-century Dutch barometer by an Italian maker.

1 Universal equinoctial sundial, silver with gold latitude scale., Signed: *Johan Martin in Augsburgh.*

2 Butterfield dial. Silver, octagonal. Signed: *Macquart & Paris.*

3 Set of drawing instruments. Late 18th century, unsigned. In tortoiseshell and silver case.

4 Walnut barometer and thermometer. Signed: *Lurasco Spinelli & Comp. fecit Amsterdam.*

5 Stone polyhedral sundial, star or lectern type. Dated 1684. Approx 5ft high.

E 6 EDINBURGH
Museum of the Royal College of Surgeons of Edinburgh

18 Nicholson Street, Edinburgh EH8 9DW

031 556 6206

The College was founded as the Incorporation of Barber-Surgeons in 1505. It contains an extensive museum with surgical, dental, pathological and historical exhibits. There are numerous medical and surgical instruments, including some associated with Alexander Wood (1725-1807), Joseph (Lord) Lister (1827-1912) and

Robert Liston (1794-1847). There is a particularly extensive collection of midwifery forceps from the mid-18th-century onwards, as well as stomach pumps, tourniquets, anaesthetic apparatus, operating knives, amputation knives, lancets, scarificators, litholabes, catheters, trepanning instruments, cupping glasses, etc. In addition there are two of the earliest hypodermic syringes. In recent years the College has acquired an extensive and important collection of dental instruments which were presented in 1964 by J Menzies Campbell. They have been catalogued in detail.
Lit: MENZIES CAMPBELL (1966)

In addition to the dental and medical instruments there are a number of microscopes, the most important being:

1 Hartsoeker-type screw-barrel microscope and compass microscope of early 18th-century date in contemporary leather case lined with calf. Microscope of gilt brass with lignum vitae focusing screw barrel. Length 1¾in. Turned wood and bone box with glass sides for viewing live objects. Ivory object holder and handle for compass microscope. Screw-barrel microscope signed: *Villette & Lieges.*

E 7 Edinburgh
Royal Observatory

Blackford Hill, Edinburgh EH9 3HJ

031 668 8100

In 1822, at the time of King George IV's visit to Edinburgh, the Observatory on Calton Hill (see *Astronomical Society of Edinburgh*) was renamed the Royal Observatory. In 1834, Thomas Henderson was appointed to the new joint post of Regius Professor of Astronomy at Edinburgh University and Director of the Royal Observatory, with the title of Astronomer Royal for Scotland. In 1846 Henderson was succeeded by the colourful Charles Piazzi Smyth, connected with whom a number of curious relics survive (see below). The smoky atmosphere of Edinburgh necessitated the move of the Observatory to a few miles to the south; in 1896 the new Royal Observatory was opened on Blackford Hill, where it remains today.

Many of the surviving instruments throw light on the Royal Observatory's history. As well as astronomical items, there are a few meteorological objects provided by the firm of Adie and Son for use at Calton Hill. There is also a fascinating collection (28-38) of 'standards' developed by Piazzi Smyth, during his work on the metrology of the Great Pyramid. He attempted to relate 'the sacred cubit' used in the building of the Pyramid to that used by Noah in the construction of the Ark and by Moses in the making of the Tabernacle.

Most historical instruments have recently been loaned to the *Royal Museum of Scotland (Chambers Street), Edinburgh.* (1986.L2)
Lit: BRUCK (1972)

1 Refracting telescope, single draw, 2in aperture. Inscribed: *Astronomical Institution Edinburgh* and signed:

Miller & Adie.

2 Refracting telescope, six draw, foreshade, 4¾in aperture. (65402)

3 Reflecting telescope, 2in aperture, mounting missing.

4 Azimuth instrument, radius 7in, signed: *Thomas Jones, London.* (51402)

5 Telescope micrometer, signed: *T. Heath, London*, dated 1736. (62401)

6 Micrometer eyepiece, signed: *Adie & Son, Edinburgh.* (62503)

7 Micrometer eyepiece, signed: *G. Calver, Chelmsford.* (61101)

8 Micrometer eyepiece, signed: *John Browning, London.* (62502)

9 Micrometer eyepiece, signed: *A. Hilger, London.* (61103)

10 Siderostat mirror, mounted in cell, diameter 16in. Signed: *Adolphe Martin.*

11 'Anomaly ruler' for drawing elliptic orbits, designed by WJ Macquorn Rankine, 1850.

12 Slide rule for lunar calculations, signed: *R. Bate, London, No.14.*

13-14 Spherometers, 4 and 5in diameter, signed *A. Hilger, London.* (66506, 7)

15 Clinometer, with improved mounting, designed by E Emslie Sang, Edinburgh, dated: *1869.* (65602)

16 Co-ordinate convertor, with superimposed rotating stereographic projections, signed: *W.B. Blaikie.*

17 Turkish quadrant, lacquered wood, with sinecal and horary scales, for latitude 40°.

18 Indian calendar dial.

19 Moon globe, plaster, with features on visible hemisphere in relief, diameter 21½in.

20 Octant, diameter 13in, divided by E. P[arson]. (61702)

21 Octant, diameter 9½in, signed: *Stebbing, Portsmouth*, divided by E. P[arson]. (71203)

22 Theodolite, diameter 12in., signed: *Troughton & Simms, London.*

23 Rotating mirror, turbine driven, for Foucault's apparatus for measuring the velocity of light, signed: *Froment, Paris.* (63102)

24 Inclinometer and dip circle, signed: *L. Casella, London, No.338.* (ROE 02243)

25 Magnetometer, signed *Carl Bamberg, Friedenau-Berlin, No.7596.* (66708)

26 Standard thermometers (13) by Adie & Son, Edinburgh; L Casella, London; Negretti & Zambra, London. Dates range from 1866 to 1877.

27 Rock thermometers (fragments) supplied by Adie and Son to the Royal Observatory, Calton Hill. Pre-1876.

28-29 Scales of the 'Sacred Cubit of Noah' (25 Pyramid inches) with MS and printed labels on wood by Piazzi Smyth.

30-31 Cylindrical rules of 10 and 25 Pyramid inches by Piazzi Smyth.

32-34 Square-section scales of the 'Sacred Cubit of Noah' with MS and printed labels on wood by Piazzi Smyth.

35-36 Square-section scales of '50 inch or two-cubit measure' with MS and printed scales on wood by Piazzi Smyth.

37 Square-section scale, wood, 'Cubit of Memphis or of the Nilometer, or the Profane Egyptian Cubit' by Piazzi Smyth.

38 Mounted thermometer, graduated with 'Great Pyramid Scale of Temperature' by Piazzi Smyth.

39 Standard scale of six inches, divided on bell metal, signed: *L. Casella, London, No.10* and dated 1868.

40 Metre scale in MS on wood, with inscription: *Model of a British copy of the French metre* by Piazzi Smyth.

41 Parallel rule of 'British Inches 1876', signed: *Adie & Son, Edinburgh.*

42 Box of drawing instruments, etc., including a standard inch engraved on glass by Adie & Son, dated: *1868*; and two scales signed: *Apps, London.* (65310)

43 Circular protractor, signed: *Elliott Bros, London.* (66505)

44 Pantograph, signed: *John Bleuler, London.* (66502)

45 Standard barometer, signed: *Negretti & Zambra, London*, numbered *808*, dated: *1895.*

46 Polished blocks of extra-dense glass made by Michael Faraday in his discovery of the plane of polarisation of light in a magnetic field. (63601)

47 Circular mirrors of vacuum deposited arsenic and platinum by William Crookes, 1877.

E 8 EDINBURGH
Royal Museum of Scotland (National Museums of Scotland)

Chambers Street, Edinburgh EH1 1JF

031 225 7534

The Museum was established in 1854 when funds and site were voted by Parliament. The foundation stone was laid in 1861 and the building opened in 1866. Originally named the Industrial Museum of Scotland, it later became the Edinburgh Museum of Science and Art, and finally the title the Royal Scottish Museum was conferred in 1904. In October 1985 it combined with the *National Museum of Antiquities of Scotland* (see E 9). The Museum, whose collections cover natural history, geology, technology, ethnography and the decorative arts, houses in the Department of Science, Technology and Working Life, one of the major collections of scientific instruments in the British Isles. Scottish makers are particularly well represented. There is an extensive range of electrical machines and accessories, a considerable quantity of early chemical apparatus and glassware, numerous microscopes, telescopes and sundials, and an interesting selection of draughtsman's and calculating instruments. The entry which follows does not attempt to list individual instruments, but is limited to describing the scope of the collection, both by source and category of instrument.

Lit: EDINBURGH (1954) for a general history of Royal Scottish Museum on its centenary.

No catalogue exists as yet of the scientific instruments in the Royal Museum of Scotland (Chambers Street), although individual items and groups have been published. See, for example, BRYDEN (1968) (Short telescopes), BRYDEN (1972B), NUTTALL (1973A) (Scottish made instruments since acquired by the Museum), ANDERSON (1978) (chemical apparatus in the Playfair Collection), NUTTALL (1979) (microscopes subsequently acquired by the Museum); CLARKE, MORRISON-LOW and SIMPSON (1989) (Scottish instruments in the Frank Collection)

I   SOURCES OF THE COLLECTION

The abbreviations used to refer to these sources in the ensuing text are indicated in parenthesis.

1 *Playfair collection (PC).* Early chemical and physical apparatus presented to the Museum by Professor Lyon Playfair on his appointment to the Chair of Chemistry at the University of Edinburgh in 1858. The collection includes such items as a pneumatic trough, a pyrometer by J Dunn, Edinburgh (fig 111); differential and air thermometers, a model metallic thermometer, a voltaic pile, a Sturgeon cell, retorts, chemical glassware ('Black's glass'), an Argand lamp, etc. The apparatus dates from the latter part of the 18th and the first half of the 19th century, and includes a number of items associated with Joseph Black (Professor 1766-1799) and Thomas Charles Hope (Professor 1795-1843). An important single item is a double-barrel air pump of Hauksbee type as modified by William Vream, probably dating from the first half of the 18th century. (fig 1)

Lit: ANDERSON (1978)

2 *Department of Natural Philosophy, University of Edinburgh (NP).* The Department possessed an extensive collection of philosophical instruments and apparatus which is listed in a manuscript catalogue apparently compiled at the appointment of Professor Forbes in 1833, and subsequently continued as new acquisitions were made. Only a small part of this material still survives. A number of instruments were lent to the Royal Scottish Museum some years ago; all remaining material, acquired by the Department before 1905, was donated in 1974. This includes apparatus associated with Sir John Leslie (Professor 1819-1832), and JD Forbes (Professor 1833-1859). Other important items include a thermometer made for the Accademia del Cimento, Prins of Amsterdam, and Alexander Wilson of Glasgow, a mountain barometer and stand by Cary, London, and a cometarium (fig 106) and a large vertical orrery by Miller of Edinburgh. There is an interesting early 18th-century equatorially mounted quadrant with sighting telescope (fig 116), a Bramah hydraulic press and a reflecting telescope by James Short.

3 *University of Edinburgh, other departments.* Items have been deposited by the Departments of Chemistry, Physiology, Geology and Mathematics.

4 *Department of Mining Engineering, Heriot Watt University, Edinburgh.* A collection of 18th- and 19th-century mining and surveying instruments, including a Bavarian 18th-century miner's dial. (fig 90)

5 *Astronomical Society of Edinburgh, Calton Hill Observatory (EAS).* The Society has placed in the Museum a six-foot mural transit circle by Troughton and Simms, 1833, as well as a seven foot deep-earth thermometer. Other material associated with the Society remains in the *Calton Hill Observatory, Edinburgh* (see E 1).

6 *Auld Collection.* A private collection, consisting chiefly of electrical and pneumatic apparatus and accessories, which was presented to the Museum in 1901.

7 *Fox Talbot (FT).* A considerable number of items, including early cameras, air pumps, electrical machines, microscopes and a camera obscura came to the Museum in 1936 from Miss Talbot. Many are known to have been used by WH Fox Talbot. Other material from the same source is in *Science Museum, London.*

8 *Admiralty.* Many of the drawing, navigation, astronomical and surveying instruments and marine compasses, chiefly of early to mid-19th century date are derived from this source.

9 *Scottish Meteorological Office.* The Office collection of meteorological instruments was deposited in the Royal Scottish Museum in 1972. The instruments were used either at the Ben Nevis Observatory (operational 1883-1904), or at other Scottish meteorological stations, and include thermometers, anemometers, hygrometers and rain gauges.

10 *Royal Observatory, Blackford Hill (RO).* The Observatory houses both a University Department of Astronomy as well as a government Research Department. A number of instruments have been placed on loan to the Museum by the government Department, but others remain in the *Royal Observatory* (see E 7).

11 *Findlay collection (FC).* An important collection of astronomical, gnomic and horological material, for many years on loan to the Royal Scottish Museum from its owner, Sir John Findlay. After Sir John's death the collection was dispersed by auction at Sotheby's in 1961 and 1962. Several items from the collection were purchased for the Royal Scottish Museum, which also retains a full photographic record.

12 *Arthur Frank Collection.* A portion of the collection of instruments formed by Mr Arthur Frank was acquired in 1979 and 1980. The majority of Mr Frank's Scottish-made instruments and microscopes are now in the Museum. *Lit:* NUTTALL (1973A and 1979); CLARKE, MORRISON-LOW and SIMPSON (1989)

II SCOPE OF THE COLLECTION

1 *Astrolabes.* A collection of seven astrolabes, including the earliest known dated European-made scientific instrument, an early Hispano-Moorish example, signed and dated 1026/7 AD, maker Muhammad b. as-Saffar, Cordoba (fig 6) (*Lit:* PLENDERLEITH (1960)). In addition there are three Persian, one Indian, and two European astrolabes – one of Fusoris type, French, *c*1400, the other Flemish, *c*1600.

2 *Astronomy.* Includes an Indo-Persian celestial globe signed by Diya' ad-Din and dated 1663 (*Lit:* SAVAGE-SMITH (1985) no. 27), and a Blaeu celestial globe, 1640 (*NP*), as well as a number of globes of lesser importance including a pocket globe by Miller of Edinburgh (fig 57). There are orreries by George Adams, W and S Jones and Miller, Edinburgh (*NP*). Miller was also the maker of a cometarium from the same source (fig 106). A tellurium was constructed by Jamieson, Hamilton, a luminarium (a card volvelle serving to provide information on solar and lunar motions for any day of the year) was designed by James Ferguson and engraved by Benjamin Cole. A unique item is a Chinese planisphere probably of 18th-century date (*Lit:* KNOBEL (1909); NEEDHAM (1959) p279 fig 108). Astronomical observation instruments (excluding portable telescopes which are separately noted) include the 6ft transit circle on loan from the *Astronomical Society of Edinburgh.* Signed by Troughton and Simms, it is dated 1833. There is a 5in equatorial by Ramsden with a 1¼in refractor, as well as a portable transit instrument by Troughton presented to Professor John Playfair in 1805. There are two reflecting telescopes made by Sir William Herschel.

3 *Barometers.* Chiefly of 19th-century date, they include instruments by Yeates and Son, Dublin; James White, Glasgow; J Hicks; George Adams; Negretti and Zambra; Ronchetti, Manchester; Richards, Paris. Edinburgh makers represented in the collection are Knie (*Lit:* BRYDEN (1972D)), Adie, Miller, Miller and Adie, Stevenson, Henderson, Bryson and Lovi. Of interest are a glycerine barometer invented by James B Jordan (the only example still operating); a sympiesometer by Adie and portable mountain barometers by Cary (*NP*); Mossey, Paris (late 18th century); Miller and Adie (as described in the *Edinburgh Encyclopedia*); and by Lawrence Buchan of Manchester. There is also a pillar barometer by Daniel Quare, London, and a barometer made by James Watt when practising as an instrument maker in Glasgow.

4 *Chemical apparatus.* In addition to the balances described below, the Royal Museum of Scotland houses a quantity of chemical apparatus derived in the main from the *Playfair Collection.* This includes chemical glassware, much of which appears to be of late 18th-century date – an alembic, retorts, flasks of various form, etc. There are pottery and fireclay retorts, crucibles, muffles and furnaces. Other items include a set of four washing bottles on a portable wooden stand, the boiler of a copper still (18/19th century), a lamp furnace, an Argand lamp, double bellows for use with a blowpipe in assay work, a Newman blowpipe, and a number of hydrometers including Richardson's hydrometer (made by Troughton), Sikes' hygrometers, hydrometers designed by JT Buchanan for the Challenger Expedition of 1872, and several sets of specific gravity beads.

5 *Circles.* Borda-type reflecting circle by Lenoir, Paris; a reflecting circle by Dollond and a repeating circle by Thomas Jones. See also *Astronomy*.

6 *Computing instruments.* The collection includes a number of unusual and important early calculators: an 'Arithmetical Jewell' of William Pratt, *c*1616, recently discovered in the Lake District (*Lit*: BRYDEN (1985)); an example of Oughtred's double horizontal dial and circles of proportion of mid-17th century date, constructed by Robert Davenport, a London-trained instrument maker who subsequently worked in Edinburgh (the earliest identified Scottish practitioner) (*Lit*: see BRYDEN (1976)); Sir Charles Cotterel's instrument for arithmetic, an arithmetical compendium of strip form Napierian rods and an abacus-type counting board (*Lit*: BRYDEN (1973)). There are two examples, one complete and dated 1701, of George Brown's 'rotula arithmetica' (*Lit*: BRYDEN (1972A)). In addition there are further examples of Napierian rods, and a modern reconstruction of Napier's promptuary. Gunter's scales, sectors, an adding machine after John Goss's design of 1812, and various 19th-century arithmometers and calculating machines. Many of the later examples are from the Department of Mathematics, University of Edinburgh.

7 *Drawing and plotting instruments.* There are sets of drawing instruments, dividers, compasses, beam compasses, protractors (including a protractor and scale by Henry Sutton dated 1660), ellipsographs and elliptic trammels. There are several pantographs, including examples of Dunn's improved pantograph, and the 'eidograph' invented by Professor W Wallace of the Department of Mathematics, University of Edinburgh, and constructed by RB Bate, London, *c*1823.

8 *Electricity.* The Museum houses an extensive collection of instruments relating to both electrostatics and to current electricity. Electrical machines include a Nairne-type medical electrical machine, *c*1785, presented by the Royal Medical Society, Edinburgh, and smaller Nairne-type machines by Keir and Charles Todd, both Edinburgh makers. A small globe electrical machine with open gearing is of a type attributed in Middleton's *New Dictionary of Arts and Sciences*, 1778, to the design of a local man, D Hutcheson. Another globe electrical machine is signed by William Lunan, Aberdeen, and can be dated to around 1827. A Cavallo type electrical machine by W and S Jones, on loan from the *Science Museum, London*, contains a very wide range of accessories including a magic picture. Unsigned machines include two plate instruments of Cuthbertson type, and another small plate machine used by WH Fox Talbot.

Accessories for use with an electrical machine include dischargers, electric eggs, thunder facades, fire houses, a voltaic cannon, a head of hair, etc. Six very large Leyden jars are from the *Royal Observatory, Edinburgh*. Electrometers include Henley's quadrant electrometer, Cuthbertson's balance electrometer, and a Peltier electrometer (*c*1860), on loan from King's College London. Other items include gold leaf electroscopes,

torsion balances (including an early example from NP), Kelvin multicellular voltmeters and a Kelvin water dropper.

The material relating to current electricity includes an experimental galvanic pile by KT Kemp of Edinburgh, 1828, and a trough type battery by the same maker. Some of Kemp's samples of liquified gases, through which he measured electrical conductivity, are also in the Collection.

9 *Magnetism.* Compasses of 18th/19th-century date include instruments by Thomas Short, Leith; Springer, Bristol; Lynch, Dublin; Bradford, London; Jas. Gray, Liverpool. There are dip sectors and dip circles of 19th-century date, one constructed by Robinson, London (*NP*). One instrument for measuring dip was used in the Polar expedition of 1875; another instrument of early to mid-19th-century date was probably used for measuring diurnal magnetic variations.

There are several Chinese geomantic compasses; also lodestones and a set of bar magnets (*NP*), the latter associated with Professor JD Forbes. Arago's apparatus for illustrating the inductive action of magnets on bodies in motion is also associated with JD Forbes (*NP*).

10 *Meteorology and oceanography.* In addition to instruments on loan from the Scottish Meteorological Office, there are several interesting barometers as well as Leslie's atmometer, a hair hygrometer by Hurter and Haas, and examples of Daniell's hygrometer. There is an interesting collection of deep-sea sounding apparatus.

11 *Metrology and chemical balances.* The oldest balance, possibly used by Professor Joseph Black (*PC*) is of 18th-century date. Later balances include one by TC Robinson on loan from the *Department of Chemistry, University of Edinburgh*, a copy of a Robinson balance signed by Adie and Son, Edinburgh (*Lit*: STOCK/BRYDEN (1972)); and examples by De Grave, London; August Oertling of Berlin and Ludwig Oertling of London; Parnell, London (*NP*, Hall's patent of 1863); Verbeeks Pechholdt, Dresden. One balance was the property of WH Fox Talbot, and a hydrostatic balance (Walker's steelyard) is on loan from the Department of Geology, University of Edinburgh.

The metrological items of prime importance are a number of standard bars, and in particular several Scottish measures. (*Lit*: PLENDERLEITH (1959)). There are chondrometers by Adie, Edinburgh, and Snart, London.

12 *Microscopes.* In addition to the microscopes from the Frank Collection, there are more than 60 instruments ranging from early English microscopes of late 17th- and early 18th-century date (including an instrument similar to Yarwell's 'double microscope', its body tube being tooled with the Royal coat of arms (*Lit*: TURNER (1966) p125) to achromatic microscopes of the 19th and 20th centuries. Major English microscope makers are represented (including a John Marshall instrument signed on the barrel), as are instruments by Edinburgh makers – Clarke (fig 77), Finlayson, Barbon and Adie (*Lit*: BRYDEN (1972C)). Several microscopes came from the Crisp Collection. The English tripod microscopes of early 18th-century date have been transferred from the *Royal Museum of Scotland (Queen Street), Edinburgh*. An Amici reflecting microscope was owned by WH Fox Talbot (fig 87).

13 *Mining.* The mining collection includes material from the *Department of Mining Engineering, Heriot Watt University, Edinburgh*, and this is the source from which the earliest item – a set of mining instruments of Bavarian origin, dated 1748 – is derived (fig 90). Other instruments include miner's dials, theodolites, and a hanging compass and level, all of 19th-century date.

14 *Navigation instruments.* These include a cross-staff stick, two backstaves, one dated 1726, octants bearing dates ranging between 1755 and 1789, as well as numerous other undated examples. Sextants include two by Ramsden, one with a cast brass frame, the other a double frame, and serial numbers *1163* and *1345* respectively. A sounding sextant for coastal survey bears the maker's name of Jones, London, mid-19th century. In addition there is a good collection of early compasses derived from the Admiralty Compass Observatory, Slough. There is also a number of late 18th- and early 19th-century instruments – circles, dip circles, transit theodolites, etc. – which were presented by the Admiralty, and formerly used in the Hydrographic Office of the Navy.

15 *Optical apparatus other than microscopes, telescopes, astronomical observation instruments.* In addition to a camera obscura, prisms, polariscopes, spectroscopes and goniometers, there are heliostats by Duboscq, Paris; Prazmowski, Paris; and Yeates and Son, Dublin (loan from *RO*).

16 *Photographic Equipment.* There is a fairly extensive collection of photographic apparatus. Very early equipment, associated with WH Fox Talbot, includes two calotype cameras, *c*1840, a dry collodion plate camera, a daguerreotype camera by A Giroux & Cie, Paris, a camera by Kemp & Co., Edinburgh, and a long focus camera with adjustable sliding front, as well as plate frames for daguerreotype cameras, lenses and eyepieces.

17 *Pneumatics.* In addition to the Vream adaptation of the Hauksbee type air pump (PC) (fig 1) there are two pumps by Edinburgh makers – J Miller (late 18th century) and W Hume (second half 19th century). Other 19th-century pumps include a modified Toepler air pump and a Geryk pump.

18 *Sundials, nocturnals, horary quadrants.* There are more than forty sundials by English, French and German makers, dating chiefly from the 18th century, but including three Nuremberg diptych dials of the early 17th century. Makers represented include L Miller and Hans Tucher of Nuremberg; Gaspar Hommer of Luneville; Bion, Delure, Chapotot, Butterfield (all of Paris); Dunod of Dusseldorf; Nairne of London and Saunders of Dublin. Hommer and Dunod are represented by universal mechanical equinoctial dials. A 12in horizontal dial by Joseph Williamson of Aberdeen, 1728, is one of the earliest signed and dated Scottish pieces. An elaborately decorated brass nocturnal from the Findlay Collection is probably a fake, but there are three wooden nocturnals of British origin, one dated 1734. There are two examples of Gunter's quadrants, one of which, from

its maker, Adie of Edinburgh, may be dated as late as the second quarter of the 19th century. One horary quadrant of late 17th-century date bears, on the reverse, a planispheric nocturnal.

19 *Surveying.* In addition to a number of 19th-century and early 20th-century theodolites and mine survey instruments, the collection contains waywisers, pedometers, surveyor's crosses and several interesting 18th-century instruments. These include a combined circumferentor/theodolite and a surveyor's level by Cole, London, a circumferentor by Benjamin Martin, a graphometer by Baradelle, Paris, and surveyor's levels by Thomas Heath (*c*1749) and Nairne. There is a water level (fig 67) and a graphometer with telescope sights signed by Lennel, Paris (fig 59); both are dated 1774, and housed in their original cases. A further interesting item is a surveyor's sextant on a pillar stand by Troughton and Simms, London. A theodolite and a seismometer (both from *NP*) are associated with Professor JD Forbes. The former was used in his glacier observations, and the latter was designed by him.

20 *Thermometers.* Chiefly of 19th- or early 20th-century date, some of them on permanent loan from the Scottish Meteorological Office. A number of items were used in the Polar expedition of 1875 or the Antarctic expedition of 1901. Among the older items are a spirit thermometer said to have been used by Professor Joseph Black, thermometers signed by Prins and Wilson (*NP*), a metallic and two differential thermometers from the Playfair collection, Leslie's aethrioscope and a 7ft deep-earth thermometer (from the *Calton Hill Observatory, Edinburgh*).

21 *Telescopes.* A representative collection of hand-held and tripod-mounted instruments, mainly of British origin. Hand-held instruments include two vellum-covered multi-draw telescopes of late 17th-century date; other instruments include early doublet achromats by J Dollond and Son and Peter Dollond. The more extensive collection of tripod-mounted instruments includes reflection telescopes by James Short (*Lit:* BRYDEN (1968)), by B.M.... (? Benjamin Martin) and Edward Nairne. One small reflecting telescope is housed in a box bearing the trade card of Edward Scarlett. A number of early 19th-century instruments were presented by the Admiralty.

22 *Regulator clocks.* An important collection, including examples by George Graham (*c*1740), John Shelton (1756, used in 1769 transit of Venus observations), William Hardy (*c*1810 for Garnett Hill Observatory, Glasgow). Significant electric clocks include three by Alexander Bain of Caithness (1841 onwards) and clocks for the *Royal Observatory, Edinburgh* by S Riefler of Munich (1911) and Will Short (his 1922 free pendulum clock).

E 9 EDINBURGH
The Royal Museum of Scotland (National Museums of Scotland)

Queen Street, Edinburgh EH2 1JD

031 225 7534

Until October 1985 this Museum had for over 200 years been the National Museum of Antiquities of Scotland. It was then amalgamated with the Royal Scottish Museum. The basis of the collection is that assembled by the Society of Antiquaries of Scotland from the end of the 18th century. This collection was conveyed to the Government in 1851. The Museum subsequently expanded greatly and has long since outgrown its premises.

The collections cover most aspects of Scotland and Scottish life from the Stone Age until recent times, and include a number of scientific instruments with Scottish associations, as well as a few others with less obvious connections.

Lit: BELL (1981) for a series of essays on antiquarianism in Scotland and the Museum's collections.

1 Rotula arithmetica. Invented by George Brown who published an account of it in Edinburgh in 1700. It permitted addition, subtraction, multiplication and division, not only of integers but also of finite and infinite decimals. It consists of two circular brass plates, one fitting inside the other, and both mounted on a wooden plate. It is dated: *Edinburgh March 6th 1700* and inscribed *William Earle of Annandale*. (NL 77)
Lit: ERSKINE(1787) p41; WILSON (1935); BRYDEN (1972A)

2 Napierian rods. Ivory rods in leather case. Unsigned. 17th century. (NL 43)

3 Napierian rods. Wooden rods in cardboard case bearing an old label on the front '.....Merchiston .....Math.......'. This set is supposed to have belonged to John Napier of Merchiston. (NL 38) (fig 99)

4 Napierian rods, cylindrical form. Six cylinders revolve on spindles in a wooden box approx 6in in length. Late 17th century. (NL 68)

5 Octant. Light coloured wood (? pearwood), dark stained, boxwood arc with diagonal scale and wooden arm. Radius approx 15in. Signed: *Tho. Erskine 1764*. (NL 41)

6 Simple microscope. Silver, made to the 1754 subscription proposal. Signed: *J. Clark fecit Edinr.*
Lit: BRYDEN (1972C)

7 Spyglass. Turned ivory in two sections, brass mounts for lenses. Early 18th century. (NL 70) (1936-771)

8 Nautical telescope. Wood and brass. Said to have belonged to a Captain Lowe of Burntisland, Fife, in the early 18th century, but the telescope appears to be of later 18th-century date. (NL 62)

9 Equinoctial or polar dial made by John Bonar, Ayr, 1622. Silurian schist. (NL 79)
Lit: CALLANDER (1910)

10 Bloud-type magnetic azimuth dial. Ivory and silvered brass. Unsigned. (NL 46)

11 Equinoctial ring dial. Brass, 4in diameter. Unsigned. Early 18th century. (NL 25)

12 Equinoctial ring dial. Brass. In addition to the usual hour scale the chapter ring is engraved on the underside with an additional scale including the number of days in each month and the equation of time. 18th century. (NL 16)

13 Horizontal garden sundial. Brass, octagonal. Signed: *Geo. Jameson fecit.* (NL 17)
Lit: BRYDEN (1972B) p16 note 77

14 Horizontal sundial and compass with paper wind-rose. In the lid an engraved paper disc with *WR* and the maker's name *I.NE. [MES]..FECIT*. 18th century. (NL 52)

15 Magnetic dial in circular brass box. Lunar volvelle, paper wind-rose. Inscribed on lid: *John Clement 1620* and: *I'll serve you at Both night & day.* (NL 54)

16 Circular astronomical compendium and equinoctial dial. Gilt brass, paper wind-rose. Incorporates calendar and saints' days, dominical letters, nocturnal and folding equinoctial sundial. Signed on equinoctial ring: *Humfrey Cole 1574*. (NL 18)
Lit: GUNTHER (1927)

17 Boxwood nocturnal. Inscribed: *William Stone 1684.* (NL 35)

18 Perpetual calendar. Brass, silvered and gilt, finely engraved. German, possibly Augsburg, early 18th century. (NL 23)

19 Perpetual calendar. Paper and glass in rectangular wooden frame. Signed: *John Gillespie Invt. Gavin & Son fct.* (NL 165)

20 Another, in circular frame. (NL 164)

21 Compass in circular case on rectangular base plate. Brass. Unsigned, 18th century. (NL 24)

22 Circumferentor. Brass. Signed: *Cole maker Fleet Street London*. Level mounted on the underside of the instrument. Two fixed sights, two mounted on revolving alidade. (NL 56)

23 Bronze mortar inscribed: *James Borthwick Apoth in Edr 1668*. (1960.2932)
Lit: ANDERSON/SIMPSON (1976) p28 no. 38

24 Astronomical longcase clock, signed: *John Scott Edinburgh* and *Andw Smith Invt et Delint*. The dial is in three parts: Greenwich mean time, solar time and sidereal time; tide predictions of Leith, Portsmouth, The Lizard, Gibraltar and The Texel (Netherlands); Moon predictions. 18th century. (NL 58)

E 10 EDINBURGH
Scottish United Services Museum

The Castle, Edinburgh EH1 2NG

031 225 7534

The Museum (which is part of the National Museums of Scotland) illustrates the history of the armed forces with particular reference to Scotland. It houses a fine library and an extensive collection of prints, and displays uniforms, head-dresses, arms and equipment, medals, portraits and models. Its large collections include a number of scientific instruments, collected primarily for their associations, but several are of interest in their own right.

1 Model of an optical telegraph, mahogany, 7⅛ x 11 x 15¼in high. The model was made by the Revd Lord George Murray (1761-1803), son of the 3rd Duke of Atholl, and was presented by the Duke of Atholl in 1936. Murray's telegraph was adopted by the Admiralty in 1796. (1936-48)(fig 92)

2 Octant. Mahogany frame and index arm radius 17¾in, boxwood arc with diagonal scale. Signed: *Made by Ino Gilbert Tower Hill London for Ino Loughty Octr 10: 1764.*

3 Octant. Ebony frame, ivory arc. Tangent and clamping screw. Signed: *Hughes London.* (L.1932.25)

4 Walker's Harpoon Log, A2.

5 Telescope, hand-held (night glass). Single draw, wooden outer tube encased in black leather. 2¾in object glass, closed length 27in. Signed: *Gregory London.* Belonged to ancestors of Admiral Sir Berkeley Milne. Third quarter 18th century. (L.1949.4)

6 Refracting telescope, two-draw tubes. Brass, wood barrel, silver mounted. Closed length approximately 3ft. Signed: *Ramsden London.* (1933.43)

7 Refracting telescope, three-draw tubes, brass, with wood barrel. 1½in object glass, closed length 9¾in. Signed: *Cox Plymouth.* Used by the sailing master on HMS Victory. (L.1930.39)

8 Telescope, single draw. Brass, with wood barrel. 2in object glass, closed length 18in. Signed: *Dollond London Day or Night.* Third quarter 18th century. Belonged to Sir David Baird. (1948.146)

9 Refracting telescope, single draw. Brass with leather-covered barrel with inset showing windrose and flags and pendants of the signal code. Belonged to Rear Admiral John McKerlie (1774-1848).

10 Refracting telescope, five draw, brass with wooden barrel. 2in object glass. Used by Sir John Campbell in the Crimean Campaign, 1859. (L.1951.27)

11 Gunner's calipers. Brass, unsigned. Early 19th century. (1938.136)

E 11 ELGIN
Elgin Museum

1 High Street, Elgin IV30 1EQ

0343 3675

The Museum is owned by the Moray Society. It contains important collections of reptile and fish fossils, prehistoric Stone and Bronze Age weapons, butterflies, etc. In addition there are several scientific instruments.

1 Small three-draw telescope. The outer tube is leather-covered with gilt tooling, the mounts are of ivory. Unsigned, but probably early 18th century.

2 Specific gravity beads. Signed: *Jas. Corte Glasgow.*

3 Geomancer's compass. Chinese.

4 Small Chinese compass, presented 1844.

5 Horizontal sundial of brass. Maker's name erased. 18th century.

6 Horizontal garden sundial. Slate, dated 1734.

7 Instrument for constructing sundials. Incomplete. Signed: *Haye a Paris.*

8 Compound monocular microscope. Signed: *Mon. E. Hart & A. Prazmowski sur Rue Bonaparte I Paris.*

9 Wimshurst machine with a battery of eight Leyden jars. Unsigned.

E 12 EXETER
Royal Albert Memorial Museum

Queen Street, Exeter EX4 3RX

0392 265858

In this Museum are collections of paintings, watercolours, ceramics, glass and silver, with an emphasis on items by Devon artists and makers. In addition natural history, ethnography and costume are represented. Archaeology and local history are housed in a branch museum – Rougemont House Museum in Castle Street.
     The Museum contains a number of interesting scientific instruments, notably microscopes, but also an early octant and a backstaff. In addition to the items listed below there is a Watson advanced petrological microscope, and several microscope lamps and condensing lenses of the second half of the 19th century.

1 Wilson screw-barrel microscope. Brass in velvet-lined fish-skin case. Unsigned. Early 18th century. (77/1962.2)

2 Wilson screw-barrel microscope. Brass in fish-skin case. Signed: *E.A. Ezekial Optician Exeter.* (75/1918)

3 Aquatic microscope screwing into lid of mahogany box. Unsigned. Early 19th century. (78/1926.2)

4 Scarlett form of Culpeper microscope. Lignum vitae eyepiece and snout. Ray-skin outer tube, green parchment inner tube with gilt tooling. Revolving object carrier. Unsigned, c1740. (78/1926.1)

5 Jones' 'improved' microscope. Signed on barrel: *W & S. Jones 30 Holborn London.* Pyramidal case. Accessories include six swivel lenses, Lieberkuhn, stage condensing lens, fishplate, *c*1800. (32/1925.1)

6 Jones' 'improved' microscope. Signed: *R. Bancks Strand London.* Rectangular box with two drawers containing Lieberkuhn, fishplate, stage condenser, a lens acting as sub-stage condenser, forceps, and six objectives. (77/1926.1)

7 Jones' 'most improved' microscope. Unsigned. Early 19th century. (84/1917)

8 Lucernal microscope. Brass and wood. Incomplete.

9 Continental drum microscope. Signed: *Nachet Opticien rue Serpente 16, Paris, c*1840.

10 Swinging substage Ross/Zentmayer binocular microscope. Signed: *W. Heath Plymouth.* Accessories include Beck and Powell & Lealand objectives, *c*1880. (61/1909)

11 Telescope. Single draw, brass with wooden outer tube. Length (closed) 10in. Unsigned. Late 18th century. (147/1911)

12 Day or night telescope. Brass with wood outer tube. Signed. *T. Harris & Son London.* (29/1928.1)

13 Backstaff. Pearwood frame. Lacks sights. Unsigned. (25/1927)

14 Backstaff. Another similar to 13.

15 Octant. Pearwood frame and index arm, radius 25in.

Diagonal scale. Unsigned. Mid-18th century.

16 Octant. Ebony frame, brass arm. Vernier and tangent screw. Signed: *J.W. Norie & Co. London.* First half of 19th century.

17 Bloud dial. Ivory. Paper compass card. Signed: *Charles Bloud le jeune A Dieppe.* (1963)

18 Universal equinoctial dial. Silvered brass in fish-skin case. Signed: *Dollond London.* (82/1930)

19 Pillar dial. Turned fruitwood with folding brass gnomon. Probably 19th century.

20 Ring dial. Brass, inscribed: *If my Master serve me well, I'll strive all others to excell.*

21 Nocturnal. Boxwood. English, *c*1700.

22 Perpetual calendar. Silver. Inscribed: *Joseph Coles Exon. c*1710. (268/1905)

23 Sector, 6in. Brass. Signed: *E. Culpeper fecit.* (101/1918)

24 Set of drawing instruments in black ray-skin case. Includes rectangular ivory protractor with diagonal scales. Signed: *Search Crown Court Soho London.* (54/1935.1)

25 Fragment of a difference engine by Charles Babbage. Presented to a meeting of the British Association in Exeter in 1869.

26 Walker's patent sounding machine and log.

27 Camera lucida. Brass, in fish-skin case. (29/1928/2)

F

F 1 FORFAR
Forfar Museum

Meffan Institute, East High Street, Forfar

0674 73232

The Museum contains collections of archaeological, geological and natural history material. It has only one instrument, presented to the Museum by John Watt, Provost of Forfar, in 1813.

1 Gregorian reflecting telescope. Brass. Signed: *Made by James Short London 1742.* (16/323 = 12)

F 2 FORT WILLIAM
West Highland Museum

Cameron Square, Fort William PH33 6AJ

0397 2169

The Museum was founded in 1922 and houses exhibits related to the history, natural history and folk life of the West Highlands. It includes tartans and military exhibits and a large number of Jacobite relics. There are three items of relevance to this survey.

1 Surveyor's level, early 19th century. Used by the engineer Thomas Telford (1757-1834).

2 Rain gauge from the Ben Nevis Observatory (established 1888, closed 1904). Most of the surviving instruments from the Observatory are now in the *Royal Museum of Scotland (Chambers Street), Edinburgh.*

3 Anamorphose: a portrait of Prince Charles Edward, the Young Pretender (1720-1788), painted on panel, 26 x 34in. The distorted picture is rectified by means of a cylindrical mirror. Mid-18th century.
Lit: AMSTERDAM (1975) item 21

G

G 1 GLASGOW
Glasgow Museums and Art Galleries

Kelvingrove, Glasgow G3 8AG

041 357 3929

The Museum and Art Gallery houses a wide variety of material, ranging from the fine and applied arts to arms and armour, archaeology, history, ethnography, technology (covering shipping, engineering and transport) and natural history. The scientific instruments are chiefly in the Department of Technology and are largely housed at a branch museum, the Museum of Transport, Kelvin Hall, Glasgow G41 2PE. The earlier and more important of these are listed below. There is also a quantity of later electrical equipment, such as Wimshurst machines, conductors, Leyden jars, electroscopes, electrometers, etc. All are late 19th-century in date, and a number are signed by Baird & Tatlock. Others came from the Glasgow firm of Thomas, Skinner & Hamilton, who, however, acted only as agents, not as makers. There are a number of other items such as Kelvin current balances, electricity meters, etc.

1 Fulton's orrery, *c*1834. A very large orrery, designed by John Fulton (1800-1853), an Ayrshire shoemaker. It was examined and approved by the Society of Arts for Scotland in 1834, and was toured by its inventor through England and Scotland in the late 1830s and 1840s.
Lit: GLASGOW (1967)

2 Solar microscope of turned wood. The box contains a label inscribed: *Sold by T. Pether, Carver, opposite Physician's Hall in Rose Street, New Town, Edinburgh.* A small box of lenses is dated 1790.

3 Gregorian reflecting telescope. Brass, in box 15½ x 5½ x 4½in. Signed: *James Short, Edinburgh 48/105* and dated 1737. (T 7455).
Lit: BRYDEN (1968)

4 Refracting telescope, 2in aperture, single-draw tube, length (closed) 26in. Tapering mahogany tube in mahogany box. Originally possessed tripod (now missing). Signed: *Blunt & Son, Cornhill, London.* Early 19th century. From the collection of James Veitch, Inchbonny, Jedburgh.

5 Equinoctial ring dial. Brass, diameter 6in. English, 18th century.

6 Backstaff. Lignum vitae with boxwood arcs. Unsigned, 18th century.

7 Octant. Radius 19in. Signed: *Made by John Brown, Gun Dock, Wapping for J. Sampson.*

8 Octant. Signed: *Made by Richd Lekeux No. 138 Wapping High Street London 1783 John Brody.*

9 Octant. Signed: *Spencer Browning & Rust London* on ivory label and initialled *S.B.R.* on arc.

10 Octant. Signed: *Spencer Browning & Rust London.*

11 Octant. Signed: *S.A. Cail Newcastle on Tyne.*

12 Block for a compass card. Signed: *Crichton Glasgow.*

13 Waywiser. Signed: *J. & W. Watkins.*

14 Drainage level. Signed: *John Gardner Glasgow fecit.*

15 Miner's compass. Signed: *James White 14 Renfield St. Glasgow.*

16 Macfarlane's calculating cylinder.

17 Sikes' hydrometer. Signed: *Bate London.*

18 Hydrometer. Signed: *Oertling London.*

19 Specific gravity beads. Signed: *A. Marnoni 34 Brunswick Place Glasgow.*

20 Balance. Signed: *De Grave & Son London 1825.*

21 Globe electrical machine. Signed: *Nairne London.* (fig 46)

22 Plate electrical machine. Diameter of plate 15in. Believed to have been made in Glasgow *c*1830.

G 2 GLASGOW
Department of Astronomy

The University, Glasgow G12 8QQ

041 339 8855

The Astronomy Department was established in 1757 with the founding of the Macfarlane Observatory, constructed to house the astronomical instruments bequeathed to the university by Alexander Macfarlane, a graduate who had emigrated to Jamaica. The manuscript packing inventory survives, but lamentably, most, if not all of the instruments – which included a large transit, an astronomical quadrant, a zenith sector and an equal altitude telescope – are lost. At one stage the Department also possessed a 10ft Herschel reflector. The Macfarlane instruments included a reflecting telescope, and the Department still houses two Short reflectors, the earlier of which may have come from the Macfarlane bequest, while the later, dated 1759, may have been acquired by the first Professor of Astronomy, Alexander Wilson. The Department also owns a fine collection of early astronomical books.
Lit: BRYDEN (1970)

1 2½in aperture Cassegrain reflecting telescope on pillar and claw stand. Altazimuth mounting. Signed: *James Short London.* (250/1103 = 9.5)
Lit: BRYDEN (1968) no. 1

2 Gregorian reflecting telescope, 3in aperture. Altazimuth mounting on pillar and claw stand. Signed: *James Short London.* (J743/29/372 = 12)
Lit: BRYDEN (1968) no. 15

G 3 GLASGOW
Department of Natural Philosophy

The University, Glasgow G12 8QQ

041 339 8855

Robert Dick, the first professor of Natural Philosophy, was appointed in 1726, and the earliest list of instruments in the possession of the department is dated 1727. Another inventory, dated 1760, was stated by the writer, John Anderson, to be a copy of a list written in 1756 'in Dr Dick's hand'. In the year in which Dick compiled his list, James Watt was appointed 'instrument maker to the University'.

John Anderson was Professor from 1757 to 1796. He provided in his will for the foundation of 'Anderson's Institute' which later became the Technical College, and has now developed into the University of Strathclyde. The next landmark for the Department was the appointment of William Thomson (later Lord Kelvin), Professor from 1846 to 1899. He was remarkable for having established the first students' physical laboratory, for his work in thermodynamics, electricity and magnetism, and for the numerous instruments he designed and patented, many of which were commercially produced by James White, the Glasgow instrument maker. Many of the instruments in the Department of Natural Philosophy are from this workshop.

A number of instruments survive from the early days of the department. Those listed by Dick must antedate 1756, and surviving items are listed in Section I. Their numbers in the Dick catalogue are given in brackets. Section II lists instruments which are mostly late 18th- or early 19th-century in date. Kelvin material is summarised in section III, but no attempt has been made to list items individually, and the publication by Green and Lloyd mentioned below should be consulted. Section IV includes items connected with James Joule, and section V items associated with the Revd John Kerr.

Lit: LLOYD (1969); GREEN/LLOYD (1970) (this lists all the Kelvin, Joule, Kerr and some of the earlier material); SWINBANK (1982) (for an account which includes details of apparatus acquired in the 18th century).

I   *ITEMS IN JOHN ANDERSON'S COPY OF DICK'S LIST*

1   Lodestone. (Dick 1B)

2   Compass in mahogany box with lid. Paper windrose. Unsigned. (Dick 2B)

3   Globe electrical machine. Development of Hauksbee type with provision for screwing on to edge of table. (Dick 5)

4   Wheel and axle. (Dick 13A)

5   Pair of cycloidal channels. (Dick 27A)

6   Model Newcomen engine (Dick 51). Now on display in *The Hunterian Museum, Glasgow* (see G 5/1).

7   Large concave mirror. (Dick 79)

8   Single and compound microscope. Signed: *Ayscough London. Fecit No. 7.* Incomplete. (Dick 84A,B)

9   Compound microscope. Fixed square section pillar on round base (similar to 'Hooke' microscope). Snout of microscope screws into arm adjustable on pillar. Red morocco outer tube, white vellum inner tube. (Dick 82)
Lit: TURNER (1966) no. 58

10   Single-draw refracting telescope. Wooden outer tube, brass mounted. Inner tube vellum covered. Signed: *Ayscough London Invt et Fecit No. 7.* (Dick 85 G)

11   Gregorian reflecting telescope. 2½in aperture, 9.4in focus. Signed: *James Short London.* (222/899 = 9.4) (Dick 86)
Lit: BRYDEN (1968) no. 14

12   Pocket camera obscura. (Dick 87A)

13   Two prisms, one of green, one of white glass. (Dick 90A)

14   Set of specific gravity beads in turned wooden box. (Dick 98)

II   *INSTRUMENTS OF LATE 18TH- OR EARLY 19TH-CENTURY DATE*

15   Gregorian reflecting telescope on pillar and claw stand. 4½in aperture. All brass. Signed: *Miller Edinburgh.*

16   Brass refracting telescope on pillar and claw stand. Signed: *T. Harris & Son London.*

17   'Museum' microscope with revolving cylinder on which specimens are mounted. Early 19th century.

18   Two hand polariscopes with revolving discs containing samples of aragonite, nitre, calcspar, topaz, borax and quartz. Early or mid-19th century.

19-20 Two Norremberg polariscopes, one signed: *Duboscq Soleil Rue de l'Odeon a Paris.*

21   Octant. Ebony frame, radius 10in. Anchor engraved on ivory degree arc. Signed: *Gardner & Neil Belfast.* This instrument belonged to Professor James Thomson, father of Lord Kelvin.

22   Octant. Ebony frame, radius 11in. Unsigned. Probably late 18th or early 19th century.

23   Sextant (125° arc). Ebony frame 14in. Vernier and sighting telescope. Signed: *D. Adams Charing Cross London. I R* engraved on arc. Late 18th century.

24   Sextant. Brass double frame with silver degree scale. Radius 8½in. Signed: *Troughton London.*

25   Cylinder electrical machine on massive wooden base. Cylinder length approximately 8½in. Probably not professionally made.

26   Plate electrical machine, De Winter type. Signed: *D. Davis Glasgow.* No. 63 on base. 1840-50.

27   Electrical discharger. Signed: *Watkins & Hill 5 Charing Cross London.* First half 19th century.

28   Plate electrical machine. Unsigned. Probably mid-19th century.

29 Voss type electrical machine. Unsigned.

30 Wimshurst machine. 14in plates. Unsigned.

31 Electric cannon.

32 Wooden double cone for use with an inclined plane.

33 Two rollers for use with an inclined plane.

34 Four-wheeled carriage for use with an inclined plane.

35 Pulleys (three).

36 Nicholson's hydrometer. Signed: *Crichton Fecit* and marked: *800 Grams Tempr 62°*, c1800.

37 Nicholson's hydrometer. Signed: *Crichton Fecit* and marked: *800 Grams Tempr 62°*. c1800.

38 Weight and bucket in box. *Apparatus for Archimedes Principle* engraved on ivory label.

39 Bimetallic thermometer. Signed: *Crichton Invt et Fecit Glasgow No.4.*
Lit: CRICHTON (1803); MIDDLETON (1966) p171

40 Stirling air engine. Working model presented by the inventor, the Revd Robert Stirling, in 1827.

41 Siren. Signed: *F. Sauernwald in Berlin No.12.*

42 Early resistance box. Consisting of 6 coils, each of '800 yards' of wire, each with simple switch and mounted on a circular wooden base. Base marked: *N.J. Holmes 1848.*

43 James Thomson's PV surface (a model of the thermodynamic surface for carbonic acid). Made by Professor James Thomson from results of Andrews' experiments.

44 Electric motor. Signed: *Froment, Paris,* supplied through Pixii, 1849 (receipts still extant).

45 Coulomb torsion balance. Signed: *Pixii.*

III *INSTRUMENTS ASSOCIATED WITH LORD KELVIN*

The Department contains very considerable quantities of instruments originated by Lord Kelvin or known to have been associated with him. Many were constructed by the firm of instrument makers set up in Glasgow by James White in 1849. Some were donated by the Kelvin-Hughes Division of Smith's Industries. Much of the material is standard commercial production of instruments originated by Kelvin, others are items he is known to have used for research, demonstrating or teaching. The following types of instruments are included:

46 Electrometers. Spring balance electrometer, absolute electrometers, quadrants electrometers, electrostatic balance, suspended vane electrometer, long range electrometer, etc.

47 Electrostatic voltmeters, vertical and multicellular types.

48 Mousemills.

49 Marine and moving coil galvanometers.

50 Astatic galvanometers including astatic mirror galvanometers, graded potential and current galvanometers. Tangent galvanometers, including lamp counters.

51 Volt balances, current balances, ampere gauge, recording voltmeter, centi-ampere balance.

52 Demonstration and marine compasses.

53 Gyrostats, precessional top.

54 Tide machine.

55 Ballistic pendulum.

56 Pitch glacier, solids floating in pitch.

57 Kelvin's standard thermometers including two signed: *Fastre aine a Paris 1850* and four from Kew Observatory dated 1852 and 1853.

58 Low-resistance bridge.

59 Apparatus for determination of the ohm, 1860.

60 Harmonic analyser – working model.

IV *APPARATUS ASSOCIATED WITH JAMES JOULE (1818-1889)*

61 Electric motor (c1840-43) apparently used in early experiments on mechanical equivalent of heat by electrical methods.

62 Electromagnet (c1840). Described by JOULE (1884-87) pp27-42.

63 Iron calorimeter. Used in experiments of friction of mercury (pre-1849). Presented by Joule to Lord Kelvin.

64 Brass calorimeter. Used by Joule in his final (1877-8) determination of the mechanical equivalent of heat. Presented by Joule to Lord Kelvin.

65 Tangent galvanometer, made c1847 for Kelvin to the specification of Joule. Signed: *J.B. Dancer Optician 15, Cross Street, Manchester.*

66 Small tangent galvanometer. Signed: *J.B. Dancer Optician Manchester.*

V *APPARATUS ASSOCIATED WITH KERR*

The Revd John Kerr, MA, LLD, FRS (1824-1907) was a close friend of Kelvin. He became lecturer in mathematics and physical science at the Free Church Training College in Glasgow, and published his paper on the Kerr electro-optic effect in 1875, and on the magneto-optic effect in 1876. The following items were presented by Kerr to the University.

67 Electro-magnet used in the discovery of the magneto-optic effect.

68 Kerr cells (2). Liquid form.

69 Kerr cell. Solid glass form.

70 Shaped glass blocks used by Kerr.

G 4 GLASGOW
Department of Natural Philosophy

University of Strathclyde, Royal College, George Street, Glasgow G1 1XQ

041 552 4400

The University of Strathclyde has developed from Anderson's Institution, founded in 1796 by John Anderson, Professor of Natural Philosophy at Glasgow University from 1757 to 1796. It was affiliated to the University of Glasgow from 1913 to 1964 and was presented with a University Charter in 1964.

The Department of Natural Philosophy houses an extensive collection of teaching apparatus from Anderson's Institution, dating from the 18th century onwards. The earliest item, Bird's octant, antedates the foundation of Anderson's Institution, and probably belonged to Anderson himself. Many others are of late 18th- or early 19th-century date when the Institution must have been building up its stock of instruments. Some items appear to be those listed in a manuscript dated 1799. The considerable quantities of mid-19th-century instruments, including Kelvin current balances, electrostatic voltmeters and a Bryson camera, have not been listed, but as much as possible of the earliest material has been included. The collection includes many items of the kind found in the King George III Collection in the *Science Museum, London*, and represents one of the more complete surviving College or University collections of teaching and demonstration apparatus.

Many instruments bear one or more inventory numbers (quoted in brackets after the description). The two inventories are, firstly, a copy of one 'subscribed by Mr. Anderson' and dated 1760, in which a distinction is made between apparatus which belonged to the University of Glasgow and that which belonged to Anderson himself, and secondly, one dated 1799, and therefore drawn up three years after Anderson's death. Another inventory, referred to by Anderson in his will, no longer survives.
Lit: MUIR (1950) (Chapter III discusses the inventories and some of the surviving instruments); CAMPBELL (1980) for a comprehensive catalogue written after this survey had been undertaken.

1 Model of a Gregorian reflecting telescope on wooden pillar and tripod stand. Octagonal wooden tube, one side removable. 4in speculum, tube length 23in. The secondary mirror is missing. (Inv: 256)

2 Model of a Newtonian reflecting telescope on wooden pillar and tripod stand. Octagonal wooden tube, one side removable for demonstration purposes. 4in speculum, tube length 31½in. (Inv: 257)

3 Model of a Herschelian reflecting telescope. Octagonal wooden barrel. 4½ in speculum, tube length 39½in. (Inv: 255)

4 Facsimile of Galileo's telescope. Bears inscription 'This instrument a copy or facsimile of the telescope constructed by the illustrious Galileo, the original being presented in the Public Museum of Florence.... Presented to the Anderson University by John Hart'. Probably early 19th century.

5 Large solar microscope. Brass, mounted on varnished wood stand. Tube length 8½in, mirror size, 3¾ x 10in. Signed: *Adams London.*

6 Small solar microscope. All wood except for brass cog wheels. Plate 6 x 6in. Tube length 6½in. Unsigned. (Inv: 11 painted on instrument; 261 on paper label)

7 Quadrant. Radius 15½in. Ebony frame, brass degree arc. In the form of a quarter circle, the arc graduated in 90°. Only the frame survives, but there are fixed points for an arm, as well as a point of attachment on the right hand outer radius. (Inv: 2)

8 Octant. Mahogany frame, radius 20in. Brass arm with vernier. Small sighting tube fixed along one edge of the octant frame. Signed: *J. Bird London.*

9 Surveyor's octant. Ebony (or ebonised wood) frame, ivory arc, wooden arm with brass sleeve. Radius 18in. The octant is attached to a pillar and tripod stand by means of a ball joint, permitting it to be tilted at any angle, and revolved around its vertical axis, with its angle of revolution being read from a small pointer moving over a brass scale at the bottom of the pillar. The tripod legs are attached to a circular base containing a compass and equipped with three adjusting screws. (Inv: 1)

10 Sextant. Signed: *M. Walker Glasgow Greenock and Liverpool.*

11 Orrery. Planetarium only. Brass, diameter of drum 8½in, mounted on pillar and tripod stand. Signed: *W. & S. Jones London.*

12 Orrery. Planetarium and tellurium. Brass, as no. 10. Signed: *W. & S. Jones London.*

13 Armillary trigonometer. Brass and wooden. Consists of a vertical semicircle, diameter 12¾in, graduated in degrees with calendar and divisions into days. One half of the circle is hinged and can be pivoted around its vertical axis, sliding as it does so over a horizontal semicircle marked with degrees and wind directions. Within the first vertical semicircle is a second one, again hinged and graduated from 90 – 0 – 90°. A second horizontal semicircle within the first and at right angles to it is graduated in hours. Both are attached to an inset circular board which can be swivelled around its central point. Inscribed: *Alexr Wilson M D Fecit.* Alexander Wilson, Professor of Practical Astronomy at the University of Glasgow was a contemporary of Anderson and known for his research on Sun spots. The instrument is described by FERGUSON (1767) as the invention of Mungo Murray the shipwright.

14 'Large Nonius scale of wood'. (Inv: 38)

15 Models of a sucking and a forcing pump. Mounted side by side on a mahogany stand. Signed: *W. & S. Jones No. 30 Holborn London.* Early 19th century.

16 Double-barrel air pump. Table model, mahogany and brass. Base 12¼ x 17in. Maker's label missing. Early 19th century.

17 Air pump. Floor standing model, legs in the form of flat pilasters with carved capitals, mahogany veneered and inlaid, support the table for the receiver. The double barrels are mounted between the legs of the table, and a barometer is set into one of the legs. Ht 42½in. Signed: *James Crichton Glasgow No. 2.*

18 Small cylinder electrical machine. First quarter 19th century.

19 Clarke's magneto-electric machine. Signed: *E.M. Clarke Inventor and Manufacturer 23 Lowther Arcades London.*

20 Flicker tubes. Five glass rods with spiral lines of metal spangles. A sixth glass rod in the centre carries a pivoted pair of brass balls.

21 Three flicker panes. One with the name Anderson.

22 Musical glass.

23 French bells apparatus.

24 Gold leaf electroscope.

25 Considerable quantities of other electrical accessories – insulating stool, conductors, head of hair, etc.

26 Lodestone, brass mounted. 18th century.

27 Dip circle. Circular brass base with three levelling screws. Circular mahogany needle case of 14½in with glass sides. The needle supported on friction wheels. Signed: *Crichton Glasgow*. Dip needle of the type designed by the Revd Mitchell and constructed by Nairne.

28 Magnetic compass in square mahogany frame 14⅝ x 14⅝in. Signed: *Crichton Glasgow*.

29 Comus's magnetic box.

30 Pyrometer for measuring the difference in the thermal expansion of metals. Bars of two different metals are connected to pointers moving in front of a divided arc. Wooden base, length 17½in.

31 Wedgwood pyrometer.

32 Guinea and half guinea balance in a box.

33 Small steelyard, length 6¾in.

34 Larger steelyard, length 16in.

35 Small red-painted balance on wooden stand. Copper pans. Another small balance, unpainted.

36 Pulley frame. Turned frame. Turned columns. From upper edge of frame are suspended a series of weights and pulleys.

37 Apparatus for illustrating plane, circular and elliptical vibrations. Maker's label missing.

38 Model of leaning tower of Pisa. Painted wood. Ht 16in.

39 Inclined plane. Baseboard mounted on four turned legs. The inclinable plane can be adjusted against an arc. The roller is equipped with a pull bar to which a cord is attached that passes over an adjustable pulley at the end of the inclined plane. Length 43in. (Inv: 43)

40 Dissected cone. Wood, ht 11½in. (Inv: 204)

41 Wooden stand for rolling cone, and wooden conic rhombus. (Inv: 10 (202))

42 Apparatus to illustrate the decomposition of white light into colours by a prism. The colours of the spectrum are indicated by coloured threads. On mahogany base, length 21in.

43 Achromatic prism. Three wedge-shaped prisms in a brass frame. To illustrate achromatic lenses. Mounted on a brass stand. Ht 4¾in.

44 Hydrometer. Wood, in a turned lignum vitae case. Length of case 4¾in. 18th century. (Inv: 22) (fig 63)

45 Set of communicating glass vessels of different shapes and capacities to demonstrate the constancy of water level.

46 Polariscope (Norremberg). Signed: *E.M. Clarke Optician Strand London.*

G 5 GLASGOW
Hunterian Museum

University of Glasgow, Glasgow G12 8QQ

041 330 4221

The Museum was founded in 1807 to house the astronomical and other collections founded by the surgeon William Hunter in London. The present Museum collections now include geological, archaeological, historical and ethnographic material, as well as well as coins, books and manuscripts and the University fine art collections. It houses a few items of relevance to this survey, of which the following are the most important.

1 Model of the Newcomen engine, formerly in the Department of Natural Philosophy, University of Glasgow and listed in the 1760 copy of Dick's 1756 inventory (see G 3.6). This scale model was repaired by James Watt in 1763-4, and as a result of his attempts to repair it, he conceived the idea of the separate condenser. It seems very probable that the model was made by Sisson; it had certainly been sent to him for repair before Watt worked on it. (fig 93).
Lit: DICKINSON/JENKINS (1927); LAW (1969) figs 6-8

2 George Adams' 'variable' microscope, ht 18in. Signed: *George Adams at No. 60 Fleet Street London*. From the Old College, Glasgow. Belonged to William Hunter.

3 Achromatic microscope. Ht 18in. Signed: *Jas Smith London*. With nose-piece adjustment and mechanical stage. Belonged to Lord Lister.
Lit: NUTTALL (1973B)

4 Electromagnetic clock by Alexander Bain, *c*1845.

5 Case containing four lancets made by Savigny & Co. Used by Edward Jenner (1749-1823).

6 Six surgical instruments, formerly the property of Lord Lister: two amputating knives, saw, scalpel, hernia knife, and saw used in cases of skull fracture.

G 6 GLOUCESTER
City Museum and Art Gallery

Brunswick Road, Gloucester GL1 1HP

0452 24131

The Museum and Art Gallery contain collections relating to local history, archaeology, natural history, geology and numismatics. There is a fine collection of English walnut furniture, also considerable collections of glass, ceramics and silver. A branch museum, Bishop Hooper's Lodging, houses local bygones and industrial, agricultural and local history collections.

The main museum has three instruments, in addition to an extensive and important collection of domestic barometers, most of which were bequeathed to the museum in 1963. The barometers include examples of most major types by such makers as Daniel Quare, London; Houghton, Farnworth near Warrington; Smith, London; James Gatty, London; Dollond, London; Burton, London; Edwards and Hunter, London; Knie, Edinburgh; Bapt. Ronchetti & Co., Manchester; Charles Orme, Ashby de la Zouche; D. Copoduro, Cirencester; Joshua Springer, Bristol; Dom. Sala, London; Cary, London.
Lit: Many of the above barometers are described and illustrated in GOODISON (1977) and BANFIELD (1976)

1 Oval horizontal dial of silver for latitude 44°35′. The dial bears a monogram beneath a ducal crown, possibly that of Amadeus II, Duke of Savoy (1680-1730). The back of the instrument is engraved: *TURIN Lat. 44°35′*, and signed *J. Rowley London*. The engraving on the instrument is of very high quality. (fig 34)

2 Brass equinoctial ring dial. Probably *c*1700.

3 Ring sundial. Brass, diameter 1½in. Probably late 17th or early 18th century.

G 7 GREENOCK
The McLean Museum

15 Kelly Street, Greenock PA16 8JH

0475 23741

The Museum, which also incorporates a picture gallery, covers natural history, geology, and shipping exhibits. In view of the fact that James Watt (1736-1819) was born at Greenock, it is not surprising that there are items associated with him in the Museum. There are several interesting models of ships and engines.

1 Balance. Presented to Greenock Philosophical Society in 1863 as having belonged to James Watt. Steel beam, 11½in long, brass pans, glazed mahogany case. The balance is suspended from a fixing point in the top of the case. Signed: *Crichton Glasgow*.

2 James Watt's patent letter-copying machine.

3 Reflecting circle. Diameter 10½in. Signed: *Troughton London 54. c*1800.

4 Octant. Ebonised frame, radius 12in. One telescope, tangent screw. Signed: *Lyon Greenock*.

5 Octant. Ebony frame. Radius 11in. Signed: *Chas. Jones Liverpool*.

6 Sector. Brass. Length (closed) 6½in. Signed: *Butterfield Paris*.

7 Model of surface condensing engine. Made by John Gray of Irvine, 1838.

H

H 1 HAWICK
Hawick Museum and the Scott Gallery

Wilton Lodge Park, Hawick TD9 7JL

0450 73457

The Museum collections concentrate on Hawick and the Borders – geology, natural history, local industry and history. An important section consists of machinery associated with the hosiery industry. The more important of the scientific instruments also have local connections, one a refracting telescope, having belonged to a local doctor, and the others to the astronomer and instrument maker Gideon Scott (1765-1833).
Lit: McKIE (1965-6)

1 Newtonian reflecting telescope. Tube of sheet iron, length 23¾in, mirror 3⁷⁄₁₀in diameter. Known to have belonged to Scott and almost certainly made by him.

2 Refracting telescope. Wooden outer tube, brass fittings, three-draw tubes. Length 11½ in (closed), maximum extension 31½in. 1¼ in aperture. Unsigned. Owned by Dr John Leyden (d. 1811).

3 Octant. Ebony frame, radius 14in. Signed (on brass arm): *W. & S. Jones 30 Holborn London*. Anchor engraved on arc, *c*1810. Known to have belonged to Gideon Scott.

4-5 Two Chinese geomantic compasses.

H 2 HELSTON
Helston Folk Museum

The Old Butter Market, Market Place, Helston TR13 8ST

03265 564027

This is a folk museum covering all aspects of life in the Lizard Peninsula. The collection contains only one scientific instrument, but one of some interest.

1 'Opake Solar microscope' as published by Benjamin Martin (1744), in wooden box bearing the trade label of *Field, optical mathematical and philosophical Instrument Maker, No.74 Cornhill London (Late Apprentice to Mr. Nairne)*. The microscope is complete and is accompanied by an edition of G Adams, *Essays on the Microscope*.

H 3 HERTFORD
Hertford Museum

18 Bull Plain, Hertford SG14 1DT

0992 582686

The Museum contains collections relating to local history, geology, archaeology, and natural history. Though not numerous, there are some scientific instruments. In addition to the items listed below, there are several optical toys such as zoetropes and stereoscopes, several cameras, an early mechanical light patented by Henry Berry in 1824, various lamps, early typewriters, telephones and telegraph machines, as well as barometers, clocks and scales and measures.

1 Camera obscura by William Storer (British Patent 1183 (1778)). Green baize covered box with domed lid and carrying handle. Dimensions 15½ x 9 x 10in. A paper label inside the lid bears the text: *By the King's Patent the Royal Delineator: to be had of no-one but W. Storer.* (fig 18)

2 Culpeper microscope of brass, ht 10½in. Signed: *Mackenzie Cheapside London. c*1820.

3 Nuremberg tripod microscope, wood and cardboard, ht 14in. Signed: *I.M.* (Junker, Magdeburg).

4 Compound monocular microscope, signed: *Dancer Optician Manchester*. Tripod foot, single pillar, Lister limb. *c*1840-50.

5 Sikes' hydrometer. Gilt brass. Signed: *Bate London.*

6 Plate electrical machine. The plate supported in a rectangular wooden frame, 31in high. Maker's label missing. The prime conductor is supported by a glass rod projecting horizontally from the frame of the machine.

7 Cylinder electrical machine. Stated by donor to have been constructed by his father, *c*1840. Rectangular base, cushion on insulated glass stand. The cushion of the cylinder can be adjusted by means of a slider fixed in position by a wooden screw. Prime conductor on a separate insulated stand with turned wooden base. Cylinder length 12in, diameter 7in.

8 Read's electrical machine. The cylinder is mounted on a vertical axis supported by a curved metal arm. Cylinder diameter 5¼in, length approx 7in. A pulley at the lower end of the axis is connected by a string to a wheel mounted horizontally on the base.

9 Accessories to these machines, including two Henley pith-ball electrometers, one signed *Ebsworth London*; an insulating stand (possibly belongs to no. 7); two Leyden jars, pith figures; jointed dischargers; Lane electrometer;

medical directors; voltaic cannon, etc.

10 Spinner to show the behaviour of bodies where the equilibrium position is governed both by gravitation and central forces. It consists of a wooden tripod foot surmounted by a pillar to which is affixed a hand-operated driving wheel. A horizontal arm is mounted on the pillar. Objects are suspended by a string from the end of the arm and are rotated by means of the wheel and pulleys. Believed to have been used by Sir David Brewster for experimenting with centripetal light, 1856.
Lit: MORRISON-LOW (1984) 92-93

11 Marine barometer. Signed: *Cary London.*

12 Globe and orrery believed to have been made by Jeremiah Cleeve of Hitchin, *c*1810. Both are mounted on similar 12-sided veneered wooden bases with ball feet, and surmounted by wood framed glazed covers. At some stage globe and orrery have been linked by means of an arm and a clockwork mechanism. The globe, a celestial globe, is inscribed: *A new celestial globe by J. Newton 1801* and an examination of the base in which it is mounted reveals the four turned legs of a late 18th- or early 19th-century globe stand, around which the veneered 12-sided base has been constructed. The base plate of the orrery has been over-painted at some stage and inscribed: *Renovated by Newton and Co. Fleet St. London*; it would appear to be of late 18th-century date.

H 4 HOLYWOOD
Ulster Folk and Transport Museum

Cultra Manor, Holywood, County Down BT18 0EU

0232 428428

The Ulster Folk and Transport Museum is a National Museum established by Act of Parliament in 1958. Its objects are the collection, study and display of material depicting traditional life in Ulster and the history of transport in Ireland.

The Transport Museum is at present divided between premises in Witham Street, Belfast (Tel: 0232 51519) and at Cultra Manor adjacent to the Ulster Folk Museum. It contains transport material of all kinds – steam locomotives and railway carriages, trams, coaches and horsedrawn carriages, early cars, vans, motor bicycles and pedal cycles, fire appliances, etc.

In addition to the items listed below the collections contain clocks, domestic barometers and instruments connected with rail transport – telegraph repeaters, signalling appliances, etc.

1 Waywiser, signed: *Richard Spear, Dublin.*

2 Surveyor's mercury level. Two ivory rods float in mercury contained in a black-painted pine case with a brass carrying handle. (cf. self-adjusting level for engineers, invented by Thomas Parker before 1832, at *Royal Artillary Institution, Woolwich, London.*)

3 Pair of globes by Newton & Son, 1857.

4 Jacot tool – watch repairer's tool for renovating pivots. Inscribed: *Tour a pivoter A Vis de Rappel Qualite Superieure.*

H 5 HUDDERSFIELD
Tolson Memorial Museum

Ravensknowle Park, Wakefield Road, Huddersfield HD5 8DJ

0484 530591

The Museum, part of Kirklees Metropolitain Borough Libraries and Museums Service, contains collections of local interest, covering such fields as geology, botany, natural history, bygones, toys, and local industries. In addition to those scientific instruments listed below there are several 19th-century microscopes including a monocular microscope by Pillischer, binocular microscopes and a group of clockmaker's tools in a reconstructed workshop.

1 Brass Culpeper microscope. Late 18th century. On loan.

2 'Solar opaque' microscope of Benjamin Martin. Signed: *Wm Harris Optician to His Majesty 22 Cornhill London.*

3 Jones' 'Improved' microscope. Signed: *W. & S. Jones 30 Holborn. London.*

4 Octant. Mahogany frame, radius 14in. Brass arm engraved with foliage. Maker's label missing, *c*1780.

5 Waywiser. Signed: *Dollond London.*

6 Miner's dial. Unusual construction. Signed: *G. Davis Leeds.*

7 Miner's dial. Signed: *Davis Leeds.*

8 Anemometer. Signed: *Davis Derby, Biram's Patent 225.*

9 Clockmaker's wheel-cutting engine. All brass. Signed: *William Terry Fecit 1774.*
Lit: HAMER (1924)

10 Clockmaker's wheel-cutting engine. Late 18th century. Unsigned.

11 Bloud-type ivory sundial. Unsigned.

H 6 HULL
Town Docks Museum

Queen Victoria Square, Hull, HU1 3DX

0482 222737/8

The maritime collections were long housed in a branch museum at Pickering Park on the western outskirts of Hull, now closed, but since 1975 a new maritime museum has been opened in the former dock authority offices in the city centre. This contains a display of Hull's whaling industry and a fisheries exhibition. There are collections devoted to the Humber estuary, inland waterways, the development of merchant shipping, shipbuilding, marine engineering and navigation.

1 Nautical telescope. Single draw tube, ten-sided outer tube of wood. Signed: *J. Gilbert Tower Hill London.*

2 Nautical telescope, single draw tube, the outer tube cardboard covered, length 19in. Signed: *Made by Jas. Chapman, St. Catherine's London Day & Night.*

3 Telescope. Wooden tube single brass draw tube. Inscribed: *Cox, Plymouth Dock Impd. Ship Telescope.* Belonged to Capt. Hornby of the Whaler *Birnie.*

4 Telescope. Leather cover, single brass draw tube. Signed: *JOHN STONES, HULL Night and Day.* Presented by Sir John Ross to Capt. Peter McBride of Hull.

5 Telescope. Red wooden tube and single brass extension. Inscribed: *Capt. Hotham R.N.* Presented to Capt. Richard Harrison during the Ashanti Wars.

6 Nautical telescope by Preston, 108 Minories, London. Single draw tube, outer tube of wood. Length closed 20in, object glass 2½in diameter. Telescope signed: *Worthington & Allen, London.*

7 Terrestrial globe. Diameter approx 10in. Column stand on turned wooden base. Signed: *Thos. Malby.*

8 Terrestrial globe. Diameter approx 5in. By Newton, 1842.

9 Celestial globe. Floor-standing model. Signed: *W. & S. Jones.*

10 Terrestrial globe with celestial gores lining case, diameter 3in. Unsigned. Early 18th century.

11 Terrestrial globe with celestial gores lining fish-skin case, diameter 3in. Signed: *C. Price fecit.*

12 Orrery by W and S Jones. With planetarium and tellurium. Engraved paper on wood, movable parts of brass. Diameter of base plate 13in.

13 Diptych dial, engraved paper on wood. Signed: *Verfertigt von David Beringer.* German, late 18th century.

14 Diptych dial. Coloured and engraved paper on wood. German, late 18th century. Unsigned.

15 Diptych dial. Mirror mounted in lid and framed with embossed metal. Paper windrose. Probably English, 19th century.

16 Equinoctial dial, brass, silvered and gilt. 18th century.

17 Equinoctial dial, brass, silvered and gilt, in leather case. Signed: *Johann Martin in Augsberg 48.*

18 Magnetic azimuth dial. Circular brass case with an engraved paper scale. Signed: *Henry Sutton fecit.*

19 Universal equinoctial ring dial. Brass, diameter 3¼in. On mahogany stand. 18th century.
Lit: HOLBROOK (1974) no. 28

20 Universal equinoctial ring dial. Brass, diameter 4in. Unsigned: early 18th century.

21 Universal equinoctial ring dial. Brass, diameter 6¼in. Signed: *Heath & Wing London.* Mid-18th century.

22 Universal equinoctial ring dial. Gilt brass, diameter 13¼in. Sliding suspension ring. Signed: *R. Glynne fecit.*

First half 18th century.

23 Circular horizontal dial. Bird gnomon, brass, diameter 3½in. Dated 1773.

24 Magnetic dial. Turned wooden box. 19th century.

25 Magnetic dial (floating compass rose), 18th century.

26 Inclined dial. Brass, partly silvered, bubble levels and adjusting screws. Signed: *Bradford 136 Minories London.*

27 Inclining dial. Brass, unsigned, late 18th century.

28 Nocturnal. Boxwood, length 3½in. Inscribed: *For Both Bears.* Late 17th/early 18th century.

29 Gunter's quadrant, with planispheric nocturnal on reverse, brass, radius approx 3¾in. Signed: *Joha Prujean Fecit Oxo.*

30 Horary quadrant, radius 3¼in. Brass. Showing equal hours with straight hour lines and shadow square. Six stars are listed and their positions marked in the zodiac. On the reverse, a revolving disc with the names of the months indicated by their initial letters. An hour scale is engraved on the face of the quadrant surrounding the disc. A pointer revolves over both scales. One pin-hole sight survives. ? Late 16th century.
Lit: HOLBROOK (1974) no. 2

31 Octant. Mahogany frame, brass arm and scale. Vernier, fixing and tangent screws. Debased rococo decoration on the arm, and the date *1775.* Inscribed *Klaas Dirkzsen Hoek.*

32 Octant. Mahogany frame and arm, the lower part of the arm encased in brass, radius 15¾in. Ivory scale and vernier. Fixing screw. Inscribed on ivory label: *Thos. Machell 1788 Gregory & Son London.*

33 Octant. Mahogany frame, engraved brass arm, ivory scale and vernier, radius 18in. Fixing screw. Inscribed on ivory label: *Cutht Broderick.* Probably third quarter 18th century.

34 Octant. Ebonised mahogany frame, radius 16in. Signed: *A. Wellington Crown Court Soho London.* Late 18th century.

35 Octant. Ebony frame, radius 11⅞in. Vernier and tangent screw. Marked: *SBR* on ivory arc. Inscribed: *Northern Hull.* Early 19th century.

36 Octant. Ebony frame, radius 13¾in. Tangent and fixed screw. One telescope. Marked: *SBR* on arc.

37 Octant. Ebony frame, radius 11in. Vernier, tangent screw. Inscribed on ivory label: *The True Love Hull.* (This boat, a whaler, operated from 1780 to 1859.)

38 Octant. Ebony frame, radius 12⅛in. Tangent and fixing screw, 3 and 2 shades. Marked: *SBR* on arc.

39 Octant. Ebony frame, radius 10in. Fixing and tangent screw, telescope, 4 and 3 shades. Signed: *Thompson London.* Label in box: *Bristol Observatory and Chronometer Dept. Estd. 1825 Edwin Langford 53 Quay.*

40 Octant. Ebony frame, radius 10in. Tangent and fixing screw, signed: *W.F. Cannon London.*

41 Octant. Ebony frame, radius 10in. Tangent and fixing screw, 3 shades, no telescope. Anchor engraved on arc. Signed on ivory label: *Thos. Harris & Son London.*

42 Octant. Ebony frame, radius 10in. Tangent and fixing screw, no telescope, 3 and 2 shades. Signed: *G. Berry & Son W. Hartlepool.*

43 Octant. Ebony frame, radius 11in. Unsigned but *SBR* on arc.

44 Octant. Brass frame, radius 7⅝in. Ivory scale and vernier, fixing screw, magnifier, telescope and 3 shades. Arc graduated to 115°. Signed: *Geo. Christian Liverpool.*

45 Sextant, 8in radius, brass index arm and frame, silver scale, arc graduated to 150°, vernier and magnifier, 4 and 3 shades. Label in box: *Thos. Helmsley & Son, King Street, Tower Hill, London.*

46 Octant, 12in radius, brass index arm, ebony frame, ivory scale, arc graduated to 90°, 3 and 2 shades. Ivory plate inscribed: *Heron, Greenock.*

47 Sextant, mahogany frame and grip, brass arm and scale, radius 16in. Fixing and tangent screw. Arc graduated to 120°. Central zero vernier, *c*1780.

48 Sextant. Brass frame, silver scale, radius 5in. Magnifier and telescope. Signed: *Potter Poultry.*

49 Quintant. Ebony frame, radius 8in. Four and three shades, vernier, microscope and telescope. Signed: *Bennet Cork.* Label in box: *H.G. Blair & Co. 53 James St. Cardiff.*

50 Wooden log reel for early ship's log (triangular piece of wood).

51 Walker's Harpoon Log A.1.

52 Craven's screw propeller pitchometer. Brass. Marked: *Edwin Craven Maker Hull No. 130.*

53 Instrument for indicating ship's roll. Consists of mahogany frame of roughly triangular shape with brass pointer moving over a brass scale.

54 Traverse board. Wood, length 14½ x 10in. 8 main wind directions, 32 points, 8 equidistant holes in each. Speed table: four sets of holes numbered 1 – 12 and four from 1 – 9. The threads for the bunches of pegs survive, but the pegs are missing.

55 Traverse board. Painted wood, 13½ x 8½in. 32 points with eight equidistant holes in each. Speed table consists of four sets of holes marked 1 – 12,¼,½,¾. Set of eight pegs in central bunch. (fig 136)
Lit: HOLBROOK (1974) no. 36

56 Boat's compass. Brass bowl hung in gimbals, engraved paper compass card. The marker's name is illegible but the address, *Queen St. Hull,* is clear. Late 18th century.

57 Boat's compass, 4in diameter card, brass bowl in gimbals in 6½in square wooden box. Signed *T.S. Negus & Co., New York.*

58 Compass in rectangular wooden box 3⁵⁄₁₆in square, an engraved compass card inside lid. Signed: *Springer Maker Bristol*. Late 18th century.

59 Not allocated.

60 Compass in rectangular wooden box with lid. Mahogany, 6in square. Early 19th century.

61 Compass. Signed: *Yeates & Son Dublin*. Second decade 19th century.

62 Chinese geomantic compass.

63 'Palinurus' compass deviation instrument signed: *Reynold & Son, 32 Crutched Friars, London*. Used to assess effect of iron on compass behaviour.

64 Graphometer. Brass. Fixed sights on diameter, moveable sights on alidade. Inset compass supported by figures blowing into sea shells and terminating in foliage. French, second half 17th century. (fig 58)
Lit: HOLBROOK (1974) no. 71

65 Lodestone. Brass mounted, *c*1700?

66 Thermometer in black fish-skin case. Signed: *Luisetti London*. Label in case stating this was Captain Cook's thermometer.

67 Callipers for measuring inside and outside diameters. Length 35in. Brass with steel points. Signed: *Troughton & Simms London*.

68 Aneroid barometer, signed: *Dollond*. Presentation instrument from the Shipwrecked Fishermen's and Mariners' Royal Benevolent Society. In wooden case.

H 7 HUTTON-LE-HOLE
Rydale Folk Museum

Hutton-le-Hole, N Yorkshire YO6 6UA

07515 367

The Museum, founded and administered by the Crosland Foundation, is housed in an early 18th-century Rydale farmhouse. The basis of the collection was assembled by the late RW Crosland, but this has been supplemented by much additional material. The collection includes prehistoric and Roman antiquities as well as covering aspects of everyday life and crafts. In the Museum grounds are a series of reconstructed buildings including a medieval house and an Elizabethan glass furnace.

There are only three instruments, but they are of considerable local interest, since they were used by the surveyor Joseph Ford (fl. 1745) for designing watercourses to supply the villages of the southern edge of the North Yorkshire moor.
Lit: McDONNELL (1963) includes an account of Joseph Ford's work and illustrates his level (p215, pl. Xc).

1 Surveyor's cross. Brass with slit and hair sights. Unsigned.

2 Theodolite. Compass in base. Spirit level mounted below telescope. Rack and pinion adjustment against vertical semicircle, as well as in the horizontal plane. Signed: *G. Adams London*.

3 Level. Telescope combined with bubble level mounted on four levelling screws. Mahogany tripod stand. The instrument appears to have been assembled from parts of other instruments.

I

I 1 INVERNESS
Inverness Museum and Art Gallery

Castle Wynd, Inverness IV2 3ED

0463 237114

The Museum's displays cover the social and natural history, archaeology and geology of the Highlands. There are a few scientific instruments; one, an electrical machine, is of considerable interest.

1 Plate electrical machine. Insulated base consisting of three solid glass pillars supported on four legged stool. 24in clear glass plate supported by glass pillars. Prime conductor missing. Overall ht 58¾in. Signed: *J. Cuthbertson London 1799*.
Lit: HACKMAN (1973)

2 Compound microscope, signed: *R. & J. Beck London 1854*.

3 Refracting telescope, hand-held. A 'military and naval' telescope patented by Cater Rand, 29 January 1799 (British Patent 2289). The telescope is inscribed: *C. Rand's Patent made and sold by Wm. Watkins St. James's Street London*. A brass plate with a crest is inscribed: *By the King's Royal Patent*. (02.838)

4 Wimshurst machine. (968.21)

I 2 INVERURIE
Inverurie Museum

Public Library Building, The Square, Inverurie, Aberdeenshire

0779 77778

The Museum, which now forms part of the North East of Scotland Museums Service, based on the *Peterhead Arbuthnot Museum, Peterhead* (to which enquiries should be addressed), has collections of archaeological material from the surrounding area, natural history and geological collections, as well as containing bygones, coins and curios. It contains only one scientific instrument.

1 Cylinder electrical machine similar to Nairne's but unsigned. Rectangular oak base, partially cut away to take the slides on which are mounted the prime conductor and the conductor with the friction cushion on green glass pillars. These can be adjusted as wished, and held in position by means of wooden screws. The cylinder is supported on green glass pillars. Length of cylinder 11in. Length of base 22in. Late 18th century.

I 3 IPSWICH
Ipswich Museum

High Street, Ipswich IP1 3QH

0473 213761/2

The collections are divided between two buildings: Christchurch Mansion which houses furniture, paintings, bygones, ceramics, etc., and the Museum in the High Street which contains geology, archaeology, natural history and ethnography. No conscious attempt has been made to collect instruments, but there are a number of items of considerable interest. In addition to the instruments listed below there are several fine domestic barometers, notably by De Salis, Leyden (combined barometer and thermometer), Nairne and Blunt, and Ferrari, Ipswich. There are also such items as weights and scales, photographic material, smoke jacks, clockwork weight jacks, dental and medical instruments, globes and engineering models.

1 Compass microscopes. Brass with ivory handle in fish-skin case. With lenses and ivory slides. Unsigned. 18th century. (1935.1055)

2 Compound monocular microscope signed: *E. Hartnack & A. Prazmowski Rue Bonaparte Paris.*

3 Compound monocular microscope. Signed: *Andw Ross London no. 173, c*1830-40. (1914.16)

4 Refracting telescope on pillar and claw stand. The tube supported behind by a steadying rod. Signed: *Banks 144 Strand London.* In mahogany box with accessories. Early 19th century.

5 Gregorian reflecting telescope. 24in focus. Altazimuth mounting, tripod stand missing. Signed: *James Short London.* (37/1037 = 24)

6 Octant. Ebony frame, brass arm, ivory scale. Ivory label inscribed: *Made by Jno Gilbert Tower Hill London for Robt Burkinshaw Janry 22nd 1783.*

7 Set of drawing instruments in ray-skin case. Protractor and diagonal scale signed: *F. Watkins London.* Second quarter 18th century.

8 Compass in circular turned wooden box. Paper compass card signed: *D. Adams Charing Cross London.* Early 19th century.

9 Wollaston's camera lucida with instructions for use. (1942.106)

10-11 Two chondrometers, one signed: *J. Long.* (R.1934.34, 963.96)

12 Cup anemometer. Signed: *Negretti & Zambra.*

13 Chemical balance. Signed: *L. Oertling London.*

14 Plate electrical machine. Late 19th-century version of French plate machine. (1927.101)

15 Plate electrical machine. Cuthbertson type. Prime conductor missing. 19th century. Both electrical machines are said to have belonged to the Ipswich Scientific Society.

K

K 1 KEDLESTON
Kedleston Hall

Kedleston, Derby (The Viscount Scarsdale)

0332 842191

The house was built by Robert Adam between 1757 and 1765. Among the magnificent collections in the house are a few scientific instruments, part of what was probably once a much larger collection. The earlier items can be dated to the first decades of the 18th century, and to judge from their date might have been acquired by Sir Nathaniel Curzon, 2nd baronet (d. 1718), or by his son (1676-1727), later 3rd baronet. Other instruments date from shortly before or after the middle of the 18th century. There is no known documentary evidence relating their acquisition, but the few items which survive suggest that they may be remnants of an important collection.

1 Fragment of Marshall-type microscope. Only the outer tube of green vellum with gold tooling, the vertical arm, and part of the box base survive. The inner tube and eyepiece are missing. A mirror is mounted on the base, and was presumably added at a later date. Early 18th century.

2 Martin's drum microscope. Incomplete. Unsigned.

3 Solar and scroll microscope. Signed *F. Watkins London.* In mahogany box. Incomplete, the screw-barrel microscope is missing.

4 Compass, for use with a plane table. Octagonal wooden case, with rectangular projection on one side. Paper compass card, signed: *Made by George Adams in Fleet Street London.*

5 Circumferentor. With two fixed sights and two sights mounted on the alidade; each with slits and threads. Signed: *J. Bird London.*

6 Waywiser for use on a carriage. Brass, in heavy wooden case. Signed: *Richard Glynne*.

7 Equinoctial ring dial. Brass, diameter 6in. Unsigned, 18th century.

8 Gunner's quadrant. A graduated quadrant, one of whose sides is prolonged as a long arm. Unsigned. Probably 18th century.

K 2 KENDAL
Kendal Museum

Station Road, Kendal LA8 6BT

0539 21374

The collection was originally based on that of the Kendal Literary and Scientific Society founded in 1835. The chemist John Dalton (1766-1844) and the geologist Adam Sedgwick (1785-1873) were members in the early days. The Society acquired the Todhunter Collection (formed at the end of the 18th century) consisting of fossils, minerals, etc., and added to it in various ways. The Museum was presented to the Corporation in 1910 and moved to its present home in 1913. The collection concentrates on material of local interest, foreign butterflies, moths, birds and animals.
Lit: A brief history of the collection appears in KENDAL (1973).

1 Gunter's quadrant. Boxwood, radius 4⅜in. Stamped with date *1700*. Reverse incompletely engraved.

2 Nocturnal. Boxwood, length 10in, diameter 4¾in. For use with Great Bear only. 17th century.

3 Octant. Ebony frame, 10in radius. Ivory arc, vernier and label. Tangent and fixing screw. Signed: *Rennison & Son North Shields*.

4 Bulb cistern barometer in roughly shaped wooden frame. Presented by Revd E Hawkes. The barometer is said to have been used by John Dalton in his annual ascents of Helvellyn. The donor recorded that after 48 ascents Dalton remarked 'that he had not discovered anything to add to his work on meteorology'.

5 Compound monocular microscope. Mid-19th century.

K 3 KESWICK
Keswick Museum and Art Gallery

Fitz Park, Keswick CA12 4NF

0596 73263

The Museum contains material illustrating the zoology, botany, geology, mineralogy and archaeology of the Lake District, as well as literary material such as original Southey, Wordsworth and Walpole manuscripts. From the point of view of this study, the items of primary interest are a series of instruments which belonged to, and were partially made, or adapted, by Jonathan Otley (1766-1856), a prominent geologist, cartographer, clock repairer, and surveyor (nos. 2-8). Otley lived in Keswick for most of his life, and was the first to describe the rock of formations of the Lake District (information from Mr Norman Gandy, the Curator). An interesting forerunner was the Museum of Peter Crosthwaite, opened in Keswick in 1780, which contained a number of scientific instruments. Most of the material in Crosthwaite's museum was sold by auction in 1870 (a copy of the catalogue is in the British Library), but one instrument, an octant, survives in the Fitz Park Museum (no. 1).
Lit: Transactions of the Cumberland Association for the Advancement of Literature and Science, Part III (1877-8); CARLISLE (1878) (on Peter Crosthwaite); *Idem*, Part II (18876-7) (on Jonathan Otley).

1 Octant. Mahogany frame and arm, boxwood arc, diagonal scale, radius 18in. Signed on the arc: *B. Martin London*. Inscribed on an ivory label: *QUADRANT OF Commander Peter Crosthwaite*. Third quarter 18th century.

2 Rain gauge. Belonged to, and probably constructed by, Jonathan Otley. It consists of a box 33in high which is surmounted by a funnel. the funnel leads into a jug inside the box. Beneath the jug is a glass measuring cylinder.

3 Graphometer. Diameter 9in. Brass, graduated from 0 to 180. Hair and slit sights.

4 Circular protractor. Diameter 9in. Graduated from 0 to 180. Probably constructed by Jonathan Otley.

5 Circumferentor. Diameter 5in. Folding sights on alidade with centrally mounted compass. The compass has an engraved paper windrose and a screw top (probably an adaptation of a magnetic dial).

6 Clinometer. Bubble level fitted in a rectangular wooden member 12 x 1 x 1in. On one side is mounted a brass rule fitted with sights and an arc graduated into 18 units. The toothed edge to the arc is manipulated by means of a gear wheel equipped with an ivory knob. A series of scales are engraved on the brass rule.

7 Cistern barometer in wooden case with lid, and paper scale graduated from 25 to 31in. Silvered sliding sleeve on tube graduated from 1 to 10. Signed: *J. Otley Keswick*.

8 Clockmaker's wheel-cutting engine. Brass plate, diameter 12in. Four bowed legs, box frame, cutter frame adjusting screw mounted in extension to main frame at rear of engine. Lancashire type, probably late 18th century. (See CROM, p98, fig 95.)

K 4 KILMARNOCK
Dick Institute

Elmbank Avenue, Kilmarnock KA1 3BU

0563 26401

The Museum contains collections of geological, ornithological, archaeological and ethnological material, as well as arms (Scottish basket-hilted swords, and small arms), incunabula, a children's museum, and paintings by English and Scottish artists. There are a small number of scientific instruments, listed below, as well as veterinary and surgical instruments, gold balances and apothecaries' scales.

1   Gregorian reflecting telescopes, 9½in aperture, focal length 58in. Wooden, metal bound tube on heavy tripod with casters, equatorial mounting. One eyepiece micrometer is signed: *Thomas Jones London*. Constructed by Andrew Barclay (1814-1900) and presented to the Museum by his grandson.

2   2½in. Gregorian reflecting telescope, signed: *James Short Edinburgh 1736 44/96*. The telescope is contained in a wooden box with a sliding lid into which the telescope can be screwed.

3   Backstaff. Pearwood, radius of 30 arc 24in. Signed: *Made by Geo. McEvoy Temple Bar Dublin*.

4   Octant. Ebony frame, radius 15in. Brass arm. Inscribed *Robert Auld* on ivory label.

5   Octant. Mahogany frame radius 18in, brass arm ormanented with engraving in rococo style. Signed: *Spencer Browning & Rust London*.

6   Octant. Ebony frame, brass arm, radius 12in. Three shades, vernier. Signed: *Della Torre & Co. London*. Early 19th century.

7   Sextant. Ebony frame, radius 11in. Inscribed: *W. Holliwell ...* [of Liverpool, fl. 1830-50].

8   Equinoctial sundial of silurian schist, by John Bonar, Ayr, 1632.

K5 KING'S LYNN
Lynn Museum

Saturday Market Place, King's Lynn PE30 1HY

0553 763044

The Museum contains collections of natural history, archaeology and folk material relating primarily to north-west Norfolk. In addition, there are coins, costumes, prints, glass and ceramics. There is a small number of scientific instruments, chiefly navigational, as is appropriate, since King's Lynn, situated on the Ouse, was once one of England's busiest ports.

1   Horizontal dial in circular brass box with lid. Paper compass card mounted inside lid and in base of the compass box. Folding gnomon for one latitude. Unsigned. 18th century.

2   Equinoctial ring dial. Brass, unsigned, 18th century.

3   Refracting telescope. Octagonal ring with ray-skin outer tube assembled in three sections. Octagonal wooden case. Early 18th century.

4   Octant. Mahogany frame and arm, boxwood arc, diagonal scale. Incomplete, index glass missing. Radius 22in. Mid-18th century.

5   Octant. Ebony frame, radius 15¼in. Brass arm with engraved rococo ornament. Vernier with zero to right. Signed: *I Gaitskell 313 Wapping London for Thadeus Coffin 1785*.

6   Octant. Mahogany frame, brass arm, radius 14¾in. Vernier zero to right, engraved anchor in centre of ivory degree arc. Signed: *[?] London*.

7   Octant. Ebony frame, radius 15in. Signed: *Spencer Browning & Rust*.

8   Octant. Ebony frame, radius 15in. Unsigned, but *UBK* engraved on arc.

9   Waywiser, large 12-spoked wheel, iron construction, recording mechanism and dial in box mounted at centre of wheel.

10   Specific gravity beads. Label in box of: *P. Kassino Glassblower St. Mary's Wynd Edinburgh*.

11   Chondrometer signed: *Corcoran & Co. 1826*.

12   Sikes' hydrometer in box. Signed: *Sikes No. 40390 p512*.

13   Saccharometer signed: *J. Long 20 Little Tower St. London*. Thermometer signed: *Dring & Fage*.

L

L 1 LACOCK
Fox Talbot Museum of Photography

Lacock, Chippenham SN15 2LG

024 973 459

The Museum, administered by the National Trust, and housed in a converted 16th-century barn, is devoted to the life and work of William Henry Fox Talbot FRS 1800-1877), the inventor of the negative/positive process of photography, who lived at Lacock Abbey from 1827 until his death. The museum contains documents, original patents, notebooks and copies of Fox Talbot's early pictures taken by the calotype process, as well as a number of instruments and cameras.

The greater part of Fox Talbot's extensive collection of instruments was dispersed by Miss Talbot in the 1930s before the idea of a Fox Talbot Museum at Lacock had been conceived, between the *Science Museum, London*, the *Royal Photographic Society*, now in Bath, and the Royal Scottish Museum in Edinburgh, now the *Royal Museum of Scotland (Chambers Street), Edinburgh*. As a result, the Museum at Lacock contains relatively few of his instruments (marked FT in the entries below). The two

cameras known to have belonged to him are on loan from the Royal Photographic Society (indicated RPS). Several other items on loan from a private collection (marked L) illustrate instrument types known to have been used by Fox Talbot. A number of later cameras are also on display. The non-photographic material is indication of Fox Talbot's wide interests in science.
Lit: LONDON – NATIONAL TRUST (1976B)

1 Compound microscopes. Brass, signed: *Vincent Jacques Louis Chevalier, Quai de l'Horloge no. 69, Paris.* Early 19th century.

2 Solar microscope. Brass, signed: *W. & S. Jones 30 Holborn London.* (L)

3 Solar microscope. Brass, incomplete (lacks lens). Unsigned. (FT)

4 Part of polarising microscope with prism for viewing mineral specimens. (FT)

5-6 Portable camera obscuras. Mahogany boxes. (L)

7 'Mousetrap' camera (small experimental camera), *c*1835-9. (RPS)

8 Calotype camera, mahogany box. Either constructed locally, or possibly obtained from Andrew Ross. (RPS)

9 Daguerreotype camera (L)

10 Part of a camera lucida (FT)

11 Camera lucida ((L)

12 Apparatus to illustrate persistence of vision. Home or locally constructed, two gear systems and revolving discs with different cut-outs. (FT)

13 Cylinder electrical machine with simple crank handle. Cushion mounted on insulated glass stem attached to the base and its pressure adjusted by means of a screw. The prime conductor is on a separate support with turn wooden base. Ivory label inset in base of machine: *Watkins & Hill Charing Cross London.* (FT)

14 Small double-barrel air pump, table model. Brass on mahogany base. Signed: *Watkins & Hill 50 Charing Cross London.* (FT)

15 Double-barrel air pump. Table model. Signed: *Andw Ross 33 Regent Street Piccadilly.* (Probably the pump listed in a bill from Andrew Ross, dated 1839). (FT)

16 Vacuum jar. Partly lined with foil. (FT)

17 Four cells mounted together in a wooden box.

18 Several electromagnets (FT)

19 Small balance, brass and iron. (FT)

20 Instrument (probably home-made) for inducing four liquids to flow into one channel. (FT)

L 2 LEAMINGTON SPA
Warwick District Council Art Gallery and Museum

Avenue Road, Leamington Spa CV31 3PP

0926 26559

Administered by Warwick District Council, the collections include paintings and watercolours, pottery, porcelain and glass. There are two scientific instruments, in addition to some early X-ray apparatus described as having been 'in use at the Warneford Hospital (Leamington) one week after Rontgen's paper was published'.

1 Sector. Brass, length 4¾in. The rule folds in to one arm of the sector, which is engraved with scale *L in. Gradium* and *Lini Fortifica*. The friction leaf is inscribed with tables relating to the construction of fortifications. Signed: *Antho. Sneewins & Delft Fecit.* (fig 119)

2 Proportional compass. Brass with steel points, length 7⅝in. Signed: *N. Witham fecit.* Early 18th century. (fig 110)

L 3 LEEDS
City Museum

Calverly Street, Leeds LS1 3AA

0532 462632

The City Museum was originally established in 1820 by the Leeds Philosophical and Literary Society. It was relocated in the Municipal Buildings after the 1939-45 war. The collections illustrate nearly every aspect of natural history, ethnography and archaeology. In part the collections relate to the Yorkshire region, though large sections of the collection have been gathered from all over the world.

There are few scientific instruments, but they do include three items of late 18th- or early 19th-century date which were formerly believed to have belonged to the chemist Joseph Priestley. However, more recently it has been suggested that these items (nos. 1-3) belonged to Joseph Dawson, founder of the Lomar Ironworks at Bradford (d. 1813), who is known to have constructed scientific instruments.
Lit: LEEDS – LEEDS CITY MUSEUMS (1974)

1 Twin plate electrical machine. Incomplete, cushions missing. 1780s, possibly modified in the late 1790s.

2 Conductor. Large tinfoil-covered wooden ball on glass stem. 1780-90.

3 Chemical chest. Mahogany chest with four drawers, the upper two of which are filled with jars of chemicals. The lower drawers contain various empty glass bottles, a test tube, a cast iron cylindrical vessel with narrow neck, an iron L-tube, a number of coarse clay crucibles, and five small pots stamped: *Wedgwood.* An ivory label on the box is inscribed: *Accum Chemist Old Compton Street Soho London.* Early 19th century.

4 Double-barrel air pump of Nairne type. Mahogany base, brass cylinders. Signed: *W. Ladd & Co. 11 & 12 Beak Street Regent Street London.* A brass single-barrel compressor pump is mounted on the base at one side and is signed: *W. Ladd & Co. London.* With brass Magdeburg hemispheres. Second half 19th century.

5 Orrery. Wooden base covered with engraved and coloured paper. Brass mechanism. Signed: *Newton & Son Chancery Lane London.* Early 19th century.

6 Refracting telescope. Signed: *S. & B. Solomons 39 Albemarle Street London.* Latter part of 19th century.

L 4 LEICESTER
Leicestershire Museum and Art Gallery

New Walk, Leicester LE1 6TD

0533 554100

The collections are housed in a central Museum and Art Gallery in New Walk and in a series of branch museums, notably Newarke Houses Museum (social history, costume, clocks, hosiery industry), Belgrave Hall, and Jewry Wall Museum (containing archaeology and particularly Roman collections). The central Museum, as well as containing the art collections, covers biology, geology, etc. The majority of the scientific instruments are housed in the Newarke Houses Museum and come under the care of the Keeper of Antiquities. Most of them have local connections. In addition to items listed below, the Museum contains a considerable collection of clocks and watches by Leicestershire makers (*Lit:* DANIELL 1951 and 1975). In addition there is an extensive collection of weights and measures including coin balances, apothecaries' balances, wool weights, etc., as well as numerous spectacles. There are several garden sundials and domestic barometers by makers such as S Mason, Dublin; J Deacon, Leicester; S Deacon, Barton-in-the-Beans (Leics.); Ramsden, London; Charles Orme, Ashby de la Zouche; Shaw, Leicester.

1 Microscope. The body of a Culpeper microscope (Matthew Loft form) with ray-skin outer tube and typical octagonal base with drawer has been adapted by the addition of a later stage and rack focusing with fine adjustment by means of a long screw. (309'1962)

2 Aquatic microscope. Late 18th century. (10.0.S.2)

3 Nuremberg tripod microscope. Branded *I.M.* on the base. (0.S.269.51)

4 Improved compound microscope. Signed: *Crichton, 112 Leadenhall St. London.* In mahogany box. Rack focusing, folding tripod feet. (131.1956)

5 Drum microscope. Brass, unsigned, *c*1840. (477'1955)

6 Spirit level with open sights for screwing on to a tripod stand. Inscribed: *Improved by Wilm Stenson Junr* (544'1954/2). This instrument, together with nos. 7-9, belonged to William Stenson, a Leicestershire surveyor employed in surveying the Leicester – Swannington railway in the early 1830s.

7 Miner's theodolite (Lean's dial) in original box with instruction leaflet. Signed: *William Wilton Mathematical Instrument Maker St. Day Cornwall* (provenance as no. 6). (544'1954/3)

8 Miner's dial with folding sights and suspended degree arc with sights. Signed: *W. & S. Jones London. Sold by Whitehurst & Son Derby.* In box labelled: *Wm. Stenson Whitwick Collieries Nr. Leicester* (provenance as no. 6). (544'1954/1)

9 Compass. Brass with silvered dial. Signed: *J. Davis Derby* (provenance as no. 6). (544'1954/1)

10 Waywiser. Signed: *John Crossley of Ashby de la Zouche*, *c*1800. (120'48)

11 Waywiser. Small wheel. Signed: *Willm. Watkins London.* (182'37)

12 Brass rule. Signed: *Cha. Orme of Ashby de la Zouche 1731.*

13 Standard yard in wooden box. Signed: *W. & T. Avery London and Birmingham.* (4.IL.1957)

14 Orrery. Wooden base, covered with engraved and coloured paper. Brass mechanical parts. Inscribed: *New portable orrery by W. Jones. Made by W. & S. Jones 135 Holborn London 1794.* (53'1958)

15 Cube sundial. Painted wood, brass gnomon. Signed: *Samuel Heyricke fecit 1687.* (1531'1885)

16 Vertical dial of slate for a South Wall, 80 x 40in. Originally fixed to the wall of the Revd William Pearson's observatory at South Kilwork, Leicestershire in 1834. (462'1960)

17 Equinoctial dial with spirit level in base. Signed: *F.L. West 35 Cockspur St. London.* (307'1962)

18 Excise officer's slide rule. Signed: *P. Corson Best Box Rule.* (10'1950/78)

19 Excise officer's slide rule. Length 18in. Signed: (under one slide) *Edwd Roberts Maker in Grocers Alley in Old Jewry London.* Inscribed in ink: *John Woodward 1755* and *J. Philips 1790 Bilson.* (142'1956)

20 Dicas hydrometer. Signed: *Dicas Patentee Liverpool.* (25'32)

21 Hydrometer. Signed: *Minaretti Leicester.* (280'1959)

22 Chondrometer. Signed: *Nicholl 16, Aldersgate St. London.*

23 Air pump. Double-barrel, table model with elevated stage for receiver plate. Signed: *Ronchetti, 1 St. Anns Place Manchester.* (12'1958)

24 Biram's patent anemometer. Diameter 14in. Signed: *Biram's Patent Davis Derby 946.* (553'54)

25 Dividers. Iron, with scale. Probably 18th century. (OS 106'51)

26 Large terrestrial globe, painted, in original wooden stand. Made in Rome, 1688. (132'1950)

L 5 LETCHWORTH
Letchworth Museum and Art Gallery

Broadway, Letchworth SG6 3PF

0462 685647

The collection includes archaeological and natural history material relating to North Hertfordshire and the history of the first Garden City. There are only two scientific instruments in the collection, both of them microscopes

1   Jones's 'most improved' microscope. Incomplete, the arm into which the microscope body should be screwed is missing. Signed: *S. Bithray Successor to J. Smith Royal Exchange. London*. Late 18th-century type, constructed *c*1830.

2   Culpeper tripod microscope. Brass on round wooden base. Rack focusing. Height 14½in. Contained in a pyramidal oak box with a drawer in the base containing accessories. Unsigned. Late 18th or early 19th century.

L 6 LINCOLN
Museum of Lincolnshire Life

The Old Barracks, Burton Road, Lincoln LN1 3LY

0522 28448

The Museum contains locally orientated collections with an emphasis on material illustrating life in Lincolnshire during the last two hundred years. It possesses a small number of instruments as well as a number of garden sundials.

1   Brass Culpeper-type microscope in pyramidal case. Unsigned. Late 18th or early 19th century.

2   Compound monocular microscope. Signed: *A. Ross London 1800*.

3   Orrery. Engraved paper on wooden base, brass mechanical parts. Dated 1784 and signed: *W. & S. Jones, 135 Holborn London*. Later (?) globe signed: *Bardin, London*. (23.63)

4   Gunter's quadrant. Wood with one surviving brass sight. Inscribed: *Movin William Young 1705*.

5   Case of instruments for restoring the apparently drowned – includes bellows for inflating lungs. Kept for many years at the Old Lord Nelson Inn, High Bridge, Lincoln. Early 19th century. (10819.06). (See similar example, fig 72)

L 7 LIVERPOOL
Department of Physics, University of Liverpool

Oliver Lodge Laboratory, Oxford St, Liverpool L69 3BX

051 794 2000

The Department of Physics houses, as well as a quantity of late 19th-century equipment, a small number of earlier instruments. Items 1-5 came to the Department by gift, but the source is unknown.

1   Quadrant on stand. Brass frame of 12½in. radius. Signed: *J. Bird London*.

2   Transit theodolite on stand. 15in vertical circles, telescope of 1¾in aperture. Signed: *Troughton London*.

3   Cassegrain reflecting telescope on pillar and claw stand. 4in aperture, signed: *James Short London 1738 25*. *Lit*: BRYDEN (1968) p4, no. 2

4   Refracting telescope on pillar and claw stand, 2½in aperture. Signed: *Ramsden London*. Late 18th century.

5   Spectroscope, signed: *J. Browning*.

6   Time assimilator – Sidereal/Meantime. Signed: *Ackland, Hatton Garden London*.

7   Wimshurst machine, plate diameter 15½in.

L 8 LIVERPOOL
Liverpool Museum

William Brown Street, Liverpool L3 8EN

051 207 0001

The Museum, now part of the National Museums and Galleries on Merseyside, was founded in 1851. The gift of the Joseph Mayer Collections in 1867 formed the basis of the important and wide-ranging collections covering the decorative arts (ceramics, metalwork and horology, textiles and costume, woodwork and musical instruments, medieval art), archaeology, technology, transport, navigation and astronomy. There is a large collection of watches, long-case clocks, turret clocks and chronometers. There is also a considerable number of scientific instruments. Although these are distributed throughout several departments, the majority comes under the care of the Departments of Physical Science and Decorative Art. Of great interest is an extensive collection of 18th- and 19th-century clock and watchmaking tools (see no. 72 below).

In addition to the instruments listed here, the collection contains photographic equipment, domestic barometers, tide measuring and predicting apparatus, and more recent instruments related to space rocketry. Some of the navigation instruments are now to be found in *Merseyside Maritime Museum* on Albert Dock, Liverpool, opened in 1986 as a constituent part of the *National Museums and Galleries on Merseyside*.

Lit: No catalogue exists of the scientific instruments, but handlists have been issued of both the watch and clock collections. These are now out of print.

In the case of loan material – from the *Science Museum, London,* or the *National Maritime Museum, London* – both the accession number of the lending institution and that allocated by the National Galleries and Museums on Merseyside are given. Accession numbers commencing M... indicate that the item comes from the collection of Joseph Mayer.

MICROSCOPES, TELESCOPES, ASTRONOMICAL DEMONSTRATION APPARATUS

1 Culpeper microscope. Brass. Signed: *J. Blunt London.* (CELL.2;66)

2 Ross/Zentmayer combined binocular and monocular microscope. Signed: *Ross London.* Serial No. 5406. (OFF.33)

3 Cary-type microscope. Unsigned. (OFF.20)

4 Heliostat for microscope. Engraved *Presented to Microscopical Society of Liverpool by Thos. Higgin F.R.S. Huyton 5th Feb. 1886.* (1967.150)

5 Transit instrument. Lent by the Jeremiah Horrocks and Wilfred Hall Observatory, Preston. (DIS.3;15)

6 Gregorian reflecting telescope, 2½in aperture. Unsigned, 18th century. (DIS.3;14)

7 Gregorian reflecting telescope, 2½in aperture. Signed *James Short Edinburgh 1738 77/165.* (60.56)

8 Gregorian reflecting telescope, 3in aperture. Signed: *Jo. Jackson London.* (DIS.3;13)

9 Gregorian reflecting telescope. 4in aperture. Signed: *Bate London.* Finder telescope, barrel supported between trunnions with elevating and traversing mechanism. Pillar and claw stand. (1968-248)

10 Transit instrument. Signed: *Troughton & Simms 1845.* Originally commissioned for the Liverpool Observatory. From the Institute of Coastal Oceanography and Tides. (DIS.1;5)

11 Speculum mirror for Lassell, 24in diameter, *c*1846. In use at Greenwich 1883-92. Loan from *National Maritime Museum London.* (DIS.3;3)
Lit: LONDON MARITIME MUSEUM (1971) inv. 35-10

12 Weight-driven governor for equatorial telescope. Loan from *Science Museum, London.* (1967.149.5)

13 Graduated circle with suspension ring and alidade. Loan from the Jeremiah Horrocks and Wilfred Hall Observatory, Preston. (DIS.3;27)

14 Vertical circle with sighting telescope. Loan from Jeremiah Horrocks and Wilfred Hall Observatory, Preston. (DIS. 3;27)

15 Quadrant on stand supported on three screw feet. Attached to upper edge of quadrant, a Newtonian reflecting telescope. An inclinometer? Signed: *W. & S. Jones 30 Holborn London.* Loan from Jeremiah Horrocks and Wilfred Hall Observatory, Preston. (CELL.2;46)

16 Heliostat. Loan from *Science Museum, London.* (SM.1887.33) (1967.149.4)

17 Equatorial refractor (demonstration model). Signed on name plate: *Horne & Thornthwaite. London.* Loan from Jeremiah Horrocks & Wilfred Hall Observatory, Preston. (CELL.2;67)

18 Equatorial telescope, 8¾in objective signed: *Merz,* stand signed: *Troughton and Simms, 1845.* Commissioned for the Liverpool Observatory. (DIS.1;3)

19 Reflecting telescope by Sir Howard Grubb (1881), 24in aperture. Made for WE Wilson. Loan from University College, London. (BAR11)

20 Equatorial refracting telescope, 1797, 4½in aperture. The 'Huddart' telescope. Loan from *Science Museum, London.* (SM.1918-70) (DIS.3;1)

21 Reflecting telescope by AA Common, 36in aperture. Loan from *Science Museum, London.* (SM.1913-657) (1969.61)

22 Celestial globe on tripod stand with turned wooden legs. Diameter 5in. Signed: *J. & W. Newton.* (1967.273.3)

23 Armillary sphere. Brass, lacking stand. Possibly 15/16th century.

24 Armillary sphere on bird's foot stand. Loan from *National Maritime Museum, London.* (Greenwich SP.21,52/6) (OFF.31)

25 Orrery. Wooden base covered with engraved paper. Inscribed: *New portable Orrery, made and sold by W. & S. Jones 30 Holborn London.* Globe signed: *Bardin.* Early 19th century. (60.220)

26 Orrery. Signed: *Newton & Co. Opticians to HM the Queen, 3 Fleet Street London.* (60.254)

27 Orrery. Base only. Signed: *Bancks 441 Strand London.* (CUP.2;4)

28 Orrery on stand. *c*1870. Signed: *Philip Son & Nephew, Liverpool.* (1962.269)

29 Planets and radius arms of orrery (base missing). Unsigned. (PART.6)

30 Equatorium. French, *c*1600. (7.1.69/39M) (fig 54)
Lit: NORTH (1969)

31 Persian astrolabe. 17th century. (1976.308)

SUNDIALS, NOCTURNALS, PERPETUAL CALENDARS

32 Compass dial. Brass with screw lid. Coloured paper compass card. Unsigned. ? Late 17th century. (25-1-83-4)

33 Compass dial (part only, compass box missing, plate and folding gnomon survive). Brass. Signed: *Gio. Savoi F. Firenze 1565.* (10-4-84-42)

34 Compass dial. Brass cased, with printed paper compass card, *c*1780. (23-11-82-1)

35 Horizontal dial. Fixed gnomon, two adjustable feet, silvered brass. In original (?) chamois-lined tooled leather box. Signed: *Koch Wien.* 18th century. (10-4-84-36)

36 Horizontal string-gnomon dial. Trapezoidal form, gilt brass with copper compass box. Folding gnomon-support. Can be adjusted for used in latitude 45, 48, and 50°. Signed: *Prague fecit Erasm. Habermal.* (23.11.82.2)
Lit: ECKHARDT (1976-77)

37 Horizontal string-gnomon dial. Octagonal, brass, partly gilt. Folding gnomon-support and plummet. Signed: *Nicolaus Rugendas in Augsp.* (M.402)

38 Butterfield dial. Silver. Signed: *Butterfield A Paris.* In original case. (10.4.84.41)

39 Butterfield dial. Silver. Signed: *Lefebure A Paris.* In original case. (10-4-84-45)

40 Diptych dial. Ivory. Signed: *Leonhart Miller*, dated 1652. (10-4-84-40)

41 Diptych dial. Ivory. Signed and dated: *Leonhart Miller 1635.* (M.404)

42 Diptych dial. Wood, unsigned. Probably English. (25.1-83-3)

43 Diptych dial. Wood. Signed: *Stockert a Bavaria.* (10-4-84-46)

44 Diptych dial, covered with engraved and coloured paper in the manner of Beringer. German, probably Nuremberg, late 18th century. (10-4-84-47)

45 Equinoctial ring dial. Silvered brass. Signed: *Nicholas Renard La Rochelle.* (M.406)

46 Equinoctial ring dial. Brass. Signed: *Cadot a Paris 1740.* (10-4-84-44)

47 Equinoctial ring dial. Brass. Signed: *Tho. Mann fecit 1673.* (M.407)

48 Universal equinoctial dial. Silver and gilt brass, octagonal. Signed: *Johann Willebrand in Augsburg.*

49 Sundial, 'Patent Helio-Chronometer'. Signed: *Pilkington & Gibbs Ltd. Preston.* (1967.276)

50 Crucifix dial. Brass with red inlay. Probably German or central European, late 17th century. (M.401)

51 Cube dial. Engraved and coloured paper on wood with brass gnomons. Signed: *D. Beringer, c1780.* (4-1-83-1) (fig 39)

52 Combined altitude dial and nocturnal incorporating a small compass for a horizontal sundial and a lunar volvelle. Gilt brass, unsigned. Late 16th century, possibly French.

53 Nocturnal, gilt brass. Italian, late 17th century. (19-4-84-43)

54 Perpetual calendar. Fire gilt brass with a high copper content. Octagonal, 7½ x 8in. Probably German, early 18th century. (M.410) (fig 109)

55 Perpetual calendar, silver diameter 1⅛in. English, late 17th century. (M.405)

56 Horary quadrant. Engraved paper mounted on wood. Radius 10¾in. Signed: *Henr Sutton Londini fecit 1658 lat. 51°32'.* (DIS.3;23)

SURVEYING AND NAVIGATION

57 Compass in gimbals. Signed: *Carlo Fiorini Bologna 1738.* (10.4.84.39)

58 Circumferentor in circular brass box with lid. Lugs for sights. Signed: *Thos. Cave Dublin Fecit.* (8.2.83.1)

59 Surveyor's level. Bubble level mounted below the telescope. Length 9½in. Signed: *Martin London.* (1967.316)

60 Engineer's level (CELL.1;6)

61 Backstaff. Lignum vitae with boxwood arcs. Replacement vanes. Signed: *John Gilbert London. c1740.* Loan from *National Maritime Museum, London.*

62 Octant. Ebony frame, radius 18in. Signed: *Gregory London* and inscribed with name of owner (on arm): *Tho. Thomason.* Fire-damaged. (OFF.18)

63 Octant. Ebony frame. Signed: *Jones & Gray Liverpool.* Early 19th century. (1970.126)

64 Sextant. Signed: *Heath & Co. Crayford. London. No.4148.* (59.58-2)

65 Box sextant. Signed: *Barton Linnard Ltd. 1918. No.5377.* (OFF.20)

66 Box sextant. Brass, diameter 2½in. Unsigned. Loan from Jeremiah Horrocks and Wilfred Hall Observatory, Preston. (DIS.3;16)

67 Sextant collimating apparatus (1968.405.2)

68 Quintant. Brass frame, signed: *Warden Amselem Gerrard & Co. Canning Place Liverpool.*

69 Station pointer. Brass. Signed: *Newman Exeter Change London.* From the Jeremiah Horrocks and Wilfred Hall Observatory, Preston. (DIS.3;26)

70 Drainage level. Signed: *Gardner & Co. 21 Buchanan Street, Glasgow, 1850.* (1967.316.2)

71 Transit theodolite. Signed: *Chadburn Ltd. Liverpool.* (1967.281.1)

CLOCK- AND WATCH-MAKING TOOLS

72 The Museums possess extensive collections of 18th- and 19th-century tools and other effects, principally from various Prescot workshops, notably that of Joseph Preston and Sons and the Lancashire Watch Co., as well as the collection of the late Dr DS Torrens of Dublin. The complete contents of two watch-case making workshops are also held: those of RJ Oliver of Clerkenwell and E O'Shaughnessy of Liverpool. Of 18th-century date are various hand and measuring tools, a number of which were made for Peter Stubs of Warrington, and including such items as screw plates, spring winders, two fusee engines and a wheel-cutting engine of c1770, signed: *Wyke and Green Liverpool.*
Lit: CROM (1970) p110, fig 105

73 Lodestone, brass mounted, with a suspension ring. Probably 18th century.

74 Barometer. Signed: *J. Hicks 8,9,10 Hatton Garden London, c1870.* From the Bidston Observatory. (1968.405.4)

75 Mercury barometer. Engraved *Troughton & Simms London*. (CELL.2;77)

76 Barometer from the Liverpool Observatory and Tidal Institute. Loan from *Science Museum, London*. (CELL.1;4)

77 Vacuum jar and press with single-barrel air pump. Maximum dimensions of base 15 x 14in. Height of vacuum jar 11in. Signed *Bate London*. (WORK8) (fig 3)

78 Double-barrel air pump. Signed: *W. & S. Jones No.30 Holborn London*. (WORK9)

79 Double-barrel air pump. Ivory nameplate engraved *Adie 55 Bold St. Liverpool*. (WORK2)

80 Cylinder electrical machine on base 26 x 13½in. Cylinder approx 11½in. long. Signed: *R. Adie Liverpool*.

81 Wimshurst machine. Many accessories. Diameter of plate 17in. (1967/273.1)

L 9 LONGLEAT
Longleat House

Longleat, Nr Warminster, Wiltshire (The Marquess of Bath)

0985 844551

The house was begun in 1567 by Sir John Thynne. His descendant was created the first Viscount Weymouth by Charles II; the first Marquess of Bath was created by George III. The scientific instruments at Longleat date mostly from the 18th and early 19th centuries, and are among the most important still to be found in the hands of the family who acquired them. They are housed in the Bishop Ken Library, named after the lifelong friend of the first Viscount Weymouth (1640-1714) who lived there for a period of twenty years.

The earliest of the scientific instruments at Longleat must have been acquired by the first Viscount Weymouth since they belong to the years around 1700 (nos. 1, 2, 4, 10 and 11). Other instruments (nos. 7, 8, 9 and 12) seem to be datable to the years around 1720-40. With the exception of one small telescope there are no instruments from the third quarter of the century, but there was another active period of acquiring instruments during the latter years of the 18th century and the early years of the 19th century. In addition to nos. 6, 13, 16, 17 and 18 three pairs of globes were acquired, to judge by their publication dates, between 1785 (by Bardin) and around 1824 (Cary, sold by Dollond). The third pair, by Cary, dates from around 1800. There is certainty regarding the date of acquisition of only one item, a reflecting telescope by Adams which is accompanied by a letter dated 1811 from the instrument maker to Lord Bath (from the date, this would have been the second Marquess who died in 1837).

Lit: HEYWOOD (1954); HEYWOOD (n.d.). This catalogue was used extensively in the compilation of these entries. There are copies at Longleat and in the Museum of the History of Science, Oxford. Where instruments have been listed in the catalogue, the number of the entry, identified by the prefix H, will be given here.

1 Air pump, Hauksbee-type. Walnut stand, brass barrels. A stool with carved stretchers carried the double-barrel pump operated by rack and crank handle and equipped with an elevated plate. Overall height 41in, *c*1700-1720.

2 Compound microscope of Marshall type, early 18th century. Brown leather tube with gold stamping. Body mounted on square pillar inclinable by means of a ball and socket joint. Octagonal wooden base with drawer. Complete except for eyepiece. Ht 20in. (H.2)

3 Adams Universal Compound microscope. Limb hinged by means of a compass joint to the top of a heavy turned pillar standing on a folding tripod foot. The narrow snout of the microscope tube screws into a brass ring at the end of a sliding arm. Mahogany box with numerous accessories. Unsigned, late 18th century. (H.3)

4 Telescope. Nine-draw, with vellum-covered cardboard tubes. The outer tube is mottled green and red with gold stampings. Wooden eyepiece holder. Lenses and objective lens holder and cover are replacements. Diameter of outer tube 3¼in, extended length 90in. Unsigned. *c*1680. (H.4)

5 Gregorian reflecting telescope, pillar and tripod stand. Tube diameter 2½in, length 17¼in. Unsigned, *c*1760-80. (H.5)

6 Gregorian reflecting telescope, pillar and tripod stand with altazimuth mounting. Slow motion drives for horizontal and vertical adjustment operated by means of wooden handles. Tube diameter 5in, length 34½in. Signed: *Adams London*. The box contains a letter dated 1811 from 'Mr Adams' (presumably Dudley Adams) to Lord Bath, in which reference is made to the delivery of the telescope. (H.6)

7 Waywiser. T-handle with brass dial mounted beneath. Painted wood, iron-clad wheel, diameter 31½in. Signed: *Made by Tho. Wright Instrument Maker to his Majesty*. First half 18th century. (H.9)

8 Set of mathematical instruments in velvet-lined wooden box, 18¼ x 12¾ x 4½in. The contents include: 12in sector, elliptic trammel, parallel rules with protractor, dividers with interchangeable ink and pencil legs and extension bar, 12in beam compasses, gunner's callipers, 12in scale (including a protractor scale), 6in protractor, 6in square, dividers with three legs and an adjustable fourth leg, small dividers, four ivory boxes with pallets and ink powders, five bottles for inks. Signed on gunner's callipers and protractor: *T. Heath London* and on sector: *Tho. Heath fecit No. 34*. (H.7)

9 Universal equinoctial ring dial mounted on a stand with adjusting screws, compass and graduations of 'watch faster' and 'watch slower'. Ht approx 14½in. Signed: *T. Heath fecit*. (H.8)

10 One vane of a cross staff. Boxwood. (Listed under H.10)

11 Backstaff. Lignum vitae with boxwood arcs. Complete with vanes including a Flamsteed glass. Maker's nameplate missing. 1700. (H.19)

12 Azimuth compass in gimbals in a mahogany box, 14½ x 14½in. Diameter of compass 11¾in. Fitted with alidade and diagonal scale marked on broad flat ring mounted on top of the compass bowl. Signed: *T. Heath London*.

13 Sextant mounted on an azimuth compass, diameter 2¾in. The azimuth compass is fitted with a prismatic sight with two shades, and with a sight with cross-wire. Signed: *Thomas Jones 62 Charing Cross London*. First half 19th century.

14 Compass in gimbals. Signed: *Spencer Browning and Co., 111 Minories London*. First half 19th century.

15 Compass in gimbals. Signed: *S. Fay & Son, Cowes, I. of W.*. First half 19th century.

16 Six's maximum and minimum thermometer. Signed: *Geo. Adams Fleet Street*. (H.12)

17 Double or multiple tube barometer. Signed: *Made by John Cerutty at the Tuns Lodging House Bath*. Late 18th century. (H.11)

18 Chondrometer. Signed: *Fraser London 27 3 New Bond St., c1800*. (H.13)

19 Sikes' hydrometer. Signed: *Buss 33 Hatton Garden London*. (H.14)

20 Sikes' hydrometer. Signature as no. 19. (H.14)

21 Gauging rod, folds in four. Signed: *Pastorelli & Rapkin 48 Hatton Garden London*.

22 Lactodensometer. Signed: *Pastorelli & Rapkin 48 Hatton Garden London*.

L 10 LUDLOW
Ludlow Museum

County Museum Office, Old Street, Ludlow SY8 1NW

0584 873857

The Museum houses an important collection of fossils, as well as local archaeological material, arms and armour.

There are a few scientific instruments.

1 Backstaff. Frame of lignum vitae with boxwood arcs. Inscribed: *Willm. Legg December 20th 1748*.

2 Waywiser. Mahogany with brass dial. Signed: *Heath & Wing*, mid-18th century).

3 Compound monocular microscope with Lister limb supported between two pillars. Signed: *Davis Derby*. c1840.

4 Compound monocular microscope, signed: *Powell & Lealand 170 Euston Road, London*. No.1 model, c1870.

L 11 LUTON
Wernher Collection

Luton Hoo, Luton (Sir Harold Wernher, Bt., GCVO)

0582 22955

The mansion was designed by Robert Adam, but much of the interior was destroyed by a fire in 1836. It was remodelled internally in the French 18th-century manner in the early 20th century. The collections include paintings, porcelain, tapestries and furniture, ivories, metalwork and jewels. With the exception of a floor-standing wheel barometer by John Whitehurst of Derby (according to Goodison (1977) it is by John Whitehurst I (1713-1788)), there is only one scientific instrument.

1 Astronomical compendium in the form of a rectangular box with a sliding lid containing two hinged and two loose leaves. Gilt copper with silver and translucent enamel. Signed: *Christophorus Schisler Augustanus fecit* and dated: *1553*. Combines compass, horizontal and equinoctial sundials, nocturnal, and perpetual calendar.
Lit: BOBINGER (1966) p299

London

Lo 1 LONDON
Royal Armouries, HM Tower of London

London EC3N 4AB

071 480 6359

The Armouries are housed in the White Tower (commenced by William the Conqueror) and the New Armouries, of late 17th-century date. Armouries existed at the Tower from earliest times, but the present collection took shape during the reigns of Henry VIII and his successors. There are large collections of European armour from the 16th century onwards, as well as important material of earlier date. In recent years the scope has been extended to make the Armouries a national museum of European arms and armour. Oriental material is also included in the collection.

1 Gunner's quadrant made for Duke Julius of Brunswick and Luneburg, 1585. The quadrant is in the form of an axe with a hammer face balancing the axe blade. The staff bears scales giving the diameter of lead, iron, stone and slag shot for a given weight. When the staff is inserted in the bore of the gun a pointer pivoting across a calibrated quadrant etched on the axe blade

marks the degree elevation. Length 52in. The quadrant is inscribed in Gothic script: *Princeps Julius Dux Brunswigensis et Luneborgensis me fieri fecit Henricopoli aliis in serviendo consumor 1585.* (Inv. XIV.19) (fig 62)
Lit: BLACKMORE (1976)

2 Commander's baton. Blued steel damascened in gold, bearing arithmetical tables as an aid to drawing up large bodies of troops. The baton was made for Vincenzo Gonzago, *c*1590. (VIII.75)

3 Gunner's callipers. Length (closed) 7in. Brass with steel interior points. The opening of the arms can be read in degrees or inches from scales engraved around the swivel point. The arms are engraved with an inch scale, scales giving the weight of a cubic foot of various materials (stone, lead, etc.), scales of chords, etc. Signed: *Edm. Culpeper Fecit.* Early 18th century. (XIX, 222) (fig 60A)
Lit: BLACKMORE (1976)

4 Two gunsmith's gauges. In a black ray-skin case. One is calibre gauge for smooth-bored small-arms barrels and inscribed: *Col. Borgards Gages for Small Armes.* Signed: *W. Deane Fecit.* The second gauge consists of three perforated plates linked by a compass joint. The perforations correspond to the diameter of lead bullets varying in size from 3 to 48 to the pound avoirdupois. Colonel Borgard was the founder of the Royal Regiment of Artillery. Another instrument devised by him is in the Museum of Artillery, the Rotunda, Woolwich (see *Museum of Artillery, London*).

5 Gunner's callipers. Brass. From the Watervliet Arsenal, USA, 19th century. (fig 60B)

Lo 2 LONDON
British Museum

Great Russell Street, London WC1B 3DG

071 636 1555

The Museum owes its origin to the bequest to the nation of Sir Hans Sloane, subject to a payment of £20,000, of his library and collection – which, incidentally, includes one of the museum's finest instruments, a late 13th-century English astrolabe. In 1753 an Act of Parliament authorised the provision of a building to house the collection.

Funds were raised by public lottery and a body of Trustees appointed. Subsequently Montagu House in Bloomsbury was acquired, and the Museum opened its doors in 1759. Its collections were wide-ranging, covering natural history, coins, drawings and other items. By 1823 the Museum had outgrown its premises, and Sir Robert Smirke was engaged to design new premises, which, while additions have since been made, are substantially unchanged. In the latter part of the 19th century, as the Museum's collections again outgrew their premises, the Natural History collections were rehoused in South Kensington. The ethnographic collections are housed at Burlington Gardens in the Museum of Mankind. The scientific instruments in the Museum are in two

Departments – Medieval and Later Antiquities, and Oriental Antiquities. The former Department contains the large group, but the latter includes several of outstanding importance. The Department of Western Asiatic Antiquities possesses fragments of three Babylonian planispheres of an astrological nature.

Related material occurs in the Department of Coins and Medals, which houses numerous coin balances, while the Department of Egyptian Antiquities contains fragments of water clocks, measuring rods, balances, etc. Similar material can be found in the Departments of Greek and Roman and of Prehistoric and Romano-British Antiquities.
Lit: MILLER (1974); LONDON – BRITISH MUSEUM (1976) for accounts of the history of the Museum and content of its various departments.

I DEPARTMENT OF MEDIEVAL AND LATER ANTIQUITIES

Contains all scientific instruments of European origin (with the exception of one Hispano-Arab astrolabe which is housed in Department of Oriental Antiquities), as well as an exceedingly important collection of clocks and watches (*Lit:* TAIT (1968, 1983)). The clock and watch collection originated in the 19th century with two major bequests, but it has been augmented – indeed has more than quadrupled in size – as a result of the acquisition, in 1958, with the help of Mr Gilbert Edgar CBE, of nearly 2000 items from the Ilbert collection. In addition to the groups of instruments listed below, the Department houses a collection of sandglasses, several of which may be of 16th/17th-century date, a number of weights and measures, and one or two coin balances. Because of the size of these collections and because of the published catalogue (below), this entry provides only a summary of the holdings.

Major sources of the instrument collection, which includes some items from the Bernal and Spitzer collections, are:

   a AW Franks. Franks was Keeper of the Department for many years. During his lifetime he presented numerous items, including instruments; additional material was bequeathed on his death in 1897.

   b Octavius Morgan bequest, 1888. A bequest of important material, including more that 30 instruments or sundials.

   c Max Rosenheim

   d Maurice Rosenheim bequest, 1922

   e Lt. Col. GB Croft Lyons bequest, 1928

   f Ilbert Collection

Of more than 400 instruments in the Department, about one half are sundials, ranging in date from the 15th to the 19th century with a preponderance of early examples. The remainder cover surveying, gunnery, astronomy and mining, as well as including calculating, drawing and optical instruments. Many of the instruments are of great interest both from a scientific and from a decorative point of view. In the following paragraphs an attempt has been made to pinpoint makers and the most important instruments, but in addition to the signed examples there are frequently others which are unsigned. Unless very early in date, or of unique design, no specific reference is made to these.

The large collection of sundials is discussed under a number of main headings: altitude dials with gnomon, azimuth and analemmatic dials, diptych dials, equinoctial dials, horizontal and multiple dials.

A catalogue of the Western scientific instruments has been published (WARD 1981), but refer also to TURNER (1982).

1 *Astrolabes.* An important collection of more than 20 European astrolabes, ranging from two of Hispano-Moorish origin, of *c*1195 and 1350 through to instruments of the 16th and early 17th century. One of the finest instruments, an English astrolabe of late 13th-century date, derives from the collection of Sir Hans Sloane. Other makers represented include Blakene, 1342; Falconi, 1507; Euphrosynus Vulparia of Florence, 1525; Georg Hartmann; Humphrey Cole; Johann Anton Lynden; Adrian Descrolieres, Lieges; Philippe Danfrie; Aegidius Cuniet; Tobias Volckmer; Cornelius Vinchz (two astrolabes dated 1599 and 1600); Adriano, Turin; and Bastien le Seney, London, dated 1545, an astrolabe constructed for King Henry VIII. The latest instrument is inscribed Richard Melbourne, and dated 1631. Most of the astrolabes are of the normal planispheric type requiring a separate plate for use in each latitude. That by Richard Melbourne is, however, based on the universal orthographic projection of Roias, while the astrolabes by Humphrey Cole and Descrolieres both employ the universal stereographic projection of Gemma Frisius.

2 *Astronomical ring.* An unsigned example (attributed to Walter Arsenius of Louvain, *c*1570)) of a rare Renaissance instrument.

3 *Clock and watchmaker's tools.* The department houses a collection of tools, including a number of small items on loan from the Antiquarian Horological Society. There are two examples of clockmaker's wheel-cutting engines, one of *c*1780, while the other, with a wooden hand-wheel, in inscribed: *Knight's Horologium Machina 1787*. There are examples of Jacot tools, fusee tools, a watchmaker's turn with a bow, an early 19th-century watchmaker's mandrel by H Leyland of Prescott, Lancs, a balance wheel engine by John Wyke of Liverpool, as well as smaller instruments such as callipers, depth tools, mainspring winders, trammels for painting clock dials, etc.

4 *Compendia.* More than 20 instruments, a particularly fine collection, of which the most complex is by Johann Anton Lynden, Heilbronn (dated 1596). A compendium with a bias towards surveying is signed by Tobias Volckmer (fig 12). Other instruments are signed by Schissler, Melchior Reichle, Tobias Klieber, James Kynuyn of London (with the coat of arms of Robert Devereux, second Earl of Neukirck, dated 1606).

5 *Computing instruments.* A range of instruments including a 17th-century English abacus, an example of

William Pratt's Arithmetical Jewel (1616), several sets of Napierian rods, a range of sectors including an early example dated 1625 (maker *EB*) and other instruments by Antony Thompson of London, dated 1665, Butterfield of Paris, Cadot of Paris, Domenicus Lusuerg of Rome, Andreas Conrad of Ulm and Spear of Dublin. Slide rules are mainly for gauging and brewing: one is dated 1777 and was constructed by Walter Field, another is signed by Rix, Salisbury Court, London. Less usual items are a circular slide rule of *c*1800 and a slide rule with an *Index for the Time of General Labour as appointed by the Rules of the Prison*, printed for the Prison Discipline Society by RB Bate. (See also *Gunnery instruments* for computing devices used by gunners.)

6 *Dials, altitude.* Several ring dials of 16th-18th-century date, the most important of which is by Humphrey Cole, 1575. Pillar dials are unsigned, save for a mid-19th century example signed by Henry Robert of Paris. A vertical disc dial of Spanish origin and late 16th-century date bears the medal bust of King Philip II of Spain. There are two examples of scaphe dials, one signed by Bartholomew, Abbot of Aldersbach, dated 1554. (See also *Quadrants* for horary quadrants.)

7 *Dials, azimuth and analemmatic.* There are several examples of Bloud-type dials, including instruments signed by Cha. Bloud of Dieppe and by Jacques Benecal, also of Dieppe. A horizontal and analemmatic dial is signed by Thomas Tuttell of London.

8 *Dials, diptych.* A form favoured by the Nuremberg sundial makers, whose work is represented by dials by George Hartmann, Reinmann (5), Hans Tucher (4), Trouschel (3), Leonhard Miller (2), Nikolaus Miller, Conrad Karner, Albreicht Karner and Michael Lesel. There are instruments by C Schissler of Augsburg, Ulrich Schniep of Munich and Marcus Purman of Munich. One dial is signed *LR* and dated 1631, another, of 18th/19th-century date, is by Stockert. An important Italian instrument is signed by Carolus Platus, Rome, 1593. There are one or two unsigned diptych dials of French 17th century origin.

9 *Dials, equinoctial.* These include an English dial of late 15th century date and instruments of the late 16th century by such makers as Christopher Schissler and Cornelius van den Eedt. Most Augsburg makers of late 17th and 18th century date are represented, among them, N Rugendas, J Willebrand, JG Vogler, LT Muller, L Grassl and JN Schretteger. An unusual form with a wide equatorial ring and a wire gnomon is signed Joh. Paul Kraus and was engraved by JG Gutwein. A spherical equinoctial dial dates from the early 17th century. There are several mechanical universal dials of differing form by Michael Bergauer of Innsbruck, and GI Wisenpaintner of Eichstatt, as well as instruments of English origin by William Deane and Ben Workman. Of four examples of crescent dials three are of German origin, one signed by Phillipp Peffenhauser of Augsburg. The fourth instrument is signed by Johann Diepholth of The Hague.

Universal equinoctial ring dials range in date from a dial signed by Elias Allen of *c*1650 to one signed by W and S Jones of *c*1800. Other instruments are signed by E Culpeper, R Glynne, Johann Willebrand and GF Brander.

One unsigned dial of mid-18th century date appears to be of Spanish/Mexican origin.

10 *Dials, horizontal.* Include garden dials, portable dials with fixed or folding gnomon, Butterfield dials, string gnomon dials (including a particularly early example dated 1453) and pin gnomon dials. There are several examples of inclining dials, mainly by English makers who particularly favoured them.

Interesting forms are a dial incorporated in an ivory globe, whose two halves screw together in the middle, and others in lute form, in casket form, and in the form of a pommel surmounting a mace. There are several examples of rings which include a string-gnomon dials. The plate of one dial is mounted a Cardan suspension.

Makers represented include Hans Graf, 1563, C Whitwell, Carolus Platus, Hieronymus Vulparia, Ulrich Schniep, Erasmus Habermel, Andreas Plieninger, F Resel, N Rugendas, Gary, Bion and Butterfield, I Bloud, T Wright, and D Bolton.

A horizontal dial including an engraved map on the polar projection is signed by Johann Anton Ostravsky and dated 1719, while a dial of honestone bears the signature of Conrad Schneid, Canonicus Regular Collegii Wetten Husani. A fine garden sundial dated 1734 bears the coat of arms and inscription associating it with Anton, Abbot of Stift Admont in Austria.

There are inclining dials by Thomas Tompion (two examples, one of gold), R Glynne, T Wright, Fraser, and Jones, while two French instruments are signed respectively by Menant, Paris and by Julien le Roy as inventor and Jacques Le Maire as maker. Another form of universal horizontal dial, by JM Hager, Braunschweig, dated 1705, has a gnomon raised by means of a spring to a pre-set angle.

11 *Dials, multiple.* These include an early example of a crucifix dial, dated 1541, as well as several examples of crucifix or polyhedral dials of later 16th-century date. Signed examples are by Melchior Reichle, 1569, and MS, 1553. A star-shaped dial, probably of Nuremberg origin, is signed MF and may be dated *c*1550. Other dials are signed by Schniep, Hans Tucher, P Reinmann and D Beringer. Unsigned 16th/17th-century examples are Italian or German in origin.

12 *Drawing instruments.* The outstanding item in this category is a set of gilt brass instruments in a case, made by Barthelmewe Mewsum of London, *c*1570. In addition to rules, folding squares, protractors and dividers, there are several instruments listed under surveying and gunnery which included as one of their functions that of drawing on paper measurements made in the field. A pair of drawing compasses is signed by John Hill and dated 1680, folding squares are signed by F Rousselot, Butterfield, and Meunier, all of Paris.

13 *Globes, armillary spheres and orreries.* These include a gilt-copper Ptolemaic armillary sphere of South German origin (early 17th century) and an English 18th-century hand orrery. A celestial globe of brass is signed by P Petit and dated 1659, while a parcel-gilt silver globe goblet is also French and dated 1569. At the time of writing a second globe goblet, the Kentwell Hall globe goblet, was on loan to the Museum. Of English or

German origin, it can be dated *c*1580. A clock of *c*1550 by J de la Garde of Blois includes an armillary sphere.

14 *Gunnery instruments.* A range of 16th- to 18th-century instruments. This includes several 16th-century gunner's levels, an instrument by GZ (Georg Zorn) dated 1619, and two gunner's compasses and squares by the English maker Humphrey Cole, both dated 1575. There are several examples of gunner's callipers, and gunner's rules of 16th/17th-century date, engraved with scales to enable gunners to calculate the weight of cannon balls from their diameter and material. One instrument is signed by Christoph Trechsler, 1620; another is attributed to Erasmus Habermel. (See also *Quadrants.*)

15 *Mining instruments.* There is a typical Tyrolean 17th/18th-century set of miner's instruments with hanging compass, protractor, rule and folding square, and a miner's theodolite by Tobias Volckmer. One of the compendia, also by Volckmer, includes a cord and reel with clinometer and could have served as a mining instrument.

16 *Nocturnals.* Include a highly important early nocturnal by Hans Dorn of Vienna, dated 1491. This example includes a volvelle for the conversion of equal and unequal hours (listed in GUNTHER (1932) no. 250 as an astrolabe). Other instruments are combined with volvelles for the phases of the moon, tide computers, sundials, horary quadrants, etc. One nocturnal, in compendium form, also includes a surveying compass. The earlier instruments are of brass and of continental origin (French, Italian, German and Dutch), but there are three examples of early 18th-century English boxwood nocturnals.

17 *Optical instruments.* There is a small group of seven instruments. These are a prism reputedly owned by Sir Isaac Newton, an early Galilean spyglass whose case is signed by Heinrich Stolle, Prague, and an early refracting telescope of early 17th-century Italian origin which formed part of the foundation collection of the British Museum. There are three compound microscopes, one of late 17th-century Dutch origin, a Culpeper microscope by Benjamin Martin, and a microscope used by Sir David Brewster until 1838, which contains the first garnet lens. A final and very curious item, of French origin and 19th-century date, combines a simple microscope and a series of dials. It is inscribed *microscope phakometre* and *epanaphoroscope phlatergometre* and was probably designed as a spoof.
Lit: WARD (1981) no. 450; JONES (1990) no. 79

18 *Pedometers.* Include instruments ranging from one of *c*1580, probably by T Ruckert of Augsburg, to 17th-century instruments signed by Johann M Landeck, Nuremberg and MT Hager (*c*1690). Later examples are by Spencer and Perkins, Ralph Gout, Dollond, Bregeut and Payne and Co. A waywiser movement is signed: *Penn London* (*c*1800).

19 *Perpetual calendars.* The earliest item is Samuel Norland's perpetual calender, of *c*1650. There are several 18th-century examples (makers include Jacob Senebier and John Boddington) and two examples combine a perpetual calender with a tide computer. Perpetual calender token discs are numerous, by a wide range of English makers, but one or two continental examples are also represented.

20 *Quadrants (portable).* Include two 14th-century English quadrants, one a *quadrans vetus*, the other a horary quadrant, dated 1399, and engraved with the rebus of King Richard II. There are Italian horary quadrants of 15th-century date and of *c*1600. A highly important English instrument was made by Thomas Gemini for King Edward VI and bears the initials of William Buckley and Sir John Cheke. There are simple horary quadrants by John Browne (mid-17th century), Henry Sutton and George Hooper. An instrument by Erasmus Habermel combines an horary quadrant and a dialling instrument, while a large instrument by Tobias Volckmer combines a sinecal and an horary quadrant. A quadrant by Henry Sutton bears a planisphere on the reverse with an orthographic projection. Gunter's quadrants are signed by Abraham Watlington, 1644, Robert Gray, 1709, and W and S Jones, *c*1800. There are two examples of gunner's quadrants, one signed: *Adams, London.*

21 *Surveying instruments.* These range from the 16th to the 18th century in date. Important, although in an incomplete state, is the brass staff of a cross staff by Gaulterus Arsenius of Louvain, signed and dated 1571. Other late 16th-century instruments include a surveyor's compass and folding rule by Erasmus Habermel, an Italian circumferentor, a German surveyor's quadrant, and a surveyor's compass dated 1594 by Jacob Ramminger, Stuttgart. A surveyor's cross dated 1562 is signed: *Eieronymo de Antonio Buzzefo.* 17th-century instruments include graphometers by P Sevin and Nicholas Bion, and a German circumferentor signed *HG 1667.* From the 18th century, as well as an Italian circumferentor by Cesare Costa, there are several unsigned examples. A compass for use with a plane table is by Giovanni Riva, Venice.

22 *Miscellaneous.* An instrument in the form of a torquetum signed by Jacobus Lusuerg in Rome and dated 1688 was presumably intended primarily as an instrument for the construction of sundials, although it could also have been used as a sundial, or have fulfilled other functions of the medieval torquetum. A guinea scale bears the signature of the London instrument maker, Benjamin Martin, while a hydrostatic balance is signed: *G.F. Brander fecit Aug. Vind* and is of mid-18th-century date. There are thermometers by Breguet and Urbain Jurgensen, the latter dated 1843, as well as domestic barometers by Whitehurst of Derby, J Patrick, P Salvade and G Rossi of Norwich. A windvane is of South German origin, and dates from the late 16th century.

II *DEPARTMENT OF ORIENTAL ANTIQUITIES*

The collections cover the cultures of the whole of Asia from neolithic times until the 19th century. The only exceptions are the ancient civilisations of the Near East and of Persia before the founding of Islam in AD 622.

The Department houses a number of important instruments from the Islamic world, notably globes, astrolabes and quadrants, but also a few sundials and compasses from China and Japan.

1 Syrian astrolabe. Unsigned, 13th century. (80, 3-8, 1)
Lit: GUNTHER (1932) no. 105

2 Hispano-Arab astrolabe. Engraved with both Latin and Arabic characters. Unsigned, 13th – 14th century. (OA+371)

3 Egyptian astrolabe, by 'Abd Al-Karim al-Misri of Cairo, AH 633 (1235/6 AD). (55, 7-9, 1)
Lit: GUNTHER (1932) no. 104; MAYER (1956) p30, pl. XIIb

4 Syrian astrolabe. Single plate, one side engraved with a universal stereographic projection, the *Saphaea Arzachelis.* By Ibrahim ad-Dimashqi, Damascus, AH 669 (1270 AD). (90, 3-15, 3)
Lit: GUNTHER (1932) no. 78

5 Syrian astrolabe. Made by Auhad b. Muhammad B. Muhammad B. Jamal al-Auhadi in AH 890 (1485 AD) for Taj ad-danla wa-d-dunya wa-d-din Jan Ali. (64, 12-21,1)
Lit: GUNTHER (1932) no. 78

6 Indian astrolabe. Made by Muhammad Muqim b. Isa of Lahore in AH 1070 (1959/60 AD). (93, 6-16,4)
Lit: GUNTHER (1932) no. 78

7 Persian astrolabe. By Muhammad Mahdi al-Khadim, Yezd, 17th century. From the collection of CJ Rich. (86, 3-17, 1)
Lit: GUNTHER (1932) no. 9; MAYER (1956) p71, pl. XVII

8 Persian astrolabe, constructed by 'Abd al-'Ali d. Muhammad Rafi al Juzi and decorated by his brother, Muhammad Baqir, for Shah Husain as-Safawi, AH 1124 (1712/13 AD). (OA+369)
Lit: GUNTHER (1932) no. 33; MAYER (1956) p27; MICHEL (1967) pp180, 184

9 North African astrolabe. Unsigned, probably 18th century.
Lit: Possibly this is the one described in GUNTHER (1932) no. 133

10 Astrolabe. The front and back of an astrolabe mounted on a circular gilt brass box incorporating a clockwork mechanism of uncertain function. Of unknown origin and date. (88, 12-1, 169)

11 Syro-Egyptian quadrant made by Muhammad b. Ahmad al-Mizzi, AH 727 (1326-7 AD). (95, 11-16, 1)
Lit: MAYER (1956) p61, pl. VIb (signature)

12 Syrian quadrant by Muhammad b. Ahmad al-Mizzi, AH 734 (1333-4 AD). (88, 12-1, 276)
Lit: MAYER (1956) p62, pl. VIc

13 Syrian quadrant by Ali b. ash-Shihab, AH 735 (1334-5 AD) for Shaikh Shams ad-din b. Sa'id, chief Muezzin of the Damascus mosque of the Umayyads. Engraved by Muhammad b. al-Ghazali. (62, 12-27, 1)
Lit: GUNTHER (1932) vol. I, p234; MAYER (1956) p44

14 Quadrant. Brass, once gilt. Engraving of poor quality. Date and origin uncertain. (1921, 6-25, 2)

15 Celestial globe made by Muhammad b. Hilal al-Munajjim, Mosul, Mesopotamia, AH 674 (1275-6 AD). (71, 3-1, 1)
Lit: GUNTHER (1932) vol. I, p238; MAYER (1956) p68; PINDER-WILSON (1975); SAVAGE-SMITH (1985), p219-220

16 Persian globe made by Muhammad b. Ja'far b. 'Umar, AH 834 (1430-31 AD). (26, 3-23, 1)
Lit: MAYER (1956) p44, pl. XIIa; SAVAGE-SMITH (1985) p249, Fig. 10

17 Compass in flat circular ivory box with lid. By Bairam b. Ilyas, AH 990 (1582 AD). (1921, 6-25, 1)
Lit: MAYER (1956) p44, pl. XVIa

18 Miniature compass in nielloed case. ?Turkish, 19th century. (AF165 Franks Coll)

19 Horizontal sundial. Circular brass case with lid. Persian, 18th century. (90, 3-15, 4)

20 Astrological or magical instrument. Silver disk with suspension ring. Engraved with magical formulae. ?Turkish. Uncertain date. (1923, 2-3, 2)

21 Prognostication table made by Mohammad b. Khutlukh at Mosul, Mesopotamia, AH 639 (1241-2 AD). (88, 5-26, 1)
Lit: MAYER (1956) p69

22 Protractor. Brass. Turkish, 18th or early 19th century. (OA+1270)

23 Still head and several alembics of glass. Syro-Egyptian, ?18th century AD. (1934, 3-16, 2, etc)

Lo 3 LONDON
British Museum (Natural History)

Cromwell Road, South Kensington, London SW7 5BD

071 938 9123

Originally part of the British Museum at Bloomsbury, increased congestion on the Bloomsbury site resulted in a new site being found for the Natural History collections at South Kensington. This is now known as the Natural History Museum. The building was erected in 1873 – 80, but has subsequently been extended. It houses the national collection of animals and plants, extinct and existing, and of the rocks and minerals which make up the earth's crust. Displays also cover evolution and other biological topics. A few of the instruments belong to the years before 1880 when the collections were still housed at Bloomsbury, but most were acquired after the move to South Kensington. Nearly all can be dated from old Museum registers.

Instruments constructed up to the end of the 1880s have been listed. However, in several cases the sequence continues into the early years of the 20th century, i.e. goniometers, spectroscopes, refractometers, microscopes, dichroscopes, polariscopes and balances. Dating from the second decade of the present century are examples of crystal-grinding apparatus by Fuess and by Swift. Many of the later instruments, although not included in this list, are of interest in illustrating developments in optical crystallography in the last decades of the 19th and the early years of the 20th century. All instruments listed below, with the exception of those indicated by (B), i.e. Department of Botany, are housed in the Department of Mineralogy. Dates in parentheses indicate the year of acquisition.
Lit: STEARN (1981) and GUNTHER (198?) for general histories; WHITEHEAD (1981) for a current guide.

1 Balance signed: *Robinson and Barrow, London.* Belonged to MHNS Maskelyne (1823-1911) and was probably used by him in the laboratory at the Old Ashmolean Museum, Oxford. Purchased for the Natural History Museum via JR Gregory (1881) (on loan to the *Science Museum London*).

2 Balance, double cone type. Signed: *Thomas Jones, 62, Charing Cross, London* (1859).
Lit: STOCK (1973)

3 Balance by Oertling, London. (1860)

4-5 Two balances by Oertling, London. (1887)

6 Balance, large size, with specific gravity pans, by Oertling, London. (1889)

7 Balance by Sartorius, Gottingen. From Murray collection. (Date of acquisition not recorded)

8 Contact goniometer by J Sax. (1863)

9 Contact, goniometer by Lenoir, Vienna. (?c1850)

10 Contact goniometer by Powell & Lealand. (c1863)

11 Contact goniometer by Lenoir, Vienna. (1882)

12-13 Two contact goniometers by Elliott Bros. (1887)

14 Optical axial angle goniometer by Troughton & Simms. (1860)

15 Vertical circle goniometer by Powell & Lealand. (1865)

16-18 Three goniometers by Fuess. (1881, 1882, 1883)

19 Goniometer by Elliott Bros, London. (1885)

20-21 Two direct vision spectroscopes. (1885)

22-23 Two simple dissecting microscopes, one incomplete, c1800. (B)

24-25 Two polariscopes by Lenoir, Vienna. (1859)

26 Polariscope by Schneider, Vienna. (1881)

27-28 Two polariscopes by Fuess. (1887, 1890)

29 Polariscope by Lenoir, Vienna. Bought from Stevens Sale Rooms. (1888)

30-31 Two students' heliostats by Spence & Co., Dublin. (1881)

32 Heliostat, weight-driven, by CJ Klaftenberger. Accessories by GW Watson. (1874)

33-35 Three dichroscopes by Fuess. (1881, 1890, 1894)

36 Locked leather box, stamped *Sir Joseph Banks.* Believed to contain a solar microscope of c1800. (B)

37 Compound microscope, by Powell & Lealand, London, *c*1860. (1863)

38 Compound microscope with monocular and Wenham binocular bodies and case of accessories by Smith, Beck and Beck. 1850s or 60s. (B)

39 Binocular dissecting microscope (Stevenson stereoscopic) by Baker, London, *c*1870-80. (B)

40-44 Five students' microscopes by Swift. (1885)

45 Compound microscopes by Nachet. (1886)

Lo 4 LONDON
British Optical Association

British College of Optometrists, 10 Knaresborough Place, London SW5 0TG

071 373 7765

The British Optical Association was founded in 1895. JH Sutcliffe, appointed Secretary shortly after its foundation, planned to establish an optical library and museum, although only in 1926, after two or three changes of headquarters, did the Association find space which could be devoted to a Museum. The first Museum, in Clifford's Inn, included spectacles, spectacle cases, small ophthalmic instruments, and a collection of glass eyes illustrating pathological conditions; subsequently it expanded rapidly, mostly by gifts from members. In 1934 the Association moved to premises in Brook Street, and the Museum was set up in a room on the first floor. As a result of intensive effort, many gifts and some purchases, the Museum grew rapidly in the years after 1934, and for the specialised field which it covers is probably the most important existing collection.

The scope of the collection ranges from spectacles of 15th- to 20th-century date, hand magnifiers, lorgnettes, and a collection of spyglasses of outstanding decorative interest. In addition there is a collection of optical fans, that is, they incorporate spyglasses. This rare collection is the only one of its kind in England. There are spectacle cases of all types and periods, as well as a collection of material illustrating the development of optics, and especially of spectacles, including coins and medals, paintings, engravings, etc. The library contains an important collection of books on optics, dating from the 16th century to the present day. The Museum houses a collection of optician's instruments – ophthalmoscopes, lens triers, optometers, refractometers, keratometers, perimeters and retinoscopes. Many of these instruments, although not normally on display in the Museum, form an important part of its collection. The range of the collection falls in the main outside the scope of this handbook, although in its own field it must be regarded as of outstanding importance. A few items, however, can be included. In addition to these listed there are several other miniature telescopes, primarily of interest from the point of view of their decoration, and examples of polemoscopes or jealousy glasses of late 18th- or early 19th-century date.

Lit: SUTCLIFFE/MITCHELL/CHITTELL (1932); MITCHELL(1975)

1 Small telescope. Four draw-tubes, red ray-skin outer tube, brass mounts. Possibly Dutch, *c*1700.

2 Small telescope. ray-skin covered outer tube. Three green vellum covered draw-tubes, early 18th century.

3 Small telescope, brass, japanned outer tube with chinoiserie decoration, 18th century.

Lo 5 LONDON
Bruce Castle Museum

Lordship Lane, Tottenham, London N17 8NU

081 808 8772

The Museum combines local history and postal history collections as well as housing the Museum of the Middlesex Regiment. Instruments in the collections are limited to waywisers and one pedometer.

1 Waywiser. Iron-bound wheel, diameter 12in. The dial is mounted on the handle immediately above the wheel which is divided to measure up to 4 miles, probably mid-18th century.

2 Waywiser. Iron-bound wooden wheel, diameter 31¾in. Brass dial mounted on the handle above the wheel, divided to measure up to 12 miles. Signed on dial: *Fras. Watkins London*. Probably Francis Watkins II, *c*1830.

3 Waywiser. Iron wheel, diameter 23in. Dial mounted on the handle immediately above the wheel. Signed: *Elliott Bros. London*. Mid-19th century.

4 Waywiser. Brass wheel, diameter 5¾in. Dial mounted in centre of wheel with four subsidiary dials, and signed: *S. Johnson's Patent Indicator and Surveying Instrument. Callaghan Optician New Bond St London*, second half 19th century.

5 Pedometer. Brass, in watch form. Signed on dial: *Spencer & Perkins London*, late 18th century.

Lo 6 LONDON
Cuming Museum

155-157 Walworth Road, Southwark, London SE17 1RS

071 703 3324/5529/6514 ext 32

The collections relate primarily to the Southwark district, and include Roman, medieval and later archaeological material and bygones. There is also a subsidiary collection specialising in superstitions from the London area. There are few scientific instruments, and such as they are, are associated either with Michael Faraday, who was born locally, or with the founder of the Museum, Richard Cuming.

1 Set of sixteen steel 'buttons' (alloys of steel with rhodium, platinum, silver, nickel, etc.) produced by

Faraday in 1820 when working with Stodart on alloys of steel.

2 Wooden tripod microscope resembling a Nuremberg tripod microscope. Dated 1792 and said to have been made by Richard Cuming.

3 Cylinder electrical machine used by Richard Cuming. The machine, whose base measures 16 x 13½in, is contained in a wooden box dated 1799 on the underside of the lid. The cylinder is broken, the T-shaped prime conductor fits on to a separate square base whose glass column no longer survives. The collecting comb is roughly cut out of sheet brass. Accessories include a discharger, a Henley pith-ball electrometer, electric bells, pith figures, knitted silks and linen cloths, Leyden jars, and numerous other smaller items, many of which are included in a handwritten list of contents pasted inside the box, or on a sheet of paper in a box containing smaller items.

4 Magneto-electric machine in obelisk-shaped, mahogany-veneered box, 15in in height. Two coils, large horseshoe magnet. Signed on ivory label: *J. Newman 122 Regent Street London*. Third decade 19th century. Said to have been used by Michael Faraday. On permanent loan from Miss L Faraday.

Lo 7 LONDON
Galton Laboratory, University of London

Department of Human Genetics and Biometry, University College London, Wolfson House, 4 Stephenson Way, London NW1 2HE

071 387 7050 ext 7411

Francis Galton's anthropometric researches led directly to the establishment of the Biometric Laboratory at University College. The laboratory houses a collection of his apparatus and personalia. Of the items listed below, nos. 1-3 are associated with Francis Galton (1822-1911); other Galton material is in the *Science Museum, London*.
Lit: THODAY (1976)

1 Head callipers used by Francis Galton in his anthropometric laboratory for taking head measurements.

2 Box used for composite photography.

3 Quincunx. Apparatus for affording physical illustration of the action of the Law of Error or of dispersion, *c*1876.
Lit: LONDON (1876) p14, no. 49

4 'Eidouranion' or transparent orrery, *c*1820. Belonged to Samuel Tertius Galton, father of Francis Galton.
Lit: BIRMINGHAM (1966) p39, no. 111

Lo 8 LONDON
Gunnersbury Park Museum

Gunnersbury Park, London W3 8LQ

081 992 1612

The Museum, housed in a mansion which formerly belonged to the Rothschild family, contains collections relating to local archaeology and history, social history, typography and local industries. A most important single exhibit is the iron printing press invented by Charles, third Earl of Stanhope (1753-1816) and dating from 1804. The only scientific instrument is the electrical machine listed below.

1 Cylinder electrical machine. Base 18 x 10in, curving outwards on one long side to take the arm supporting the friction cushion. The pressure of the cushion against the cylinder can be adjusted by means of a slot at right angles to the arm held by a wooden screw. Length of cylinder 10½in, diameter 6in. Prime conductor with insulated glass stem is mounted on a separate circular wooden base. Similar to the Cavallo machine illustrated in REES (1819), vol.2 of plates, plate IX, fig 11. The main difference lies in the method of adjusting the friction cushion. *c*1775.

Lo 9 LONDON
Horniman Museum

London Road, Forest Hill, London SE23 3PQ

081 699 1872/2339/4911

The original collections were accumulated by Mr FJ Horniman and were exhibited in Surrey House which was demolished in 1898 to make room for the existing building. Building and collection were presented to the London County Council in 1901; later it was administered by the Greater London Council, and from 1990 has been run by a Board of Trustees with funding from the Museums and Galleries Commission.

The Museum contains important ethnological collections ranging from Stone, Bronze and Iron Age implements to weapons, tools, decorative art, magic, religion, musical instruments, etc. In addition there are wide-ranging natural history collections.

The ethnology collections include simple time measurement instruments, such as water clocks, measuring sticks, shepherd's dials and calendar sticks. In additions there are a few items within the scope of this survey, some of which (nos. 1-4) are of considerable interest.

1 Islamic astrolabe by Ahmad b. Muhammad b. Ibrahim, Fez, AH 1115 (AD 1703). (26:32)
Lit: GIBBS/HENDERSON/PRICE (1973) p38

2 Gunter quadrant and almanac. Boxwood, circular, with brass sights. One side (with the sights) is engraved with the horary quadrant inscribed in a heart-shaped figure and a perpetual almanac. The quadrant is marked: *Latitude 51°32'* and this side also bears the (maker's?)

initials *W.H. 1654* (Walter Hayes?).The reverse of the quadrant bears calendar scales, as well as scales for ascertaining the rising, amplitude, declination, or right ascension of the Sun, or the length of day at any time of the year. A brass rule pivoted at the centre of the instrument serves to make readings from the scales. (31:39)

3 Astronomical compendium. Gilt brass, circular, including a nocturnal (outside of lid), list of latitudes and hour lines for a horizontal dial marked for different latitudes. The underside of the dial is inscribed: *VIGILATE QVIA NESCITIS QVA HORA DOMINUS VESTER VENTURUS SIT.* Signed: *Humfrey Cole 1590* and inscribed: *This dial everall for England and Scotland is for all contres y[e] lie east or west from there.* The gnomon for the horizontal dial is lacking. (31.183 A)
Lit: HOLBROOK (1974) no. 31, plate 18

4 Horizontal string-gnomon dial. Ivory, octagonal, of diptych form, with silvered hour ring engraved with the initials: *R.G.* (? Ralph Greatorex). A coloured engraved map of England is inset in the underside of the lid, and a paper compass card in the base of the compass box. English, mid-17th century. (31.187 A) (fig 36)
Lit: HOLBROOK (1974) no. 34, plate 10

5 Upper leaf of an ivory diptych dial. Nuremberg, 17th century. (27.438)

6 Magnetic dial in circular wooden case. Signed: *Ebsworth Fleet Street London.* (36.22)

7 Magnetic dial mounted on ivory stand. Signed: *C. Essex & Co London.* (27.417)

8 Compass in turned wooden box. Paper compass-card signed: *D. Adams Charing Cross London.*

9 Thermometer and compass. Turned ivory in red leather case. Unsigned, early 19th century.

10 Equinoctial pocket dial. Chinese.
Lit: NEEDHAM (1959) p310, plate XLII (Type B)

11-14 Four diptych dials. Chinese, 19th century.

15-18 Four gemomantic compasses. Chinese, 19th century.

Lo 10 LONDON
Institute of Ophthalmology, University of London

University of London, Judd Street, London WC1

071 387 9621

The Institute of Ophthalmology was established in 1947. Since its foundation it has acquired a small but interesting collection of optical apparatus consisting primarily of ophthalmological instruments, including a series of early ophthalmoscopes, from the Helmholtz type onwards. In addition there are spectacles, trial lenses and trial frames. Medical items include a late 19th-century stethoscope, a clockwork trephine and an artificial leech. Finally, and more important for this

survey, there is a group of nine simple and compound microscopes ranging in date from the early 18th until the middle of the 19th century.

1 Simple microscope. Brass, with rotating wheel in which are mounted 20 different objects, and a swivelling holder for two objectives. The distance between object and lens is adjustable by means of a spring. ? Early 18th century. (fig 74)

2 Ivory optical compendium. Now incomplete, but probably consisted originally of a small spy glass and a simple magnifying glass and a spike, *c*1700.

3 Screw-barrel microscope. Ivory in black fish-skin case. With ivory container for mica circles and four lenses. Two ivory slides. First half 18th century.

4 Screw-barrel microscope. Brass in black fish-skin case, 18th century.

5 Culpeper microscope. The brass tripod legs supporting the body tube are spaced between those supporting the stage. Bonanni spring-stage. Black ray-skin covering to outer tube covered with green vellum ornamented with gold stampings and with ray-skin band around top, *c*1730-40.

6 Culpeper microscope, similar to 5, but without ray-skin band around top of the inner tube.

7 Culpeper microscope, late 18th/early 19th-century type. Brass in pyramidal wooden case. Signed: *T. Blunt London.*

8 Compound microscope. Similar to the 'Compendious Pocket Microscope' described by George Adams in 1771: body focusing with rack and pinion adjustment, inclinable support, folding tripod foot. Unsigned, late 18th century.

9 Compound microscope. Brass, Signed: *Parkes & Son Birmingham.* Mid-19th century.

Lo 11 LONDON
Museum of Artillery in the Rotunda

Repository Road, Woolwich, London SE18 4JJ

081 316 5402

The collection housed in the Museum of Artillery was started in 1788 by Captain (later Lieutenant General Sir William) Congreve. At the time Congreve was Superintendent of Military Machines at Woolwich, and the Royal Military Repository he established at the Warren, Woolwich (known as the Royal Arsenal since 1804) consisted of a collection of models used for instructing officers of the Royal Artillery in the handling of guns. Some of the original models were destroyed in a fire in the collection's original home in 1802, but others were added when Sir William's son, Colonel Congreve, was in charge of the Repository from 1814 onwards. After the fire the collection was housed until 1819, near, but not on, the present site. In 1819, Colonel Congreve persuaded the Prince Regent to present one of the tents designed by John Nash for the meeting of the Allied

Sovereigns in London in 1814, and this was erected in 1820, having been made more permanent by the addition of a lead roof and a central pillar.

Since the collection was first established, material has been acquired from many sources. As well as artillery equipment and models, it includes hand firearms, both European and Oriental, swords, native weapons, and some armour and ordnance from the 16th century onwards. Important from the point of view of this survey is a group of instruments, mainly, but not entirely, connected with gunnery. While the earliest items date from the 16th century, the majority are of 18th-century date. Makers such as John Rowley, George Adams, and Benjamin Cole (probably Benjamin Cole junior) are represented, all of them among the finest makers of their period. Some of the instruments appear to be unique, notably nos. 2, 5, 26, 31 and 32.

It remains to be added that many gunner's sights and levels, including sights invented by Sir William Congreve (1772-1828) and described in his *Description of Sights* (see Congreve (1819)), are not listed here and the selection of items for inclusion has been somewhat arbitrary. The collection is of great importance for anyone wishing to study gunnery instruments of the 18th or 19th centuries; they are listed XXIV 8-9 and 12-24 in KAESTLING (1970). Another Collection which, containing relatively recent material, is not listed below is the Watkin Collection. It contains instruments, including clinometers, tangent sights, recording chronographs (for measuring variations in the velocity of projectiles), range finders, etc. designed by Colonel Watkin in the last quarter of the 19th century. *Lit:* KAESTLING (1970); this lists the instruments in Class XXIV, as well as containing a survey of the history of the building and the collection. The numbers allocated to instruments in this catalogue are inserted in parentheses after the entries in the ensuing text.

1 Gunner's level. Gilt brass. Plummet level over arc divided from 0 to 15. Alidade with sights. Signed: *CTI 1598* (? Christopher Treschler, Dresden) (XXIV/63)

2 Gunner's level. Brass rule with slit sights mounted on two pairs of curved legs. A quadrant with plummet and bubble level is mounted on the rule parallel to the line of sight, another bubble level at right angles to it. Signed: *Ancell Fecit.* Probably early 18th century. (XXIV/74)

3 Gunner's level. Brass. Flat base with central pillar on which is mounted an open quadrant with plummet level and graduations from 45 – 0 – 45°. Signed: *I. Rowley fecit.* (XXIV/92)

4 As 3, with addition of alidade with open sights and bubble level mounted at apex of quadrant. Signed : *Col. Borgard Invt 1710. I. Rowley fecit.* Colonel Borgard, a Dane, was the first Colonel of the Royal Artillery. (XXIV/73)

5 Gunner's sighting and levelling instrument. Brass quadrant, 9in radius, with 90° graduations, movable over an arc graduated 26 – 0 – 26°. One radius of the quadrant is extended as a long arm with screw thread. This supports arms apparently for inserting in the bore of a gun. Supports for a (missing) sighting telescope. Signed: *F. Morgan London* (possibly Francis Morgan, fl. *c*1770). (XXIV/66)

6 Gunner's level. Brass and steel, ht 5in. Base cut away in a curve to permit placing on the barrel of a gun. Bubble level. Steel slide in central vertical axis. Four (sighting) holes under bubble level. Signed: *G. Adams Inst Maker to his Majesty Fleet Street London.* Possibly incomplete, cf. no. 7. (XXIV/212)

7 Similar to 6. To the top of the steel slide is fixed an arm with bubble level which can be pivoted between horizontal and vertical. Signed: *G. Adams Mathematical Instrument Maker to his Majesty Fleet Street London.* (XXIV/88)

8 Instrument for laying guns. Horizontal graduated semicircle, sighting telescope with bubble level. Further bubble level at right angles. Signed: *G. Adams Instrut Maker to His Majesty Fleet Street London.* Attachment for tripod with the engraved number: *1299.* (XXIV/88)

9 Gunner's level. Consisting of quadrant, one radius of which is extended as an arm for insertion in the mouth of the cannon. Vernier with tangent screw, bubble level. Length 15¾in. Signed: *Thomas Jones 62 Charing Cross London.* Probably second quarter 19th century. (XXIV/78)

10 Instrument for laying mortars. Brass. Rule with open and sights, bubble level mounted over one sight. Signed: *J. & W. Watkins Charing Cross London.* (XXIV/85)

11 Gunner's folding rule, sector-shaped, each arm with two further engraved arms which fold away. Brass, closed length 6½in. Scales relate to gunnery, one inner arm perforated with holes probably for measuring the diameter of bullets. Unsigned, but similar to instruments by Humphrey Cole, *c*1575. (XXIV/200)

12 Gunner's scale. Boxwood, maximum length 8in. Unsigned. English, probably late 17th or early 18th century. (XXIV/117)

13 Gunner's rule. Boxwood, brass mounted, folds in four. Length when folded 6⅜in. 18th century.

14 Folding scale, sector-shaped, boxwood, brass mounted, with bubble level. A graduated arc permits it to be used as a clinometer. Signed: *W. Ladd Beak St.Regent Street.* (XXIV/6)

15 As 14. Signed: *Adie Pall Mall.*

16 Gunner's callipers, brass, closed length 6⅞in. Signed: *M. Berge London.* (XXIV/202)

17 Gunner's callipers. Brass, closed length 7in. Signed: *Gilbert London* (XXIV/10)

18 Gunner's callipers. Brass, signed: *Cummins London.* (XXIV/36)

19 Gunner's callipers. Brass, signed: *M. Berge London.* (XXIV/222)

20 Vernier callipers, divided for measuring outsides and insides. Brass, length of rule 26½in. Inscribed: *Ordnance Select Committee,* signed: *Thomas Jones Charing Cross.* (XXIV/121)

21 Protractor. Semicircular. Brass, arm with vernier. Signed: *Berge late Ramsden London.* (XXIV/182)

22 Protractor. Brass. Arc of around 120°. Radius at base of arc extended as an arm. Movable arm with vernier for reading from the degree scale, chamfered edge and scale from 10 – 0 and 0 – 10. Unsigned, ? 18th century. (XXIV/183)

23 Parallel ruler. Brass, 18th century. (XXIV/181)

24 Beam compasses. Mahogany and pearwood with brass attachments. Length 40½in. Unsigned, 18th century. (XXIV/180)

25 Sector. Brass, closed length 6¾in. Signed: *Butterfield*. Ornamented with rather crude engraving. (XXIV/81)

26 Architect's sector. Mahogany with ebony, and ivory inlay engraved with scales. A mahogany and ivory arc passes through both arms of the sector. Scales include heights of doors, entablatures, heights of orders, etc. Signed: *G. Adams Fleet Street London*. (XXIV/179)

27 Pantograph. Mahogany, ivory scales, brass attachments. Signed: *Made by Geo. Adams in Fleet Street London*. Arms of unequal length, maximum 29½in. (XXIV/178)

28 Pantograph. Ebony and brass. Length of outer arms 26¾in. Brass scales inlaid in arms. Unsigned.

29 Pantograph. Brass, length 34½in. Signed: *Fraser & Son London*. Inscribed: *No.3*.

30 Graphometer with sighting telescopes. Signed: *Canivet a la Sphere Paris 1759*. Brass. (XXIV/101)

31 Clinometer. Brass quadrant. 9⅝in radius. On one face is mounted a circular dial with divisions from 0 to 90. The hand on the dial is moved by a toothed arc and gear wheel with attached weight. The quadrant is graduated from 0 to 90° and 90 to 0°, and a bubble level is mounted on one of its radii. Signed: *Cole Maker at the Orrery in fleet Street London*. (XXIV/99)

32 Levelling instrument/clinometer. Circular base 11in radius, with toothed edge and vernier. Mounted on the base above one another are two tubes of rectangular section, one of whose vertical sides is engraved with the scales of a sector. On the other two sides is engraved a diagonal scales. The upper tube carries a bubble level and supports a telescope with bubble level mounted below. One or both rectangular tubes can be inclined. Signed: *Sisson London*. (XXIV/98)

33 Engineer's quicksilver level, presented to the Royal Repository in 1832 by Thomas N Parker, the inventor. Consists of a box with carrying handle and brass inscription plaque, with two vessels containing mercury in which float ivory rods. (XXIV/184)

34 Prismatic compass. Brass, paper dial. Signed: *Troughton & Simms London*.

Lo 12 LONDON
Clockmakers' Company Collection

Guildhall Library, Aldermanbury, London EC2P 2EJ

071 606 3030 ext 1866

The collection is composed primarily of clocks and watches, and in this respect is outstanding. It originated in the early 19th century. The Clockmaker's Company founded a library of horological books and manuscripts in 1813, and shortly afterwards began to assemble a collection of important clocks and watches, watch-mechanisms, watch keys and other accessories. Over the years many important items have been acquired or presented, including a number of early clocks and watches (15th – 16th-century in date), as well as a number of chronometers, and several items by John Harrison, including his Marine Timepiece (no. 5 below). In addition there is a number of watch and clockmaker's tools.

The small collection of instruments is derived to a large extent from the collection given or bequeathed by the Revd. HL Nelthropp, Master in 1893. Consisting primarily of sundials, there are also pedometers, waywisers, an orrery and an astrolabe. The collection and Library were deposited on permanent loan to the Guildhall Library in 1873.

Lit: BAILLIE (1939); LONDON – WORSHIPFUL COMPANY OF CLOCKMAKERS (1949); CLUTTON/DANIELS (1975). Numbers in parentheses after each entry refer to this last catalogue.

1 Celestial globe of brass engraved with constellations and Chinese characters. Globe supported by stand with four dolphins. Perhaps made under the direction of Father Fernando Verbiest who designed instruments for the Peking Observatory, *c*1680. Movement probably second half 18th century, stand early 19th century. (590)

2 Orrery glass, globe etched with constellations containing the orrery turned by a pendulum movement below. The globe is supported by three male figures in bronze. French, mid-18th century. (567)

3 Astrolabe. Brass, diameter 3.5in. Dated AH 1099 (AD 1688). Signed: *Hamid ibn Muhammed mukim ibn Isa ibn Allahdad astralabi Lahouri Huma-youni*. (648)
Lit: PRICE (1955) no. 1115

4 Ring dial. Brass, with sliding collar. Late 17th century. (664)

5 Horizontal dial with hinged gnomon for one latitude (approx 52°). Brass, *c*1800. (663)

6 Horizontal sundial for a latitude of 45°N. 7in square. Compass card and needle now missing. Signed: *T. Menant Paris 1743*. (662)

7 Butterfield dial. Octagonal, silver. Signed: *Butterfield AParis*. (658)

8 Butterfield dial. Silver, octagonal. Signed: *Butterfield AParis*. (659)

9 Butterfield dial. Octagonal, silver. Signed: *P. Sevin AParis*.

10 Diptych dial. Ivory, inscribed: *J 3 K.* (Jakob Karner), Nuremberg, *c*1630. (650)

11 Universal equinoctial dial. Gilt brass. Signed: *T.W. fecit 1600.* (652)

12 Universal equinoctial dial. Brass with silvered hour ring, octagonal. Signed: *A. Chevallier Opticien Rue de la Bourse No. 1 AParis.* (656)

13 Universal equinoctial dial. Brass, circular. compass with stop for needle. Two sets of hour figures, one for southern latitudes. English, *c*1800. (657)

14 Universal equinoctial dial. Brass, octagonal. Signed: *Johann Schretteger in Augsburg.* (653)

15 Universal equinoctial dial. Brass, octagonal, incorporating a wind vane. Signed: *Lorenz Grassl in Augsburg.* (655)

16 Universal equinoctial dial. Brass, octagonal. Signed: *L. Grassl.* (924)

17 Bloud dial (magnetic azimuth). Ivory. French, second half 17th century.

18 compass with enamelled dial. Signed: *R.B. Bate 17 Poultry London.* (666)

19 Pedometer. Signed: *Fraser Bond Street London.* Silver case with the London hallmark for 1788/9. (669) (fig 108)

20-21 Pedometers. Signed: *B. Gray Clockmaker to George 2nd,* one example hallmarked 1754/5. (667-8)

22-23 Pedometers. Signed: *Spencer & Perkins, London.* Serial number: *1261* stamped on one. Late 18th century. (670-71)

24 Waywiser for attachment to carriage wheel. Signed: *Tho. Eayre Kettering.* Mid-18th century. (672) (fig 134)

25 Waywiser for attachment to carriage wheel. Signed: *J. Brockbank London.* (673)

26 Clockmaker's gauge for pivoting arbors. 18th century. (641)

27 Clockmaker's sector for determining wheel and pinion sizes and depths. Made by R Pennington, *c*1780. (640)

28 Micrometer gauge. 19th century. (642)

Lo 13 LONDON
Museum of London

London Wall, London EC2Y 5HN

071 600 3699

The new Museum of London was opened in 1977 in purpose-built accommodation in the City. It was formed by amalgamating the collections of the former London Museum and Guildhall Museum, and presents a picture of London from prehistory to the present, covering all aspects of the life, trends and occupations of its inhabitants. The Museum incorporates the Department of Urban Archaeology (for the City of London) and the

Greater London Archaeology Department. A feature of the collections is that a number of instruments have been derived from archaeological contexts. Of particular interest are the portable sundials (nos. 1 – 3 and 15 – 17 below), some of which are similar in design and execution to those found on the *Mary Rose,* now in the *Mary Rose Exhibition, Portsmouth.* Some distillation apparatus is also present from such excavations, as are pharmaceutical ceramics.

The collections include clocks (among other items, an Ellicott journeyman clock) and 200 watches of both British and Continental origin (two of which are excavated examples from a jeweller's stock buried in Cheapside in the early 17th century). Weights and measures range from excavated Roman samples to 19th-century local standards, and coin balances of 17th- to 18th-century date (including examples by Thomas and Elizabeth Hux (29.78/4) and David Wilford (A8134), both of London). There are a few barometers and a number of watch and clockmaking tools. In addition there are a dozen or so pairs of callipers, some from archaeological excavation. Scientific instruments in the collection are mainly sundials, but there is a number of other items including interesting instruments by D Wiglias and Henry Sutton (nos. 31 and 32 below). Several important items are on loan to the Museum, notably an early microscope (no. 40) and an extensive set of 17th-century surgical instruments (no. 39).

1 Horizontal dial. Circular turned wooden base, gnomon missing. Unsigned, 16th century. Found in London. (A5139)

2 Portable horizontal sundial of boxwood with brass gnomon. Small, 1¼in diameter. Unsigned, probably 16th century. Found in London. (A3891)

3 Portable horizontal dial. Circular, ivory and brass. Unsigned. ? German 16th/17th century. (A22451)

4 Four fragments of a Butterfield type dial. Late 17th/early 18th century. (A10430)

5 Horizontal plate dial. Signed: *WH,* 19th/20th century. (A12228)

6 String gnomon dial. Oval, ivory. Unsigned French or German, 16th century. (A9978)

7 Horizontal string gnomon dial. Octagonal. Unsigned, 16th century.

8 String gnomon dial. Silver, oval, major diameter 2⅛in. Original case. Inscribed: *An 1655.* English. (46.78/707) *Lit:* TANGYE

9 Equinoctial dial (no. 707). Brass, partly silvered. Octagonal with engraved floral decoration. Augsberg, early 19th century. (A21322)

10 Universal equinoctial ring dial. Signed: *Long Maker Little Tower St. London, c*1820. (A13309)

11 Universal equinoctial ring dial. Brass, diameter 3in. Signed: *Hilkiah Bedford by Holborn Circuit, c*1660-1680. (46.78/585) *Lit:* TANGYE no. 585

12 Universal equinoctial ring dial. Signed: *Tho. Tompion*

in Fleet St. Fecit. Brass, diameter 5in. (80.271/100)

13 Ring dial, unsigned, 18th century. Diameter 1½in. (10.848)

14 Horizontal dial, found on the foreshore of the River Thames. 16th century. (78.150)

15 Portable horizontal dial (incomplete). Ivory, diameter 1⅞in. From an excavation. (10.058)

16 Portable horizontal dial (incomplete). Ivory, octagonal, 1½in across. From an excavation. (Z66364)

17 Portable horizontal dial (incomplete). Wooden, diameter 1in. (78.283)

18 Upper tablet of diptych dial. Ivory, dated: *Anno 1561.* Excavated from site of Horn Tavern, south of St Paul's Cathedral. (TAV 82)

19 Diptych dial. Ivory. Excavated from Bridewell in 1978.

20 Lunar volvelle, with conversion scale between equal and unequal hours, and lunar calendar. Probably part of an astronomical compendium. 17th century. (A26828)

21 Porter's magnetic dial. Incomplete. (A11755)

22 Orrery. Engraved paper on circular wooden base, brass mechanism. Signed: *T. Blunt optician and mathematical instrument maker to His Majesty 22 Cornhill.*

23 Astronomical telescope. Pasteboard, the outer tube covered with green morocco with gold tooling, the five draw-tubes with green vellum. English, later 17th century. Closed length 32¼in, fully extended 84in. Loan from *Museum of Leathercraft, Walsall,* Inv. 156-51. (L75.2)

24 Spyglass. Ray-skin, ebony pasteboard and brass. Signed: *I. Cuff London. c*1740. (Z6189)

25 Spyglass. Silvered brass, black-painted outer tube. Royal coat of arms. In red leather case. Signed: *Adams, Fleet Street, London. c*1760. (27.196/2)
Lit: LONDON (1970) no. 272

26 Small telescope. Shagreen-covered tube. Signed: *Dollond, London.* Early 19th century. (37.59)
Lit: LONDON (1970) no. 274

27 Spyglass. Brass, signed: *G. and C Dixey, 3 Bond Street.* (A13272)
Lit: LONDON (1970) no. 273

28 Camera lucida. Early 19th century, unsigned. (A11778)

29 Camera lucida. Signed : *Cary London. c*1830. (A11777)
Lit: LONDON (1970) no. 277

30 Four lenses, each mounted in a brass cell.

31 Brass sector. Signed: *D.Wiglias 1626.* (46.78/582)
Lit: TANGYE no. 582

32 Circumferentor/Holland circle. Two fixed sights. Brass arms with compass box mounted at its centre. Compass card of paper with magnetic azimuth dial. Signed: *H Sutton fecit 1658.* (80.271/84)

33 Altazimuth theodolite. Signed: *Thomas Jones Charing Cross London.* (Loan from the *Science Museum, London*)

34 Plane table alidade. Brass, engraved with diagonal and other scales (lacks sight). Signed: *S. Saunders fecit.*

35 Napierian rods, cylindrical form. Boxwood. *c*1700. (10053)

36 Set of drawing instruments (dividers, parallel ruler, protractor, sector etc.). Signed: *Blunt London.*

37 Twelve-inch carpenter's rule (only four inches from hinge survive). 18th century. (81.468/2)

38 Single-handed chart dividers. (81.541.4)

39 Set of surgical instruments in case. Presented to the Royal College of Physicians by Dr Francis Prujean in 1653. Loan from the Royal College of Physicians, London. (L180/1)
Lit: THOMPSON (1927)

40 Compound microscope, Hooke type, with modified stand and pillar. Early 18th century. Loan from the *Wellcome Museum, London.* (L68.1)

41 Watch and clockmaking tools, including clockmaker's screwcutter, and watch and clockmaker's lathe.

42 Alembic. Pottery, broken spout. Excavated at Fenchurch Street. (17,419)
Lit: MOORHOUSE (1972)

43 Alembic. Pottery, in Raeren ware (Germany), 16th century. Excavated at Surrey Street. (A4784)
Lit: MOORHOUSE(1972)

Lo 14 LONDON
Museum of the Pharmaceutical Society of Great Britain

1 Lambeth High Street, London SE1 7JN

071 735 9141

The Society was founded in 1841. As well as possessing an extensive library, it has a Museum which contains a collection of crude drugs of animal and vegetable origin used in the 17th century (mainly presented by the Royal College of Physicians), drug jars, leech jars, mortars, medicine chests, dispensing apparatus and prints and manuscripts relating to pharmacy. In addition there is a small number of scientific instruments – mostly microscopes – which are listed below.

1 Screw-barrel microscope of brass with ivory handle, contained in a velvet-lined, fish-skin case, accompanied by a compass microscope. The screw-barrel microscope is signed: *E. Culpeper fecit.*

2 Culpeper microscope. Ray-skin outer tube edged with scalloped brass bands, highly curved legs, rectangular base. Inner tube vellum covered. Ebony moulding between brass eyepiece and inner tube. Mid-18th century.

3 Culpeper microscope. Brass, on rectangular base. Strongly curved legs terminating in a scroll at top and bottom. Unsigned. Mid-18th century.

4   Jones 'Most Improved' microscope. Signed: *W. & S. Jones, 30 Holborn London*. Incomplete, the eyepiece section having been replaced, probably in the latter part of the 19th century.

5   Cary type microscope screwing into lid of red leather box. Accompanied by leaflet: *Description of a new improved pocket compound microscope*.

6   Drum microscope in box. Signed: *Abraham Optician Bath*. Early 19th century. (fig 80)

7   Scale of chemical equivalents, engraved paper on wood, signed: *Published by W. Cary 182 Strand Jan. 1. 1814*.
Lit: WOLLASTON (1814)

Lo 15 LONDON
National Army Museum

Royal Hospital Road, Chelsea, London SW3 4HT

071 730 0717

The Museum covers the history of the British Army from the reign of Henry VII, the Indian Army until 1947 and other Commonwealth countries up to independence. Opened in 1960 at the Royal Military Academy Sandhurst, the Museum's main displays have, since November 1971, been on view in a new building adjacent to the Royal Hospital Chelsea. A detachment, including the reserve collections of the Uniform and Art departments, remain at Sandhurst. The collections include paintings, prints, drawings, uniforms, medals, weapons, personal mementos and equipment, including a number of scientific instruments. Some are associated with particular commanders or campaigns. In addition to the items listed below, the collection contains a number of more recent items, including binoculars and late 19th- and early 20th-century artillery instruments, many with personal associations.

1   Hypsometer, used by Colonel Sir Sidney Burrard, Royal Engineers. *c*1890. (7710-32)

2   Watkin mirror clinometer, signed: *Elliott Brothers, London*, owned by Captain FHW Sherwin, Royal Inniskilling Fusiliers. (6310-2-9)

3   Clinometer, signed: *Charles Baker, London*, owned by Field Marshal Frederick Roberts, Royal Artillery, *c*1860-1868. (6602-54)

4   Clinometer, signed: *Short and Mason, London*, owned by Lieutenant AWW White, Hampshire Regiment, 1899. (7408-30-11)

5   Prismatic compass, signed: *Savage and Son, Sandhurst*, used at the Royal Military College, Sandhurst, 1864. (6006-19)

6   Theodolite, signed: *Cary, 181 Strand, London*, *c*1850. Used by Earl Kitchener in Western Palestine, 1872-77. On loan from the *Royal Engineers Museum, Chatham*.

7   Gunner's callipers, brass, Irish. Signed: *Spear, 23 Capel Street, Inst Maker to His Majesty's Ordnance*. Closed length 7in. Owned by Lieutenant RJ Dacres, Royal Artillery, 1817. (6106-45)

8   Shot gauge used in the Royal Gun Factories, *c*1880. (18202-35)

9   Protractor, brass, radius 2½ in, inscribed with scale of polygons.

10   Pocket quadrant, radius 2⅛ in, signed *T. Jones, Charing Cross, London*, owned by Lieutenant-Colonel Charles Napier, Monmouthshire Light Infantry, 1813. (6310-208)

11   Sextant, signed: *Troughton and Simms*, owned by Captain R Biddulph, Royal Artillery, *c*1860. (6012-67)

12   Weldon's patent rangefinder, *c*1885. (7512-115)

13   Mathematical drawing instruments, silver and ivory, protractor signed: *T. Heath Fecit*, sector signed: *Bennett London*, early 18th century, in black leather case. Box lid bears silver plaque with arms of member of Marlborough family. Transferred from the Royal United Services Institution. (6310-302)

14   Horse-measuring gauge, owned by Lieutenant Colonel S Longhurst, Veterinary Corps, 1877. (6007-190)

15   Artillery rule, 1832. (6403-58-1)

16   Artillery set square, brass, signed: *Lasnier, Paris*. Engraved with scales of Rhineland foot and *Pied de Roy*, *c*1867. (6504-43-2)

17   Telescope, 5 draw, brass and mahogany, signed *Matthew Barge, London*. Owned by the Duke of Wellington at Waterloo (1815) and given to Sir Robert Peel. Lent by the *Royal Armouries, HM Tower of London*. (7009-33)

18   Telescope, 3 draw, signed: *Dollond, London*. Belonged to Major General Sir Henry Havelock at the Indian Mutiny (1857). Transferred from the Royal United Services Institution. (6309-71)

19   Telescope, signed: *Dollond, London Day & Night*, owned by Lieutenant-Colonel WL Beresford, The Bedfordshire Regiment, *c*1830. (5012-51)

20   Telescope, signed: *S. and B. Solomons*, owned by Field Marshal GJ Wolseley, Light Infantry, *c*1859. (5210-64-2)

21   Telescope, signed: *Thomas Jones, 62 Charing Cross, London*, early 19th century. Owned by Major-General Sir AG Woodford in *c*1859. (6310-213)

22   Telescope, brass. Inscribed: *Henry Alfonso Henri RN. Found on the Battlefield at the relief of Lucknow November 1857*.

23   Telescope, signed: *A. Ross, London*, mounted on a frame rifle stock for use with one arm only, owned by Field Marshal Lord Raglan (whose arm was amputated in 1815), *c*1850. (6310-214)

24   Telescope, signed: *Abraham, Bath*, owned by Colonel Sir John Moore, Oxfordshire Regiment, *c*1820. (6310-217)

25 Telescope, signed: *Cary, London,* owned by Major T Nicholson, Madras Native Infantry, *c*1840. (6603-13-3)

26 Telescope on stand, signed: *W and J Gilbert, London,* owned by Field Marshal Viscount Gough, *c*1850. (6605-14)

27 Telescope, signed: *C W Dixey,* owned by Captain J Laye, Rutlandshire Regiment, *c*1845. (8209-82)

28 Wheatstone automatic telegraph, used by Royal Engineers (Signals Section), *c*1890. (6005-203)

29 Photographic chemical set, owned by Field Marshal HH Kitchener, Royal Engineers, 1876. (6604-30)

30 Schmalcalder's patent compass, signed: *Schmalcalder's Patent, 82 Strand, London,* diameter 2¾in. Post 1812.

31 Marine barometer, signed *Berge London Late Ramsden.* Used by the Duke of Wellington (1769-1852) during the campaign in Spain and Portugal (1804-1814). A metal plaque dated 1832 records: *This barometer, the companion of the Duke of Wellington in the Peninsular was presented by his Grace to Commodore Sir George R. Collies RN while co-operating with the Army on the coast of Spain, and was given by his family to his friend P. Miller MD Knowle Cottage 1832.*

32 Three sets of surgical instruments, signed: *Savigny and Co., Evans and Co. London* and *Ferris and Company, Bristol* respectively.

Lo 16 LONDON
National Maritime Museum

Romney Road, Greenwich, London SE10 9NF

081 858 4422

The Museum was established by Act of Parliament in 1934, and opened in 1937. Its collections include maritime and marine paintings, prints and drawings, ships models, medals and seals, uniforms, weapons and relics, as well as an important library. Of great importance from the point of view of this survey is the extensive collection of astronomical, navigational and other scientific instruments housed in the Department of Navigation and Astronomy. The navigation instruments are exhibited chiefly in the Navigation Room, much of the astronomical material in the Old Royal Observatory, but as would be expected with a collection of this size, only a part is on display.

In the summary which follows, Section I indicates the major sources from which the collection is derived, and Section II the scope of the navigation and astronomy collections, under the headings which appear in the Greenwich Inventory. Section III refers briefly to other catagories of instruments, mainly derived from the Gabb Collection, which, since they are not related to navigation and astronomy, do not appear in the Greenwich Inventory.
Lit: (a general guide) HILL/PAGET-TOMLINSON (1958); HOWSE (1975); LONDON (1971) (References below to this inventory are made in the following way: *GREENWICH INVENTORY* Section... Reference must be made to this work for details of instruments and makers represented in the collection.)

I SOURCES OF THE COLLECTION

1 *Old Royal Observatory.* The collection contains a group of instruments used at the Old Royal Observatory from its foundation in 1675, which have been displayed, as far as is possible, in their original working settings. They include material associated with all Astronomers Royal, ranging from the Angle Clock constructed by Thomas Tompion for John Flamsteed, to an eight-foot quadrant made for Edmund Halley in 1725 by Jonathan Sisson working under the supervision of George Graham; James Bradley's zenith sector of 1727; his quadrant and telescope by John Bird; a mural circle (1812) and transit telescope (1816) by Edward Troughton; George Biddle Airy's transit circle of 1851 and more recent instruments.
Lit: GREENWICH INVENTORY Section 35

2 *Barberini Collection.* A second small but very important group of instruments formed part of a collection said to have been assembled by Cardinal Francesco Barberini (1597-1679). It consists of armillary spheres, astrolabes, sundials, surveying and mathematical instruments, dating between 1600 and 1630.
Lit: GREENWICH INVENTORY Section 34, Barberini Collection. PRICE (1954) lists these instruments as well as some others which are now in the *Science Museum, London* and the *Museum of the History of Science, Oxford.*

3 *Herschel Collection.* The Museum houses a large quantity of material – instruments, a few manuscripts and other items – which, although acquired from a number of sources are known to have belonged to Sir William Herschel and to his son Sir John.
Lit: GREENWICH INVENTORY Section 34, Herschel

4 *Harrison Collection.* Books, portraits and timekeepers relating to John Harrison, as well as Larcum Kendall's copies of Harrison's fourth timekeeper.
Lit: GREENWICH INVENTORY Section 34, Harrison

5 *Sir James Caird.* Many important items in the collection were presented by the Museum's chief benefactor, Sir James Caird, or were acquired after his death through the Caird Fund.

6 *Admiralty Compass Observatory.* A collection of several thousand compasses and associated material from this source were transferred to the National Maritime Museum in 1969. (There is also material from this source in the *Royal Museum of Scotland (Chambers Street), Edinburgh.*) A small number of items from the same source were deposited before this date, and are included in the inventory, but the great majority await cataloguing, although where the maker is known, they are listed in the index of the *Greenwich Inventory* under the maker's name.

7 *Other sources.* The National Maritime Museum preserves instruments from the Gabb, Adams, and Lord Stokes collections, as well as single instruments or small groups of items from other sources. Material on loan includes items from *St John's College, Cambridge, The Castle Museum, York, Science Museum, London,* the *Museum* *of the History of Science, Oxford* (including some instruments used in the Radcliffe Observatory, Oxford).

II SCOPE OF THE NAVIGATION AND ASTRONOMY COLLECTIONS

The cataloguing of instruments has been taken from the *Greenwich Inventory*, and the numbering of these categories corresponds with that of the relevant section in the Inventory, to which reference must be made for details of instruments and makers represented in the National Maritime Museum's collection.

1 *Armillary spheres.* More than 20 instruments including five of French or Italian origin from the Barberini collection (q.v.). The earliest, by Caspar Vopel, is dated 1543, the latest *c*1870.
Lit: WATERS (1964)

2 *Artificial horizons.* Most main types represented.

3 *Astrolabes.* One of the largest collections of these rare instruments, containing more than 20 Western astrolabes, nearly 30 oriental ones, and mariner's astrolabes. Many of the astrolabes were acquired through the generosity of Sir James Caird during the years 1935-1939. The major types of astrolabe are represented: Roias' orthographic projection, Gemma Frisius's stereographic projection and the normal stereographic projection from one of the poles to the plane of the equator.

4 *Astronomical and optical instruments (excluding portable telescopes).* Primarily instruments or accessories (lenses, furniture) associated with the Old Royal Observatory or with William or John Herschel. Included are altazimuths, equatorial sectors, transits, mural and zenith instruments, quadrants and reflecting and refracting telescopes. In addition such items as burning glasses, a camera lucida, lens polishing instruments and a lathe are listed.

5 *Backstaves.* Sixteen examples, including a specimen of Elton's quadrant.

6-7 *Charts.* In manuscripts and printed, celestial and terrestrial.

8 *Circles.* Reflecting and repeating, Borda and Troughton type.

9 *Clocks and watches.* Primarily regulators and marine chronometers, but also domestic clocks of various types, as well as watches. Particularly notable are Harrison's four marine timekeepers, for longitude determination at sea, dating from 1728-1759. The last won a prize of £20,000, offered by the British Government.
Lit: GOULD (1935)
   Many of the items in this section are on loan from the Hydrographic Department of the Ministry of Defence.

10 *Compasses.* Primarily nautical (fig 24), but also including aircraft compasses, surveying and mining compasses and compass cards. The earliest compass is of late 16th-century type. This section of the inventory, covering more than 200 items, does not include most of the compasses from the Admiralty Compass Observatory, which are in a subsequent listing.

11 *Computing instruments.* Napierian rods, Gunter scales, sectors, gunner's callipers, slide rules, tide computers, as well as other instruments able to fulfil computing functions, are included.

12 *Cross staff.* The collection contains only one example of a cross staff, a navigational instrument which rarely survives. In this case its survival is due to the fact that it is made of ivory and was part of a presentation set which was never put to practical use. (fig 26)

13 *Depth sounders.* A representative collection showing the mechanical development of the depth sounder.

14 *Dip circles.* 18th-century example by Nairne and others of 19th-century date.

15 *Drawing instruments.* Parallel rulers, protractors, station pointers, beam compasses, camera lucida, dividers, drawing instrument sets, ellipsographs, pantographs, rulers, scales, set squares.

16 *Globes.* This is probably the largest single collection in the world, much of it having been purchased by Sir James Caird, and presented in the 1930s. A list of these globes was published by WATERS (1964), but the more up-to-date list is found in the *Greenwich Inventory*.

17 *Hour glasses*

18 *Lodestones.* Estimated dates between the 16th and 18th century.

19 *Logs, ships'*

20 *Meteorological instruments.* Barometers, anemometers, thermometers, sunshine recorders, rain gauges, hygrometers.

21 *Oceanographic instruments*

22 *Octants.* Nearly 100 instruments representing all the main types.

23 *Orreries.* 23 examples, dating between 1730 (Richard Glynne) and 1890.

26 *Quadrants.* Hand quadrants for time measurement and surveying, including several Gunter's quadrants, horary quadrants, Islamic quadrants, and a mid-17th-century sinical quadrant.

27 *Radio aids,* post 1940, and *rangefinders.*

28 *Sextants.* An extensive collection, largely but not entirely by British makers, many sextants from the Adams and the Dryad Collections.

29 *Sundials.* Very extensive collection (figs 27 and 40) including numerous astronomical compendia.

30 *Surveying.* A wide range of instruments, including interesting examples of late 16th-century and early 17th-century date, some of them from the Barberini collection. Includes alidades for use with plane tables, circumferentors, clinometers, surveyor's crosses, geometric squares, graphometers, levels (18th century), recipiangles (17th century), a plane table (*c*1600), theodolites (16th-20th century), a trigonometer of Danfrie (late 16th century).

31 *Telescopes, portable.* Reflecting and refracting, floor or table mounted, 18th-19th century date, with telescopes of English manufacture particularly well represented. Also included are telescope objectives and mirrors.

32 *Telescopes, hand.* An extensive collection of hand telescopes of all types, including a refracting telescope dated 1661.

33 *Traverse boards*

III *INSTRUMENTS NOT INCLUDED IN THE NAVIGATION AND ASTRONOMY COLLECTIONS*

In addition to the instruments which have been listed in the *Greenwich Inventory*, the National Maritime Museum also possesses a number of other instruments which belong to the main groups listed below. To a large extent these are derived from the Gabb collection.

1 *Computing instruments.* These range from a spiral slide rule dated 1655, and a set of Napierian rods, to slide rules designed to fulfil various specialised purposes, dating from the 18th and early 19th centuries. They include a Coggleshall slide rule, Ewart's cattle gauge, and timber merchant's, engineer's and excise officer's slide rules.

2 *Hydrometers.* Ranging from ivory hydrometers of early 18th-century date, and a Clarke's hydrometer (*c*1755), to specific gravity beads and hydrometers of early and mid-19th century date. Also included are saccharometers, a salinometer, lactodensometer and an acetometer. (A number of hydrometers from this source have been transferred to the *Science Museum, London.*)

3 *Microscopes.* A range of microscopes, predominantly of 18th-century or early 19th-century date, and almost exclusively of English origin. Included are many of the main types of microscope, with instruments by Scarlett, Culpeper, Benjamin Martin, George Adams, Clarke of Edinburgh and Francis Watkins. Important early items are a Marshall type microscope, and a microscope of Campani type of *c*1670.

4 *Optics (excluding microscopes (III/3) and telescopes (II/31-2)).* Optical instruments include drawing aids such as a 19th-century camera obscura, a camera lucida and Claude Lorraine glasses. There are mirrors, prisms, a kaleidoscope, several jealousy glasses, anamorphoses, spyglasses and binoculars, as well as many spectacles and faceted lenses for multiplying images.

5 *Other instruments.* An aeolopile may be dated to the late 17th century, while an air pump by Edward Nairne belongs to the third quarter of the 18th century. A chemical balance in a mahogany case may be dated to *c*1790. In addition there are a number of balances (diamond scales, coin testers, money changer's scales), thermometers (19th century) and several examples of Doebereiner's lamp.

Lo 17 LONDON
Royal Botanic Gardens, Kew

Richmond TW7 3AB

081 940 1171

The Botanic Gardens date back to 1759 when Princess Augusta established a botanical garden in the grounds of her private dwelling, Kew House. Her son, King George III, who inherited her tastes, occupied Richmond Lodge, adjacent to Kew House, and the two gardens were subsequently joined together. Princess Augusta and King George III were aided in their venture, first by Sir Joseph Banks and then by the Earl of Bute. In 1841 the gardens were handed over to the State, and greatly extended in area by the addition of the gardens and parklands of two other adjacent properties. Sir William Hooker (1785-1865), the eminent botanist and previously Professor of Botany at the University of Glasgow, was appointed as the first Director in 1846. He was succeeded, on his death in 1865, by his son, Sir Joseph Dalton Hooker (1817-1911), and it is to these two directors that the Gardens owe much of their present form, as well as their reputation as the centre of the botanical world. Scientific instruments are to be found in two departments in the Gardens: the Herbarium and the Jodrell Laboratory. Those in the Herbarium are in the main associated with eminent botanists including the Hookers. Those in the Jodrell Laboratory were mostly acquired for the Laboratory during the last quarter of the 19th century (more recent material has not been listed). The locations of the instruments listed below are indicated by J (Jodrell Laboratory) or H (Herbarium).

1 Simple microscope. Silver with leather-covered box which serves as a base for the microscope pillar. Believed to be a copy, made for the botanist Robert Brown (1773-1858), of the microscope used by Louis Claude Marie Richard (1754-1821) when working on his *Demonstration Botanique au Analyse du Fruit* (Paris 1808). (H)

2 Dissecting microscope. Rack focusing, circular pillar screwing into lid of fish-skin covered box. Signed: *A. Abraham Liverpool.* Used by Dawson Turner FRS. (H)

3 Simple microscope. Brass, in mahogany box. The microscope pillar screws into the top of the box. Trade card. Signed: *Banks, no. 441, Strand London Maker to H.R.H. Prince of Wales.* Used by Sir William Hooker in describing the *British Jungermanniae* and *Musci Exotici.* Presented by Sir JD Hooker. (H)

4 Dissecting microscope. Circular pillar screwing into top of wooden box. Signed: *Bancks & Son, Insr^t Makers to his Majesty.* Used by George Bentham FRS (1800-1884) and presented by Sir JD Hooker. (H)

5 Dissecting microscope with pentagonal mahogany box. Signed on microscope: *H. Powell London,* and on a trade card in the box: *H. Powell, Mathematical & Optical Instrument Maker, 24 Clarendon Street, Somers Town, London.* Used by Professor JS Henslow, 1833. (H)

6 Dissecting microscope. Circular pillar mounted on

circular brass base, rack focusing. Unsigned, early 19th century. Said to have been made for Brown, as a copy of Richard's microscope (but cf. no. 1). Appears to be English, early 19th century.

7 Simple and compound microscope. Brass, unsigned, in wooden box with accessories, as well as bottles of dyes, tweezers, etc. Used by NE Brown (1849-1934). (H)

8 Compound microscope, Cary type. Presented 1859. (H)

9 Camera lucida. Brass, unsigned, early 19th century. (H)

10 Camera lucida for microscope, in leather box inscribed *Zeichenapparat nach Abbe*. Signed: *Carl Zeiss, Jena*. (J)

11 Dissecting microscope with camera lucida. Signed: *J. Swift & Son London*. (J)

12 Compound microscope and accessories. Signed: *Carl Zeiss Jena 19290*. Used by DH Scott (1854-1934), first Keeper of the Jodrell Laboratory.

13 Rotary microtome. Unsigned. ? Last quarter 19th century. (J)

14 Rocking microtomes. Signed: *Camb. Sc. Inst. Co. 1892*. Used by DH Scott. (J)

15 Single prism spectroscope on tripod stand. Signed: *A. Hilger London*. Acquired 1878. (J)

16 Chemical balance in glazed wooden case. Signed: *L. Oertling London*. Acquired 1878. (J)

17 Thermometer (broken). Signed: *J. Newman 122 Regent Street, London*. Used by Sir JD Hooker in the Himalayas, 1847-50. (H)

18 Mason's hygrometer, and compass in circular brass case with lid. Signed: *Thomas Jones 62 Charing Cross*. Belonged to Dr George Gardner, and used by him during his travels in Brazil and Ceylon. (H)

19 Sympiesometer. Signed: *Patent. Adie & Son Edinburgh No. 1929*. Used by Mr Hewett Cottrell Watson to determine the altitude at which British flora were found. Can be dated from maker and serial number as being *c*1843-9. (H)

20 Barometer. Kew pattern. Signed: *Casella Maker to the Admiralty and Ordnance, No. 256*. Checked against barometer at Kew Observatory in 1879. (J)

21 Aneroid barometer. Signed: *Negretti & Zambra London 5091*. (Housed in chronometer case used by Sir JD Hooker on the voyage of the *Erebus* 1839-43). (H)

22 Clinometer and compass. Consists of compass in rectangular mahogany box with lid. A quadrant is engraved on the inner side of the lid, and would have been equipped with a plumb bob. Designed by Professor Henslow and used by him in India, Morocco and Syria. Presented by Sir JD Hooker. (H)

23 Clinometer and compass. As no. 22, save that the quadrant inside the lid is of brass and professionally engraved. The compass has been equipped with a bubble

level. Signed: *Cary London*. Used by Sir JD Hooker in India, Syria and Morocco. (H)

24 Instrument used by Sir JD Hooker in the Himalayas (1847-50). Consists of brass rule with inset darkened glass panel, of varying degrees of darkness. Silver scale inset along one edge of rule. A slide with silver vernier scale is equipped with an observation slit, and a screw thread operated by a handle for fine adjustment.

Lo 18 LONDON
Royal Geographical Society

1 Kensington Gore, London SW7

071 589 5466

The Society, founded in 1830, contains an extensive collection of geographical books as well as a considerable number of maps. In addition it houses a collection of scientific instruments, many of which are associated with distinguished explorers or were used on a particular expedition. Some of the instruments are of considerable interest in their own right. The more important instruments are listed below, but some of the later compasses, telescopes, etc. have been omitted.

1 Single-draw spyglass with ray-skin outer tube. Length 3¾in. Signed: *Dollond London*.

2 Pocket terrestrial and celestial globe, inscribed on the terrestrial globe: *A new and correct globe of the earth by I. Senex F.R.S.*. Diameter 3¾in.

3 Miniature terrestrial globe, diameter 2¾in. Unsigned, *c*1800.

4 Orrery. Mechanical parts of brass, base of wood and paper. Signed: *Sold by P. and J. Dollond London*. Late 18th century.

5 Astrolabe, diameter 8⅞in. Italian, *c*1500.
Lit: GUNTHER (1932) vol. 2, p328, no. 174; PRICE (1955) no. 174

6 Butterfield dial. Brass in leather case. Signed: *Butterfield AParis*.

7 Pillar dial, boxwood and brass. Unsigned. ? 17th century.

8 Diptych dial. Wood and brass. Unsigned. English, early 19th century. Used by Sir Roderick Murchison.

9 Diptych dial. Chinese, 19th century.

10 Universal equinoctial dial. Signed: *And. Vogler* (Andreas Vogler, Augsburg). Brass, partly silvered, in leather case.

11 Equinoctial ring dial. Brass, partly silvered, 9in diameter. Signed: *C. Lincoln London*.

12 Nocturnal. Boxwood, England, 17th century.
Lit: TAYLOR/RICHEY (1962) fig 22

13 6in Gunter's quadrant. Boxwood. Signed: *W. & S. Jones 30 Holborn London*.

14 Compass, signed: *S.J. Browning Portsmouth*. Used by Sir Edward Parry on expedition to the North Pole in 1827.

15 Compass. Chinese, wood, varnished in red and black ink, diameter 9¹¹⁄₁₆in.

16 Compass. Japanese, in wooden case.

17 Kamal (position-finding instrument of Arab pilot). *Lit:* TAYLOR/RICHEY (1962) fig 16

18 Backstaff. Signed: *Made by Daniel Scatliff at New Stairs for Jonathan Cape 1727*.

19 Octant. Mahogany frame and index arm, boxwood arc with diagonal scale radius 20in. Signed: *John Urings fecit London*.

20 Octant. Ebony frame, brass arm, radius 18in. Maker's plaque missing. Third quarter 18th century.

21 Sextant. Brass with silver degree scale. Signed: *Troughton and Simms London no. 2932*.

22 Sextant. Brass with silver scales. Radius approx 7⅜in. Signed: *Henry Hughes 59 Fenchurch Street London*.

23 Pocket sextant. Signed: *Troughton and Simms London*. Mid-19th century.

24 Pocket sextant used by Sir Clements Markham in Peru. Signed: *Troughton and Simms London*.

25 Pocket sextant, signed: *Newman 122 Regent St. London*. Belonged to Charles Darwin and was used by him on the *Beagle*, 1831-6.

26 Artificial horizon, signed: *King, Optician, 2 Clare Street Bristol*.

27 Circumferentor, brass, partly silvered, diameter 12in. Signed: *Geo. Adams London*. Believed to have been used by Jeremiah Dixon in the Mason-Dixon survey of the Maryland-Pennsylvania boundary, 1763-9.

28 Theodolite, signed: *Cox, London*. Used by HH Kitchener (Earl Kitchener of Khartoum) during the 1882 survey of Cyprus.

29 Lendy's patent topograph, *c*1865. Incorporates prismatic compass, plane table, levels, clinometer.

30 Protractor and diagonal scale. Brass, 4 x 4in. Signed: *H. Sutton fecit 1660*. Contained in a copy of J Collins *Geometricall Dyalling* (London 1659).

31 Excise slide rule. Signed: *Edwd. Roberts Maker in Grocer's Alley in Old Jewry London* and inscribed: *Henry Bladen 1756*. Boxwood, length 10in.

32 Slide rule, 14in. in length, signed: *Dring & Fage Makers No. 20 Tooley St. London Bridge*.

Lo 19 LONDON
Royal Institute of British Architects

21 Portman Square, London W1H 0BE

071 580 5533

In addition to its outstanding collection of architectural designs and prints, the RIBA holds a number of scientific instruments in its Drawings Collection which have been presented from time to time by member architects. These include a number of items of considerable interest which were exhibited at the Case Heinz Gallery of the RIBA in 1982 (see HAMBLY (1982)).

1 Case of instruments of 16th- to 18th-century date, contained in a flat leather covered box lined with calf, 12¼ x 10½in. The case appears to be German, of early or mid-18th century date, and was clearly made for an existing collection of instruments. (fig 42)

  a Universal equinoctial dial. Brass, part gilt, octagonal. Signed: *LTM* (Ludwig Theodor Muller, Augsburg). Second half 18th century. This instrument is probably a later addition to the set.

  b Sector. Brass, closed length, 6⅛in. Engraved inside the arms of the sector: *Daniel Chorez F.p.*; French, by Daniel Chorez, early 17th century. Chorez constructed the sector described by Didier Henrion in his *Usage du compas de proportion* (Paris 1618).

  c Surveying or gunnery instrument. Consists of double rule inside which can be folded two smaller, pointed rules. One is graduated from 0 to 100; the other lacks graduations, but can be held at right angles to the double rule by means of a small arm. Brass, signed: *Marcus Purman Fecit Monarchio 1598*.

  d Gunner's quadrant. Gilt brass, height 2⅛in. With a sliding sight movable in the vertical plane, and a clinometer (plummet missing). Possibly German, early 17th century.

  e Gunner's quadrant. Gilt brass, with iron plummet. The graduated arc is surmounted by the figure of a stag from whose horns the plummet is suspended. On the back of the instrument a circular dial with a steel pointer is graduated from 1 to 12. Possibly German, late 17th or early 18th century.

  f Parallel ruler. Brass. Letter N impressed in one rule. 18th century.

  g Parallel ruler. Brass, length 9⁹⁄₁₀in. 18th century.

  h Set square. Brass, maximum length 4in. 17th/18th century.

i Diagonal scale. brass, length 7¼in. The back of the scale is engraved with scales comparing 'Etliche Halbe Werckschueh' – 6in scales according to the systems prevailing in the Rhineland, Vienna, Nuremberg, Augsburg, Strasbourg and Munich.

j Protractor. Brass, diameter 4in. German, probably early 18th century.

k Compasses with one adjustable and jointed leg; two pairs of dividers; jointed pencil holder for compass; tracing wheel attachment.
Lit: HAMBLY (1982) pp42, 43

2 Dividers. Wood with silver mounts and steel points. Length 12½in.

3 Set of drawing instruments in green ray-skin case. The contents include an ivory protractor and scale signed: *Watkins & Hill Charing Cross*. Boxwood scale, ivory parallel ruler, compasses, etc. Inscribed on base of case: *Thos Farnolls Pritchard Arch*. *Salop.* (Pritchard (d. 1777) was responsible for the initial design of the Iron Bridge over the River Severn at Coalbrookdale.)
Lit: HAMBLY (1982) p43

4 Set of drawing instruments in green ray-skin case with silver mounts. Incomplete, but includes square protractor with diagonal scales of ivory, signed: *Wm. Elliott Gt. Newport Street London.* 19th century.

5 Set of drawing instruments in green ray-skin case (now incomplete). Inscribed on case: *John White Esq.* Late 18th century.
Lit: HAMBLY (1982) p43

6 Architectural sector. Silver, length of arms 12in. Signed: *Made by Geo Adams Mathl Instrut Maker to His Royal Highness the Prince of Wales, Fleet Street London.* A wide ivory arc can be fixed at any point by means of fixing screws. The arc is engraved with scales relating to the proportions of the Tuscan, Doric, Ionic and Corinthian orders. This type of sector was illustrated and described by Joshua Kirby in *The Perspective of Architecture* (London 1761) (fig 120).
Lit: HAMBLY (1982) pp36, 37

7 Boxwood slide rule. Length 2ft. Equipped with two slides. Inscribed: *Fred^k A. Sheppard Patentee Stanley Great Turnstile Holborn London.* Second half 19th century.

8 Ellipsograph. Brass. Signed: *Farey London No. 22,* *c*1814. (fig 52)
Lit: HAMBLY (1982) p23

9-10 Volute compasses designed by Professor Stevens to draw volutes of the form of those on the Temple at Priene, and the Erechteum. Presented 1958 by Professor Stevens.
Lit: HAMBLY (1982) p24

11 'Teleconographe'. An instrument to permit the accurate drawing of distant and inaccessible objects, consisting of a development of the camera obscura with a prism in front of the lens which permits the enlarged image to be projected on to a sheet of paper on a drawing board. The instrument was invented by Revoil and described by Viollet-le-Duc in 1869. Signed:

Teleconographe H^y Revoil Brevete SGDG L.Lefebvre Constr 54 Rue des Tournelles, Paris.

Lo 20 LONDON
Royal Institution of Great Britain

21 Albemarle Street, London W1X 4BZ

071 409 2992

The Royal Institution was founded in 1799 by Benjamin Thompson (Count Rumford) with the object of 'diffusing the knowledge and facilitating the general introduction of useful mechanical inventions and improvements, and for teaching by courses of philosophical lectures and experiments the application of science to the common purpose of life'. It was one, but by far the most distinguished, of a number of institutions formed with the idea of extending the benefits of science to the poorer classes.

The object of educating the poorer classes was soon dropped from the activities of the Institution, but it has consistently fulfilled its other aim of diffusing scientific knowledge, and some of its professors were among the greatest names in British science, notably Humphry Davy and Michael Faraday. Other distinguished Professors were Thomas Young, John Tyndall, Lord Rayleigh, James Dewar and Sir William Bragg. As a result of the strong sense of tradition characteristic of the Royal Institution, apparatus associated with all these scientists is preserved, as well as a number of interesting items presented to, or acquired by, the Royal Institution during the course of its long history. Notably, there are several items associated with Henry Cavendish (1731-1819), including a balance, eudiometers and a thermometer calibrated by Cavendish himself in 1779.

One group of material which deserves mention, though chronologically outside this survey, is the Spottiswoode Collection, presented to the Royal Institution in 1899. (*Lit:* LONDON – ROYAL INSTITUTION (1899)). It consists mainly of electrical and optical apparatus, including electromagnets, induction coils, a Toepler machine, optical projection apparatus, spectroscopes, polariscopes, Nicol prisms, etc. dating from the 1870s. William Spottiswoode, a publisher by profession, but with a great interest in mathematics and physics, was Honorary Secretary of the Royal Institution and delivered a series of lectures there.

The instruments and apparatus in the Royal Institution are listed under the headings mechanics and mechanical models, hydrostatics, surveying, chemistry, heat, sound, electricity and magnetism. The list is preceded by details of the work of the main scientific figures at the Royal Institution, and an indication of the nature of the surviving instruments and apparatus connected with each.
Lit: JONES (1871) for history of the Royal Institution up to 1829; MARTIN (1961) for a brief history to 1950; BERMAN (1977) for a critical view of the period 1799-1844; BRAGG (1959); KING (1973) for Faraday's work and apparatus.

I ASSOCIATIONS

1 *Benjamin Thompson, Count Rumford (1753-1814).* American by birth, Rumford, arrived in London in 1798 as representative of the Elector of Bavaria. Through his interest in social welfare and the practical application of science (he studied fuel economy in heating and cooking, producing designs of stoves and cooking utensils), he proposed, and succeeded in establishing, in association with the Society of Bettering the Condition and Increasing the Comforts of the Poor, the Royal Institution. His association with it was short, however, and he moved to Paris in 1801. The Royal Institution still possesses models of some of Rumford's stoves and flues, as well as a thermometer known to have belonged to him.

2 *Thomas Young (1773-1829).* Thomas Young was Professor of Natural Philosophy from 1801 until 1803. He had studied medicine and practised much of his life as a physician. He established a wave theory of light, and deciphered the Rosetta Stone. As Professor at the Royal Institution he gave a series of lectures which were published as *A Course of Lectures on Natural Philosophy* (London 1807). Some of the models and apparatus described in this work still survive.

3 *Sir Humphry Davy (1778-1829).* Davy became Lecturer, and later Professor of Chemistry, lecturing on tanning and the chemistry of agriculture. He investigated the chemical action of the electric current, decomposing water, potash and soda, also discovering chlorine and isolating the alkaline earth elements. Later (1815) Davy was asked to investigate the causes of explosions of fire-damp in coal mines, and as a result designed the miner's safety lamp. Various pieces of apparatus connected with Davy's work have survived, notably electrolysis apparatus and voltameters, and a series of trial versions of the miner's lamp. A scale of chemical equivalents is known to have belonged to him.

4 *Michael Faraday (1791-1867).* Faraday was appointed as laboratory assistant to Sir Humphry Davy in 1813 and thereafter spent the whole of his working life at the Royal Institution, as Superintendent of the House and Laboratory from 1821, and as Fullerian Professor in Chemistry from 1833. Apparatus connected with almost every major aspect of his work survives at the Royal Institution, much of it constructed by the London instrument maker Newman. In addition to instruments and apparatus connected with most of his major discoveries in the fields of electricity and magnetism, as well as apparatus connected with his chemical research, there are the products of some of his chemical experiments, such as alloys of steel prepared during his research on steel with James Stodart. (Other samples can be found at *Cuming Museum, London* and *Science Museum, London.*) There are also specimens of optical glass produced during his research on optical glass for the Royal Society in the late 1820s.

5 *John Tyndall (1820-1893).* Appointed Professor at the Royal Institution in 1853, he succeeded Faraday as Superintendent in 1867. Initially Tyndall carried out research on the radiation and absorption of heat by gases and vapours and investigated the scattering of light by fine particles suspended in the air, as well as the effects of the dust of the atmosphere in spreading bacterial infection. Some of his apparatus – a 'blue sky tube' and an astatic sine galvanometer among other items – still survives.

6 *Sir James Dewar (1842-1923).* From 1877 Dewar held the position of Fullerian Professor concurrently with that of Professor of Natural Philosophy at the University of Cambridge. He carried out his work on the liquefaction of gases at the Royal Institution.

7 *John William Strutt, third Lord Rayleigh (1842-1919).* After resigning from the position of Cavendish Professor at Cambridge, Rayleigh became Professor of Natural Philosophy in succession to John Tyndall. Together with Sir William Ramsay he discovered argon in 1894 and succeeded in producing it in quantity at the Royal Institution in 1895. Other apparatus used by Ramsay and Rayleigh is on loan to the *Science Museum, London* from University College, London.

II INSTRUMENTS AND APPARATUS

In the ensuing entries the association of the instrument, if any, is indicated in parenthesis, and this is followed by the inventory number allocated by the Royal Institution.

MECHANICS AND MECHANICAL MODELS

1 Inclined plane; the vehicles modern. (T Young)

2 Model to demonstrate the properties of levers, signed: *B. Hooke 159 Fleet St.* (T Young) B.11

3 'Model to demonstrate fallacies of belief of some supporters of perpetual motion'. This model consists of a circular mahogany disc with projecting rim, mounted on a turned wooden stand. Eight thin mahogany strips run from the rim so that they form an octagon around the centre of the disc. Each section formed by the edges of two strips and the rim of the disc contains a silver ball. The disc revolves freely in a vertical plane around its central point. (T Young)

4 Trajectory tube model. Consists of a graduated quadrant mounted on a turned wooden stand. A brass tube is pivoted at the centre of the quadrant and can be set at any angle to the horizontal. A trigger at the base of the tube is used to fire a small ball. The apparatus demonstrates that the horizontal distance travelled by the ball varies with the angle at which it is fired. (T Young) B.12

5 Atwood's machine, signed: *Fidler London*. Early 19th century.

6 Model anchor escapement. 19th century.

7 Model crown wheel escapement. 19th century

8 Model escapement, signed: *J. Bishop 4 North Audley St.*

9 Balance. Said to have been constructed by Harrison (Thomas?) to the instructions of Henry Cavendish, late 18th century. Glazed cabinet with double doors and matching table base. Iron beam, 19½in loup. There is a replica of this in the *Science Museum, London*.

Lit: STOCK (1969) pp11-12, where earlier literature is cited.

10 Balance. Double-cone beam of Ramsden type, 17in long. Glazed mahogany case with five drawers on a matching table base. Signed: Fidler London.

11 Universal joint of Hooke type. (T Young)

12 Mahogany paddle wheel for raising water, mounted in box. (T Young) B.14

13 Parts of a centrifugal machine of Nollet type.

14 Model of self-closing gate in wooden box inscribed: *Presented to the Royal Institution by TN. Palmer 1802.* L.77

HYDROSTATICS

15 Hydrostatic balance. Brass stand in three sections with extending stem. Steel balance, three brass pans, glass bucket and weight. Ht 19in. Signed: *W. Harris Holborn London.* First half 19th century.

16 Acetometer in mahogany box containing (a) lacquered tin containing hydrate of lime; (b) thermometer signed: *J. & P. Taylor Ltd;* (c) hydrometer, 70° marked on stem, silver cup for weights; (d) boxwood slide rule; (e) two glass flasks; (f) set of weights; (g) forceps; (h) spatula. Label on box: *No. 128 Revenue Acetometer J. & P. Taylor London.* Accompanied by description and instructions for use, dated 1818: *Description of the acetometer for determining the strengths of acetic acid, invented and made for the Revenue of the United Kingdom by John and Philip Taylor, 10 Bur Court, St. Mary Axe, with directions for its use.*

17 Hydrometer, signed: *J. Ronchetti 43 Market St[t] Manchester.* Accompanied by an engraving and a complete set of weights, *c*1830-40.

18 Hand hydraulic press, signed: *J. Bramah Inv. et Fect. London.* Used by Faraday to purify frozen benzene, 1825.

PNEUMATICS

19 Air pump. Single- and double-barrel mounted at either end of a 27in long table. The elevated plate is shared by both pumps. Signed: *Newman 122 Regent Street.*

20 Air pump. Single diagonally mounted barrel on wooden base, slightly elevated receiver plate. (T Young) B.5

21 Hand pump for exhausting and compressing. (Faraday) D.114

22 Hand pump on wooden base, used by Faraday for compressing his gases in liquefying experiments. D.115

23 Transferer. Stopcock and junction on turned base. A tube branches out on each side to two bell jar tables. (T Young) B.4

24 Pressure vessel with single-barrel diagonally-mounted pump. Wooden screws and frame clamp the bell jar in position. Early 19th century.

25 Bell jar with turned wooden stopper, the mouth closed by a piece of pigskin. (T Young) B.19

26 Gas manipulation globes for use with an air pump. (Davy)

27 Glass globe mounted on wooden base with a stopcock and brass cap for supporting materials in the globe. (T Young) B.17

28 Egg-shaped glass container on wooden stand with stopcock. (T Young) B.3.

29 Baroscope. Signed: *Fraser London.* (T Young)

30 Liege barometer. 18th/19th century.

31 Portable stick barometer. Signed: *J. Newman Lisle Street London,* serial number: *323.* L.29

32 Portable stick barometer. Unsigned. Serial number: *233.* Inscribed: *Sir H. Davy, afterwards Mr. Faraday 1814 H.B.J. 1867 J. Tyndall 1873.* L.29

33 Aneroid barometer, signed: *Construit par Vidi en 1857 Bourgeois Ing[r] Opt. 27 Rue des Pyramides Paris.* L.28

SURVEYING

34 Surveyor's level. Four-screw, bubble level, telescope. Signed: *Thomas Mason Dublin.*

CHEMISTRY

35 Metallic eudiometer used by Cavendish in experiments on the composition of water, *c* 1780. Presented by Sir Humphry Davy. C.1
Lit: BARCLAY (1937) part II, p10, no. 11

36 Eudiometer. Glass with metal stopcock and fixing for lid. signed: *Newman.* (Cavendish) C.2

37-8 Two eudiometers. (Cavendish) C.3

39 Eudiometer, glass. Labelled: 'Eudiometer formerly in the possession of Cavendish'. C.4

40 Boxwood rule with scale of chemical equivalents. Signed: *J. Newman Lisle Street London.* (Davy) D.59

41 Glass globe with brass stopcock, on mahogany base, used by Faraday for the preparation of 'chloride of carbon'. Stopcock signed: *Newman.*

42 Glass globe containing crystals of 'tri-chloride of carbon'. (Faraday) D.30

43 Gas analysis cells. (Davy) D.45, D.55

44 Gas volume measures (standard units of volume). (Cavendish) C.6

45 Globe for the preparation of argon in quantity, 1895. (Rayleigh). Facsimile in *Science Museum, London* discussed by BARCLAY (1937) part II, no. 58.

HEAT

46 Thermometer (spirit in glass) made for the Accademia del Cimento. L.26

47 Alcohol thermometer calibrated by Henry Cavendish, 1779.
Lit: MIDDLETON (1966) pp138-9

48 Space thermometer, brass end-piece, boxwood scale.

Signed: *Fidler London*. (Rumford) A.12

49 Air thermometer. (Davy) D.55

50 Thermometer. Mounted on silvered backplate in wooden case. Inscribed: *Prof. Smith's [?]labl^m Kiobenavn. For F. Daniell esq. from H.C. Schumacher*. A note in the lid of the box records that the thermometer was designed by Schumacher for Daniell.

51 Jacketed thermometer. Faraday's 'second Casella' thermometer referred to in his *Diary* (see MARTIN (1932-36), vol. VII, pp367, 371).

52 Jacketed thermometer control cell used in 1855 by Faraday in his magneto-crystallic research. D.120

53 Apparatus for displaying the effect of the radiation of heat from the top of a bell jar. (Davy) D.54

54 Crookes' original radiometer. Presented to Sir James Dewar.

55 Suspended astatic sine galvanometer used by Tyndall for heat radiation experiments. E.11

56 Hot air reflection tube on stand with six jet gas burner. (Tyndall) E.46

57 Blue sky tube made by Harvey and Peake London. (Tyndall) E.9

LIGHT

58 Optometer. Ivory. (T Young) B.1

59 Brass microscope, rack focusing, folding tripod foot, signed: *Cary London*. Said to have belonged to Faraday. D.112

60 Opaque solar microscope. All brass, incomplete. Early 19th century.

61 Camera lucida to WH Wollaston's patented design. *WHW 113* on brass mount of prism. L.78

62 Steinheil spectrometer. Used by Faraday, 1861-2. D.86

63 Glass prisms. (Faraday) D.72

64 Rowland grating and refractometer plate. Presented in 1892. (Tyndall) E.39

65 Polariscope. (Faraday) D.88

SOUND

66 Siren. Signed: *F. Kercy, 12 Spanner Building St. Pancras London* (from University College London).

67 Helmholtz resonator (from University College London).

ELECTRICITY: ELECTROSTATICS

68 Vertical globe electrical machine with clamp for fixing to table top. Signed: *Fra. Watkins London*.

69 Small globe electrical machine, the globe mounted on an axis at right angles to a vertical pillar. Crank handle. The sphere was electrified by hand.

70 Small cylinder electrical machine. Believed to have been constructed by Faraday while apprenticed to a bookbinder. D.1

71 Large cylinder electrical machine. Brass plate inscribed: *Nairne's machine, c1803*. The cylinder is supported at one end by two glass pillars, at the other, by a single pillar. Two prime conductors, an insulated stool and a battery of 15 Leyden jars. The machine was used by Faraday in 1836 to observe the nature of electrical discharge.

72 Plate electrical machine. Signed: *J. Cuthbertson London* (I. Bas 12).

73 Voss electrical machine, plate diameter 16¾in. Signed: *F.E. Becker Comp. 34 Maiden Lane, London W.C.* (from University College, London).

74 Discharging sphere. (Faraday) D.21

75 Electrophorus. (Faraday) D.21

76 Flat plate condenser. Signed: *I. Newman Lisle Street London*. (Faraday) D.23

77 Harris's electrometer. Used by Faraday to show the heating effect of a magneto-electric current, 1832.
Lit: HARRIS (1827)

78 Musical glass (from University College London).

79 Electric egg. Used by Faraday to study the discharge in gases at different pressures.

80 Specific inductive capacity apparatus, the stopcock signed: *Newman*. (Faraday) D.10

81 Specific inductive capacity apparatus with glass hemisphere dielectric. Signed: *Newman*. (Faraday)

82 Mahogany moulds – gauges for the correct positioning of spheres in specific inductive capacity apparatus. (Faraday) D.11.

83 Dielectrics used with specific inductive capacity apparatus – melted sulphur, etc. (Faraday) D.13 and others.

84 Gold leaf electrometer

85 Large Coulomb torsion balance (from University College, London).

86 Galvanometer (lacks scale and original needle) used by Faraday to demonstrate electromagnetic induction in the earth's field. D.52
Lit: MARTIN (1932-36) vol. I, p382

87 Galvanometer. Vertical coil, very thick wire. Suspended magnet was polished to act as a mirror in recording deflection (Faraday's galvanometer B). D.54
Lit: MARTIN (1932-36) vol. VII, p255 ff

88 Voltaic pile. Presented by Volta to Faraday. D.2

89 Cruikshank's trough battery, used by Davy and Faraday.

90 Astatic needle galvanometer, thick wire galvanometer for low currents, 1851. (Faraday) D.53.

91 Electromagnetic rotation apparatus.

Lit: FARADAY (1859) ppIV and 14; KING (1973) fig 8

92 Earth inductor used by Faraday in experiments on the generation of electric currents by the rotation of a rectangle of copper wire in the earth's magnetic field.

93 Barlow's wheel. (Faraday) D.100

ELECTRICITY: ELECTROLYSIS

94 Troughs (5) used in decomposition of the alkalis. (Davy) D.56

95 Two V-shaped electrolysis tubes on mahogany bases. (Faraday) D.4

96 Cylindrical electrolysis apparatus. (Faraday) D.6

97 Cylindrical electrolysis apparatus with delivery tube. (Faraday) D.5

98 Electrolysis apparatus – mahogany stand, glass wedge, platinum electrodes, porcelain trough. (Davy) D.39, also D.40, D.41, D.42a-b

99 Long V-shaped electrolysis tube with platinum foil electrodes. (Faraday) D.3

100 V-shaped electrolysis tube. (Davy) D.46

101 Glass electrolysis tubes. (Davy) D.51

102 Stand for electrolysis electrodes. (Davy) D.44

103 Gas collection tubes. (Davy) D.47

104 Voltameters (two examples, one illustrated KING (1973), fig 13). (Faraday)

MAGNETISM AND ELECTROMAGNETISM

105 Lodestone and keep. Brass mounted in black fish-skin case. Inscribed in lid: *Professor Tyndall FRS with Dr. David J. Price's best regards.* (Tyndall) E.12

106 Horseshoe magnets. (Faraday) D.59

107 Bar magnets. (Faraday) D.55

108 Logemann magnet. (Faraday) D.60

109 Large Logemann compound magnet (5 core). Acquired 1852. (Faraday)

110 Small laminated Schmidt magnet with flat solenoid between the poles. Used by Faraday for demonstrations. *Lit:* KING (1973) fig 12

111 Plain helix. (Faraday)

112 Helix on wooden base for magnifying needles by discharge of battery of Leyden jars. (Faraday)

113 Paper tube solenoid. Used for mutual induction between coils. (Faraday) D.42

114 Glass tube solenoid, single helix. Tube broken. (Faraday) D.32

115 Double solenoid with iron core. Two single helices on cork cylinder. (Faraday) D.39

116 Solenoid 'B'. Wooden tube and iron core, double windings, 6 iron and six copper helices.

Lit: MARTIN (1932-36) vol. I, pp370-71

117 Small solenoid, 3in, cloth-covered. (Faraday) D.34

118 Fine wire solenoid. (Faraday) D.83

119 Short solenoid on cork. (Faraday) D.48

120 Solenoid 'C' with magnet. (Faraday) D.43

121 Paper tube solenoid 'H'. (Faraday) D.41

122 Paper tube solenoid, core of dark crystal. (Faraday) D.82

123 Solenoid on musket barrel, single helix, 'N'. D.31
Lit: MARTIN (1932-36) vol. I, p375

124 Electromagnetic induction ring, 1831. (Faraday) D.25

125 Small induction ring. (Faraday) D.27

126 Ring electromagnet, 1832. (Faraday) D.58
Lit: KING (1973) fig 10

127 Small magnet made by electric discharge. Made by Faraday and labelled by him. D.56

128 Catherine wheel coil, paper frame, iron core. (Faraday)

129 Toroidal coil. (Faraday) D.47

130 Early magneto generator, Clarke type. Signed: *Watkins & Hill 5 Charing Cross London.*

131 Early magneto generator, Saxton type. Signed: *Newman Regent London.*

132 Pole pieces used in diamagnetic research by Faraday. D.64

Lo 21 LONDON
Royal Society

6 Carlton House Terrace, London SW1Y 5AG

071 839 5561

The history of the Royal Society, prior to the granting of its Royal Charter in 1662, is a matter of some debate. In the 17th century weekly meetings at Gresham College frequently included demonstration experiments. The care of a museum of rarities was added to the duties of the Curator of Experiments. By the turn of the century, however, the Society's Repository had become sadly neglected, attracting unfavourable comment from visiting savants. In 1710 the Society moved to a house in Crane Court off Fleet Street. Later it had rooms in Burlington House, Piccadilly, before transferring to its present location. The fine library dates from the collection presented by Henry Howard (later 6th Duke of Norfolk) in 1666/7. The non-scientific works from the Arundel Collection were sold to the British Museum in 1830 so that the Royal Society library is now primarily scientific. In addition to printed works it contains important manuscripts and extensive scientific correspondence.

During the 18th and early 19th century the Society acquired a number of important instruments. It equipped

major scientific expeditions such as that of James Cook, who sailed to Tahiti specifically to observe the 1761 Transit of Venus. Almost all surviving apparatus of this class has been lent to the *Science Museum, London*. Only the items listed below are retained on the Society's premises.
Lit: HUGGINS (1906) pp128-131 lists in an appendix the instruments which the Society possessed in the early years of the 20th century. HOPPEN (1976) includes references to recent literature on the foundation of the Royal Society. See also GREW (1681), the earliest printed catalogue of 'natural and artificial rarities'.

1   Reflecting telescope, constructed by Sir Isaac Newton, 1671. Newton's first reflecting telescope, constructed in 1668, no longer survives. This, his second telescope, was presented to the Royal Society by the London firm of instrument makers, Heath & Wing, on 6 February 1776.
Lit: MILLS/TURVEY (1979); SIMPSON (1984)

2   Dividers. Steel and brass. Belonged to Sir Christopher Wren.

3-4 Pair of chronometers by John Arnold. Taken by Captain James Cook on his second and third voyages of 1772 and 1776.
Lit: HOWSE (1968-70)

Lo 22 LONDON
Science Museum

Exhibition Road, South Kensington, London SW7 2DD

071 938 8000

The Science Museum, the National Museum of Science and Industry, has collections illustrating the development of mathematics, physics and chemistry and their industrial applications, as well as the development of agriculture, engineering, medicine (see *Wellcome Museum of the History of Medicine, London*), transport and communications, mining and industries up to the present time. Essential material to illustrate these developments, where not in the collection of the Science Museum itself, is illustrated by replicas, photographs and models. The Science Museum Library is a Department of the Museum and has extensive collections of printed, manuscript and pictorial material relating to the history of science, technology and medicine. The Museum's collections are probably the largest and most comprehensive of their type in the world.

One of the results of the 1851 Great Exhibition was the establishment of the Department of Science and Art attached to the Board of Trade. The purpose of this Department was 'to increase the means of industrial education and extend the influence of Science and Art upon Product and Industry', and its plans included 'museums by which all classes might be induced to investigate those common principles of taste which may be traced in the works of excellence of all ages'. In 1857 rudimentary collections, more industrial than scientific, were housed with much more extensive decorative art collections in new iron buildings – known familiarly as the Brompton Boilers – erected on the site of the present

Victoria and Albert Museum. The combined collections were known as the South Kensington Museum.

The so-called 'Scientific Collections' were moved in 1864 to buildings which had been the refreshment rooms of the 1862 Exhibition on the present site of the Science Museum. The collections took on a wider scope with the inauguration in 1864 of a Naval and Marine Engines collection, and a Mechanical Engineering collection in 1867. In 1876 the Special Loan Collection of Scientific Apparatus was exhibited in South Kensington and subsequently many of the exhibits were acquired to form the basis of the collections of scientific instruments and apparatus. Historic instruments from several countries were also loaned (*Lit:* LONDON (1876)). In 1884 the contents of the Patent Museum established by Bennett Woodcroft were transferred to the South Kensington Museum, and all the collections developed substantially in the subsequent years. In 1909 the scientific and art collections, already separated physically by Exhibition Road, became separate institutions; the art collections retained the name Victoria and Albert Museum and the scientific and technological collections were known as the Science Museum.

In 1913 part of the old building was demolished to make way for the present East Block, completed after the 1914-18 war. The last fragment went in 1947 when the present centre block was started but the new building was not finalised until 1961, and the first part was opened in 1963. The East Block Extension was opened in 1977. The first outstation of the Museum, the *National Railway Museum* at York, was opened in 1975, and an exhibition dealing with Concorde opened in 1980 at Yeovilton in Somerset. The *National Museum of Photography, Film and Television* opened at Bradford in 1983.

Over the years sectional catalogues have been published of considerable parts of the enormous collections of the Science Museum. All but the more recent ones are now out of print and no longer comprehensive, but they are available for consultation in the Science Museum Library. The sections of the Science Museum collections which relate to this survey are described briefly below, accompanied by reference to any particularly relevant catalogues or booklets published either currently or in the past. In view of the enormous size of the Science Museum collections, it is impossible to attempt to refer, in this survey, however briefly, to either named collections (of which there are several hundred) or to individual items in the collection. The aim of the following paragraphs is merely to give the user a general idea of the kind of material included in the collection, and some indication as to its provenance where this is significant. Where they exist, catalogues have been listed under the section to which they relate, but in the case of many important sections no published catalogues are available. In other cases, although catalogues exist, much material has been acquired since their publication.
Lit: For histories of the Museum, see GREENAWAY (1951), LONDON – SCIENCE MUSEUM (1957) and FOLLETT (1978); LONDON – SURVEY (1975) pp248-52 deals with the site and buildings.

1   *Acoustics.* The collection includes 19th-century apparatus such as tuning forks, organ pipes and monochords, and also some instruments by Koenig, e.g. his vowel-sound synthesiser. There is a group of unusual musical instruments, e.g. pyrophone, musical glasses, claviola, dulcitone, bell-ringing machine and also several keyboard instruments designed to play in just intonation in several keys. A large and growing sub-collection illustrates the story of sound recording from tinfoil to stereo.

Lit: CHEW (1981) describes phonographs and gramophones from 1877 to 1914.

2 *Astronomy.* In addition to astronomical telescopes and instruments (listed under *Optical instruments*, Section 12b), the collection of astronomical instruments includes oriental and European astrolabes, early astronomical quadrants, armillary spheres (the earliest of mid-16th century date), and an important collection of orreries (figs 104 and 105) including Rowley's original orrery, once the property of successive Earls of Cork and Orrery. There are models to demonstrate phenomena such as tidal motion, the motion of comets (*cometarium*) and the Transit of Venus (mid-18th century). Celestial and terrestrial globes (fig 56) date from the 16th century onwards, and there is an example of John Russell's *Selenographia*, c1796, the earliest Moon globe.

3 *Chemistry.* The collection covers all aspects of chemistry – organic, inorganic, analytical and physical – from early times, including laboratory apparatus (chemical glassware (fig 101), balances (figs 14 and 16), hydrometers (figs 65 and 66), heating appliances, chemical autoclaves, and distillation, filtration, evaporation and desiccation apparatus), as well as samples of chemical products. There is an important group of Islamic glassware, 10th-12th century (*Lit:* ANDERSON 1983). Sources of important items of early laboratory apparatus are St Bartholomew's Hospital, the Royal Society and, slightly later, the Laboratory of the Government Chemist. The Science Museum owns a wide and representative range of balances, including instruments by makers such as Jesse Ramsden (the double cone balance of 1783), Edward Troughton, Thomas Jones, Floreiz Sartorius, Friedrich Wilhelm Breithaupt and Ludwig Oertling. There is apparatus associated with major names and discoveries in chemistry, including Henry Cavendish (a double burning glass, c1770), Michael Faraday, William Hyde Wollaston, Sir James South, William Allen Miller, Thomas Graham, George Gore, Augustus Matthiessen, William Crookes, William Ramsay, Morris William Travers and Thomas Young (on loan from University College, London), the Braggs (WH and WL) and Walter Noel Hartley (Lord Rayleigh's apparatus is in the physics collections). There are important models of large molecules prepared by Max Perutz, John Kendrew, Dorothy Hodgkin and Francis Crick and James Watson. Industrial chemistry, glass technology and metallurgy are also represented. There are samples of dyes by AG von Hofmann and William Henry Perkin (his mauveine).

Lit: BARCLAY (1937) Part I *Historical Review*, Part II *Descriptive Catalogue*; STOCK (1969) and BUCHANAN (1982) for chemical balances.

4 *Geophysics.* The geophysics collection covers instruments used to study the Earth's gravitational and magnetic fields; tidal measurement, analysis and prediction; seismology; and atmospheric electricity. The earliest instruments are those used for magnetic investigations. There is a large number of dip-circles, variation needles and force magnetometers, some portable, some from observatories. They date from the 18th century onwards and many have a known history. The extensive use of pendulums to measure relative gravitational attraction in various parts of the world is reflected in the large collection of these instruments, mostly on loan from the Royal Society and Kew Observatory. Again, most have a known history. Efficient working seismographs were developed towards the end of the 19th century; the collection includes a variety of types, from Britain, Japan and Italy, as well as a working Shaw seismograph of 1935 which forms part of the gallery display.

Lit: SHAW (1936); WARTNABY (1957); McCONNELL (1980); McCONNELL (1986A)

5 *King George III collection.* This collection, the only surviving example in England of a cabinet of philosophical instruments, probably owes its origin to Stephen Demainbray in about 1740. In 1754 Dr Demainbray became Tutor to the Royal family, and what began as his personal collection was extended under the patronage of King George III, mainly in the years before 1768, although additions continued to be made to the collection until the early years of the 19th century.

From 1769 until 1841 the collection was housed at the King's private Observatory at Kew. When the government ceased to maintain the observatory in 1841 the collection was presented, with the exception of some astronomical instruments which were sent to *The Observatory, Armagh*, Northern Ireland, and a few which remained at Kew, to King's College, London to be used for a course in 'Experimental Philosophy'. The instruments remained at King's College until 1926 when they were lent by the College Delegacy to the Science Museum. Some instruments (mostly microscopes) were separated from the main collection by Sir Frank Crisp and a few of these are to be found in *Whipple Museum, Cambridge* and *Museum of the History of Science, Oxford*; a few medical specimens survive in the *Hunterian Museum, Glasgow*.

The major part of the collection which is now in the Science Museum (some 1500 items) comprises physical and mechanical apparatus. Mechanics is represented by numerous pieces of apparatus for the demonstration of mechanical principles, many of which were to be used with the Philosophical Table (which still survives in the collection), as well as by instruments employing mechanical principles such as balances, and models of larger machines – cranes, capstans, piledrivers, etc. Thermal instruments are represented by pyrometers and thermometers, pneumatics by air pumps and accessories, hydrostatics by Pascal vases, etc. Chemistry is barely represented in this collection (fig 101), though there is an interesting balance by Sisson.

The astronomy collection is incomplete, a part of the material now being at Armagh, but it includes orreries, globes, sundials and a variety of refracting and reflecting telescopes, as well as a large mural quadrant. Other optical instruments include numerous microscopes, a camera obscura, stereoscopes, anamorphoses, a polemoscope, a polyoptric pyramid, a model eye and numerous lenses and prisms. There is also a lens-cutting machine. Surveying instruments include theodolites, clinometers and levels. There are many drawing instruments. The acoustics collection is small, including organ pipes, tuning forks, etc.

Instruments dealing with electricity and magnetism are numerous; there are lodestones, magnets, compasses and magnetic toys, several electrical machines (fig 49) with numerous accessories, as well as instruments for current electricity such as a magneto-electrical machine and galvanometers. Finally there are various models – a steam engine, a paddle wheel, models of carts and trucks and a miscellany of weapons. A manuscript catalogue is preserved in Archives Collection of the Science Museum Library. It was compiled around 1769 and lists the instruments deposited by Queen Charlotte at Kew, including many of the early items in the King George III collection.
Lit: WHIPPLE (1926); CHALDECOTT (1949 and 1951); CHEW (1968)

6 *Heat and thermal instruments.* This section contains both instruments used for measuring temperature and apparatus used in connection with investigations concerning the theory of heat. The former category contains thermometers for a wide variety of uses, including domestic, clinical, chemical, standard, self-registering, deep sea, maximum and minimum, underground and bimetallic. Pyrometers range from 18th-century instruments for measuring the expansion of solids, and for comparing the expansion of different substances when heated, to Wedgwood's pyrometer for use in measuring the temperature attained in pottery kilns, and thus a pyrometer in the modern sense of an instrument to measure temperatures beyond the range of a normal thermometer. An important group of instruments are a series of electrical resistance pyrometers designed or constructed by Hugh Longbourne Callendar at the Cavendish Laboratory in Cambridge in 1886-7.

The collection contains instruments and apparatus relating to the nature of heat, change of heat, refrigeration, very low temperatures, etc., and there is material connected with the work of Sir James Dewar (from the *Royal Institution, London*), John James Joly, James Prescott Joule, CV Boys and William Crookes. Another important item is the apparatus used by Thomas Andrews in 1863-9 and 1876 for investigating the properties of carbon dioxide.
Lit: CHALDECOTT (1954 and 1976); CHALDECOTT (1955)

7 *Magnetism and electricity.* The collection ranges from lodestones and early permanent magnets to electromagnets and induction coils. Electrical machines (figs 48 and 49) include a cylinder and a globe machine used by Joseph Priestley, Thomas Armstrong's hydro-electric machine used by Faraday (fig 50), several machines made by James Wimshurst (fig 51) and many electrical accessories. There is an extensive range of galvanometers, electrometers, and other measuring instruments including electric standards.

Lit: LONDON – BOARD OF EDUCATION (1905); STOCK/VAUGHAN (1983)

8 *Mathematics and mathematical instruments.* Includes comprehensive collections of calculating machines, slide rules, drawing instruments (fig 44), integrating instruments and mathematical models. All the machines and instruments are collected up to present day examples. Some of the items of particular note are the earliest surviving straight rule by Robert Bissaker (1654), several examples of 17th-century Napierian rods and a good selection of Victorian curve-drawing instruments. The Babbage Difference and Analytical Engines, previously in this collection, have been transferred to form the starting point of a new collection, Computing and Data Processing, which embraces both analogue and digital machines and methods.
Lit: BAXANDALL (1975)

9 *Meteorology.* All types of meteorological instruments are represented in the collection, including barometers, thermometers, sunshine recorders, hygrometers, rain gauges, evaporimeters and anemometers. Instruments have been presented or lent by the *Royal Society, London,* the Kew Committee of the Royal Society, the Royal Meteorological Society, the Astronomer Royal, the Meteorological Council and the Hydrographic Department of the Admiralty, to mention only major sources. Several of the barometers in the collection are listed or described in MIDDLETON (1964) pp452-4; other instruments are discussed in MIDDLETON (1969).
Lit: Early cataloguing efforts appear as: LONDON (1900 and 1922)

10 *Navigation.* The collection includes astronomical instruments for use at sea; backstaves (fig 13), octants (fig 102), sextants (fig 122), circles (fig 21) and artificial horizons (fig 5). Mariner's compasses, steering and azimuth compasses are included as well as Chinese geomancer's compasses. Aids to dead reckoning such as traverse boards and ships' logs are included, as well as examples of more sophisticated radio navigation aids such as hyperbolic position-finding instruments.

11 *Oceanography.* The oceanographic collection has been largely the gift of the Admiralty and the Institute of Oceanographic Sciences. It contains deep-sea thermometers, water-bottles, current meters and sounding apparatus, mostly dating from the 19th and 20th centuries. There is also a small collection of instruments used for measurements in lakes and rivers.
Lit: McCONNELL (1981 and 1983)

12 *Optical instruments.*

    a Microscopes. The Museum houses a large collection of microscopes of all periods, many from the collection of TH Court, and illustrated in CLAY/COURT (1928). For details of their acquisition, see INSLEY (1982). One of the earliest and most interesting, presented by Court, is a compound microscope of *c*1670. A microscope (by Benjamin Martin) was in the possession of Joseph Priestley. A companion to the Adams silver microscope made for King George III in the *Museum of the History of Science,*

Oxford is another made for the Prince of Wales, later King George IV. Nearly all major types of English (figs 76A and B, 78, 82-86) and some continental microscopes are represented.
Lit: PALMER/SAHIAR (1971)

b   Telescopes. The collection is extensive and includes portable and hand-held refractors (fig 127) and reflectors (fig 128), including two refractors by Guiseppe Compani of Rome and a refracting telescope by Christopher Cock (1673). From the Royal Society's collection are a number of late 17th-century objective lenses, including examples made by Christian Huygens. Additionally there is an early Gregorian telescope, reputedly constructed by John Hadley in 1726. There are numerous important major astronomical instruments, including a 7ft focus reflecting telescope constructed by William Herschel (fig 130), another by James Short, *c*1767 or 1768, on an equatorial mounting, and the equatorial refractor of its time. There is a heliometer constructed for the Radcliffe Observatory, Oxford, in the 1840s. The mirror of Herschel's 24ft focus telescope is on display as is the mirror for the Earl of Rosse's 6ft aperture, 54ft focus reflecting telescope erected at *Birr, Eire* in 1845. There are also many smaller portable astronomical instruments (figs 53 and 135).
Lit: THODAY (1971)

c   Other optical instruments. Outside the two above mentioned categories of instruments the collection contains polariscopes, ophthalmic, ancient and modern spectacles, spectroscopes (fig 126) (also in *Chemistry*) and photographic apparatus (see *Photography*). There are optical aids to drawing such as the camera obscura (including a book camera obscura which belonged to Sir Joshua Reynolds), the camera lucida (fig 17) and Cornelius Varley's Patent Graphic Telescope. Lens grinding machines are also represented. An interesting item is William Herschel's machine for grinding small mirrors. From the King George III collection come some of the few surviving anamorphoses with cylindrical and conical mirrors.

13 *Photography*. There is an extensive collection of photographic apparatus – cameras, lenses, exposure meters and other photographic accessories, as well as a range of examples illustrating different photographic techniques. Micro-photography, stereoscopy and colour photography are represented. The collection contains a considerable body of cameras and optical instruments (camera obscuras, camera lucidas, solar microscopes) used by WH Fox Talbot and presented by Miss Talbot to the Science Museum in 1937 (at the same time as other material was presented by the same donor to the *Royal Museum of Scotland (Chambers Street), Edinburgh* and the *Royal Photographic Society, Bath*. Early cameras by Giroux and Cie, Voigtlander (Daguerreotype), Noel Paymal Lerebours, Charles Louis Chevalier and John Dancer are included in the collection. The history of cine photography is also well represented, by apparatus of

Etienne-Jules Marey, Louis Jean Lumiere and RW Paul.
Lit: THOMAS (1964, 1966, 1969A and B, 1981, 1983)

14 *Pneumatics*. The core of early material is in the King George III Collection, which includes various types of pumps – single and double barrel, sucking, forcing, exhausting and compressing – as well as accessories for use with the pumps. Important items are the early Hauksbee type pump on loan from the *Royal Society, London*, and an Adams exhausting and compressing pump. There is a significant collection of later pumping machinery.
Lit: CHALDECOTT (1951) pp31-37; WESTCOTT (1932-33)

15 *Surveying and geodesy*. The collection is extensive and particularly complete for the 18th-20th centuries, although less so for the 16th-17th centuries when the majority of the instruments – surveyor's quadrants, circumferentors, graphometers – are of continental (French, German and Italian) origin. Among the 18th and 19th century instruments the most important are on loan from the Ordnance Survey, Ramsden's 3ft theodolite, *c*1790, and Colby's compensation bars of 1827/8. All were used in the triangulation of Great Britain and Ireland. On loan from the *Royal Society, London* are two items, a boning telescope and a steel chain made by Ramsden and used by General Roy for measuring the Hounslow Heath baseline in 1784. A further important source of instruments is the Admiralty from which several early 19th-century transit theodolites (fig 133) are derived, as well as marine survey instruments. Continental instruments are well represented among material of 19th-century date. There are examples of both mining and gunnery instruments (range finders).
Lit: LANCASTER-JONES (1925); WARTNABY (1968)

16 *Time measurement*. The collection includes primitive and non-mechanical time measuring devices – shadow clocks, water clocks, sand glasses and an extensive range of sundials of 16th-19th century date (figs 28-31, 33). Horary quadrants (fig 115A and B), nocturnals and astronomical compendia as well as perpetual calendars are represented. Of vital significance is a Byzantine geared calendrical dial of the 5th century, possibly Alexandrian (*Lit:* FIELD/WRIGHT 1985A and 1985B)). Mechanical clocks include clocks from Dover Castle (*c*1600) and Wells Cathedral (14th century), and a wide range of domestic clocks from the 16th century onwards (including an early pendulum example by Fromenteel), as well as regulators and chronometers. Most material is European but Japanese clocks are also represented. About 150 watches are exhibited, ranging from a stackfreed movement of the mid-16th century to quartz watches of the present day. Electric horology, from Alexander Bain's pioneer timepiece (*c*1840) to the atomic clock developed at the *National Physical Laboratory, Teddington*, is well represented.
Lit: WARD (1955, 1963, 1966, 1970, 1972 and 1973)

17 *Tools, machine*. The collection was initiated by James Nasmyth's working model of his steam hammer in 1857, and has since then grown to number approximately 300 items. Included in the collection are a wide range of machine tools for working metal and wood. The Portsmouth blockmaking machinery is a very early

example of equipment designed for mass production (*Lit:* GILBERT (1965)). Of particular interest is the sequence of dividing engines used in the manufacture of scales on precision instruments for navigation and surveying; they range from an early Ramsden-type dividing engine of *c*1780 to its 20th-century successor. Clockmaker's wheel cutting engines (fig 139) are well represented. There are several examples of copying machines and an associated collection of Industrial Metrology which includes James Watt's micrometer, Henry Maudslay's 'Lord Chancellor' test bed and Joseph Whitworth's 'Millionth' measuring machines.
Lit: GILBERT (1966 and 1975)

18 *Trade cards.* Of reference to the study of English instrument making is the collection of more than 400 trade cards and related ephemera, ranging in date from the mid-17th until the late 19th century. The majority of these items derived from the collections of TH Court and George Gabb.
Lit: CALVERT (1967)

19 *Weights and measures.* An important collection of weights, measures and balances, ranging from Greek, Roman, Egyptian and Sumerian material up to the 19th and 20th centuries. It includes money changer's balances, coin scales, apothecary's scales, jeweller's balances, assay balances, spring balances, bismars, steelyards, etc. The linear and capacity standards of King Henry VII and the standard weights of Queen Elizabeth I are among the earliest in the important sequence of pre-Imperial standard weights and measurers. A large group of Indian weights and measures collected by the East India Company forms part of the collection (*Lit:* ANDERSON (1982) pp26, 27). The Castlereagh Collection is a large collection of standard weights formed in 1818 at the instigation of Lord Castlereagh, Foreign Secretary, who instructed British consuls abroad to despatch examples to England.
Lit: SKINNER (1967); CONNOR (1987)

20 *Medicine. The Wellcome Museum of the History of Medicine, London* (q.v.) was transferred on permanent loan to the Science Museum by the Wellcome Trust in 1977. There are very large collections of physiological instruments, microscopes, surgical instruments and artefacts relating to the whole spectrum of medical technology. This collection is being actively augmented by the Science Museum. The many non-medical scientific instruments from the Wellcome Collection are being assimilated in the relevant Science Museum departments.
Lit: BRACEGIRDLE (1981); SKINNER (1986)

Lo 23 LONDON
Sir John Soane's Museum

13 Lincoln's Inn Fields, London WC2A 3BP

071 405 2107

The house, with its Museum and Library, was opened to the public on the death of Sir John Soane in 1837 under a private Act of Parliament obtained in 1833. Built to Soane's designs in 1812, the house contains his collection of paintings, Greek, Roman and Renaissance architectural fragments, casts and sculptures, Greek vases, antique gems, architectural drawings and designs, as well as an extensive library. The furnishings include a number of 18th-19th century clocks and watches, among other items, a chronometer by Thomas Mudge, 1793. Although there are no instruments as such, the collection contains the two models listed below.

1   Model pile driver

2   Model of a device for building brick sewers

Lo 24 LONDON
Society of Antiquaries of London

Burlington House, Piccadilly, London W1V 0HS

071 734 0193

The Society was founded in 1707 when a group of men agreed to meet weekly in a London tavern. The terms of reference of the society were a study of the antiquities and history of Great Britain. In 1751 the Society was granted a Charter and in 1753 acquired the lease of its first premises, at 2 Chancery Lane. In 1780 the Society was granted premises in Somerset House and remained there until, in 1875, it transferred to new accommodation at Burlington House. Since its foundation the Society has accumulated an important and interesting collection of material, much of it of an archaeological nature, but including also paintings, clocks (including a clock dated 1524 by J Zech and an eight-day regulator by J Vulliamy and B Gray, of *c*1770) and a number of items of scientific interest. Some of the more important items in the collection are described and illustrated in a Society publication by RLS Bruce Mitford which is cited below.
Lit: MITFORD (1951) and EVANS (1956) for histories of the Society and its possessions.

1   Hispano-Moorish astrolabe, 14th century. Brass, diameter 3⅞in. Plates engraved for latitudes between 32°30′ and 41°.
Lit: GUNTHER (1932) vol. 1, no. 162; PRICE (1955) no. 162

2   Perpetual calendar in turned wooden box. Paper calendar card. Signed: *Wriglesworth Easter 1688.* (Identical example in *Whipple Museum, Cambridge.*)

3   Roman dodecahedron. Excavated 1768 in Carmarthen. It has been suggested that this may have been used in surveying. For a discussion of this and other views see THOMPSON (1970).

4   Aeolipile in form of a crouching naked man with his right hand held against his brow. Found at Basingstoke, *c*1800.
Lit: ANON (1800)

Lo 25 LONDON
Victoria and Albert Museum

Cromwell Road, South Kensington, London SW7 2RL

071 938 8500

The Victoria and Albert Museum and the Science Museum, founded as the South Kensington Museum, were administratively one body until 1909 (see *Science Museum, London*). The Museum houses an outstanding collection of fine and decorative arts of all countries, periods and styles. The European collections range from early Christian times onwards, the Oriental collections include Persian, Chinese, Japanese and Indian material. The Museum contains a number of scientific instruments, collected primarily for their decorative qualities. They are mainly in the care of the Department of Metalwork, although some items (ivory sundials, for example) come under the Department of Architecture and Sculpture, while barometers (mainly of a domestic nature) belong to the Department of Furniture and Woodwork. In addition to the instruments listed below, the Museum houses a number of important early clocks and watches (*Lit:* HAYWARD (1969)). In the list which follows, Oriental instruments appear in Section A, and European items in section B.
Lit: LONDON (1975) pp97-123; COCKS (1980) for the development of the collections.

I   *ORIENTAL INSTRUMENTS*

1   Celestial globe on stand. Brass with silver points for stars. Signed: *Qasim Muhammad b. 'Isar b. al-Hadad, Lahore.* Indo-Persian, *c*1630. From the Marling Bequest. (M828-1928)
Lit: SAVAGE-SMITH (1985) p225, figs 38 and 42

2   Celestial globe on stand. Brass with silver points. Signed: *Muhammad Zaman.* Persian, AH 1051 (AD 1641/2). From the Marling Bequest. (M827-1928)
Lit: MAYER (1956) p78; SAVAGE-SMITH (1985) p226, fig 21

3   Celestial globe. Brass and silver. Signed: *Diya' ad-din Muhammed, Lahore.* Indo-Persian (Lahore), dated AH 1060 (AD 1649/50). (M507-1888)
Lit: SAVAGE-SMITH (1985) p227-8

4   Celestial globe. Brass with silver points for stars. Signed: *Diya' ad-din Muhammad, Lahore.* Indo-Persian, AH 1067 (AD 1657). (2324-1883 1.S)
Lit: SAVAGE-SMITH (1985) p228

5   Celestial globe. Brass with silver points, 8-lobed stand with five columns. Persian, 17th/18th century. (1149-1883)
Lit: SAVAGE-SMITH (1985) p257-8

6   Celestial globe. Bronze, by Muhammad Karim, dated AH 1241 (AD 1825/26). (M24-1882)
Lit: MAYER (1956) p69; SAVAGE-SMITH (1985) p255

7   Globe. Painted. N.E. India (? Orissa). (IM 499-1924)
Lit: ANDERSON (1982) p34, no. 123

8   Celestial globe. Brass. Oudh. (06475 I.S.)

9   Part of an armillary instrument. Brass disc, diameter 6⅛in, perforated and equipped with an alidade. Two broken brass rings fit at right angles around the disc. Disc inscribed: *NSL* and *Mana Yatra* (measure instrument) in Devanagari characters. W. India, 19th century. From the Royal Asiatic Society, London. (IM 413 and 413C-1924)

10   Syrian astrolabe (Saphaea Arzachelis). Brass, diameter 8in. Signed: *'Abd ar-Rahman b. Yusuf at Damascus,* AH 598 (AD 1202/3). (M504-1888)
Lit: GUNTHER (1932) no. 102; MAYER (1956) p31

11   Indian astrolabe. Brass, diameter 5in. Signed: *Diya' ad-din Muhammad, Lahore.* Dated AH 1074 (AD 1664). (419-1876)

12   Indo-Persian astrolabe. Brass, diameter 4⅜in. Signed: *Muhammad Salih Tattari* AH 1076 (AD 1665). From the Royal Asiatic Society, London. (IM 408-1924)

13   Persian astrolabe. Brass, diameter 4½in. Signed: *Khalil Muhammad b. Hasan 'Ali* and *decorated by Muhammad Mahdi al-Yazdi b. Muhammad Amin, c*1700.
Lit: MAYER (1956) p56; GUNTHER (1932) no. 24

14   Syrian or Egyptian astrolabe, 17th century. Brass, diameter 4½in. (530-1876)

15   Indian astrolabe. Signed: *Jamal ad-din of Lahore,* AH 1092 (AD 1681). (IS 30-1882)

16   Persian astrolabe. Brass, diameter 5in. Signed: *'Abd al A'Imma* the younger, AH 1127 (AD 1715) (458-1888)
Lit: MAYER (1956) p24; GUNTHER (1932) no. 34

17   Persian astrolabe. Brass, diameter 10in. Signed: *'Abd al-Ghafur in Rajab* AH 1195 (AD 1781). (Marling Bequest) (M 826-1928)
Lit: MAYER (1956), p29 (as AH 1095)

18   Persian astrolabe *c*1800. Brass, diameter 4⅝in. (26-1889)
Lit: GUNTHER (1932) no. 50

19   Arabian astrolabe. Diameter 8⅜in. No plates. Inscriptions in Arabic-Kufic characters. From Royal Asiatic Society, London. (IM 406-1924)

20   Indian astrolabe, early 19th century. Brass, diameter 3½in. Inscriptions in Hindi characters. Five plates of the same size (IPN 2467) may also have belonged to this astrolabe. From the Royal Asiatic Society, London. (IM 407-1924)

21   Indian astrolabe, 19th century. Brass, diameter 6¾in. Inscriptions in Hindi characters. From the Royal Asiatic Society London. (IM 409-1924)

22   Surveying circle. Brass with (?) Hindi inscription: Indian, 18th century. (IM 10-1915)

23   Quadrant. Square with sights along one edge. One side engraved with a sinical quadrant, the other with a circular diagram with divisions into 24 hours, the signs of the zodiac, etc. The remains of a pointer is pivoted at the centre. Indian, 17th or 18th century. (IM 11-1915)

24 Sundial. Circular with lid to top and bottom. The middle portion contains a scaphe. The instrument also includes a compass and a clinometer. ? Western India. (IM 471-1924)

25 Universal equinoctial dial. Brass, height approx 5in. Numerals are European, an inscription is in Arabic, and includes the date AH 1284 (AD 1874). (06484 IS)

26 Astrological plaque. Brass, inscribed with zodiac, talismanic signs and inscriptions. Accompanied by two sets of four dice. Persian, 19th century AD. (505a,b-1888)

27 Perpetual calendar. Brass, diameter 4½in. Inscribed: *Joutsheed Nur Mehund*. Indian, early 19th century. From the Royal Asiatic Society, London. (IM 410-1924)

28 Perpetual calendar. Ink on paper, consisting of eight superimposed rotating discs. Diameter 9⅝in. S. Indian, Tamil. Early 19th century. From the Royal Asiatic Society, London. (IM 412-1924)

29 Compass. Brass, engraved. Persian, 18th century. (574-1878)

30 Compass. Brass, engraved. Persian, 18th century. (762-1889)

31 Compass. Brass. Persian, 19th century. (M307-1887)

32 Compass. Brass, engraved. Signed: *Muhammad Hasan*. Persian, 19th century.

II EUROPEAN INSTRUMENTS

33 Celestial globe. Gilt brass with clockwork mechanism. Engraved with Imperial double eagle and signed: *Elaborabat Georgius Roll et Johannes Reinhold in Augusta Anno domini 1584*. Perhaps constructed for the Emperor Rudolph II (M246-1865).
Lit: ZINNER, p493; BOBINGER (1969); HAYWARD (1950)

34 Pair of globes on mahogany stands. Late 17th century. (HH113 and 113a; on display at Ham House, London)

35 Pair of globes, diameter 18in, on mahogany floor stands with tripod legs and compass. Celestial globe undated, terrestrial globe records discoveries up to 1843. (W 52 and 52a-1916)

36 Terrestrial globe in fish-skin case, inscribed: *Newton's New and Improved Pocket Terrestial Globe*. Dated: 1817. (W 34-1974)

37 Armillary sphere. Brass, mounted in carved mahogany stand. Height 38½in. Unsigned, c1750. (W 36-1938)

38 Astrolabe. French, c1410. Brass, with trefoil gothic rete. One of the plates is engraved for the latitude of Paris. (M 128-1923)
Lit: GUNTHER (1932) no. 190, fig 7

39 Quadrant. Brass, the front engraved with Gunter's quadrant, the reverse with revolving planispheric nocturnal. Signed: *W. Hayes fecit*. (M 341-1926)

40 Perpetual calendar. Gilt copper, dated 1667. Finely engraved, originally incorporated small clock face. (M23-1952)

41 Lunar calendar for the years 1530-1559. Brass, diameter 8¾in. Gearing around the central point is connected with a model controlling the dial for phases of the Moon. (M 1175-1893)

42 Nocturnal. Chiselled and engraved steel. ? English, ? 19th century. From the Spitzer collection. (M 702-1893)

43 Nocturnal. Applewood, diameter 4½in. length 13in. English, c1700. (21-1954)

44 Astronomical compendium. The cover engraved with the Royal arms. Incorporates a list of latitudes, an equinoctial dial (with initials RG on painted compass card), lunar volvelle, windrose, nocturnal and scale showing The Prime and Epact beginning 1617. Signed: *Elias Allen fecit*. Probably made c1616 for King James I. Transferred from the Royal United Services Institution.
Lit: BLAIR (1964)

45 Astronomical compendium. Gilt brass, circular. Includes plummet, windvane, horary quadrant, map, horizontal sundial for five latitudes, compass, lunar volvelle, nocturnal and scale for length of day and night, times of sunrise and sunset. Signed: *Christophorus Schissler me faciebat Augustae Vindelicorum Anno Domini 1561*. (M 165-1938)
Lit: ZINNER, p508; BOBINGER (1966) p305, fig 58

46 Astronomical compendium. Gilt brass, rectangular. Includes lunar volvelle and aspectarium, nocturnal, horizontal dial, table of latitudes, hour conversion diagram, and compass. Signed: *Christophorus Schissler me fecit Auguste Vindelicorum Anno Domini 1557*. (M 167-1938)
Lit: ZINNER, p506; BOBINGER (1966) p302

47 Horizontal string-gnomon dial. Gilt brass with folding gnomon support, plumb level and list of latitudes. Signed: *V.S.* (Ulrich Schniep Munich). German, 16th century. (M 166-1938)
Lit: ZINNER, p526

48 Horizontal string-gnomon dial with folding gnomon support and plumb level. Inside the lid a 'viatorium'. Gilt brass. Signed: *VS 1586*. (M 641-1965)

49 Butterfield dial. Silver in fish-skin case. (388 and 388A-1906)

50 Butterfield dial. Silver in fish-skin case. Signed: *Butterfield AParis*. (M 879-1882)

51 Horizontal table dial on three screw feet. Signed: *Johan Koch fecit Vienae*. Mid-17th century. (M 125-1953)

52 Horizontal table dial. Brass, with folding gnomon support and plummet level. Signed: *Franz Antoni Knitl Linz fecit*. Austrian, first half 18th century. (20-1879)

53 Horizontal dial on four feet. Gilt brass, rectangular. The upper side with compass and horizontal sundial, the underside engraved with scales of Epacts, Golden Numbers, etc. Signed: *Jacob Marquart von Augspurg*, dated 1567. (9035-1863)
Lit: ZINNER, p438; BOBINGER (1966) p266, pl. 205

54 Horizontal string-gnomon dial. Ivory, with paper

compass rose. French, *c*1800. (loan from *Science Museum, London* 1938-346)

55 Analemmatic dial and horizontal dial. Brass. Signed: *Tho. Tuttell Charing X Londini fecit.*

56 Bloud (magnetic azimuth) dial. Ivory. Signed: *Fait par Gabriel Bloud a Dieppe.* (48-1894)

57 Bloud dial. Ivory and silvered brass, 4¾ x 3½in. Signed: *Fait par Henri Robert Marseille.* Coat of arms engraved on lid.

58 Bloud dial. Ivory, unsigned. (A 52-1923)

59 Diptych dial. Gilt-brass, rectangular, the lid engraved with the Fugger-Kirchberg Weissenhorn coat of arms. Signed: *Christophorus Schissler Senior faciebat Augusta Anno 1581.* (M 86-1921)
Lit: ZINNER, p514; BOBINGER (1966) p319, pls. 21, 226

60 Universal equinoctial dial. Engraved brass, with red lacquer filling. Rectangular. Folding equinoctial plate. Inscribed: *JHB* (in monogram). Possibly Johann Hertel Braunschweig, German, second half 17th century. (M 164-1938)

61 Crucifix dial. Gilt brass. South German, *c*1560. (M 258-1864)

62 Drawing instruments in green ray-skin case. Ivory rule, other instruments (compasses) of brass and steel. Unsigned, late 18th century. (M 137-1906)

63 Protractor combined with set square. Brass, 3in. x 4¾in. The protractor is also engraved with a scale of polygons. Signed: *Domenicus Lusuerg F. Romae 1701.* (M 217-1939)

64 Set of instruments of brass in a gold-stamped flat leather case, 10 x 7¾in (fig 43). The contents include:

  a Set square and protractor. Signed: *Petrus Galland Fecit Romae sub Signo Pulcrae virtutis In via Coronaria.*

  b Compass in heavy brass circular case with turned lid.

  c Pair of compasses with steel points, one point is interchangeable with a pencil holder and other accessories.

  d Pair of pincers. Brass with steel points.

  e Folding rule. Ungraduated.

  f Folding plumb-bob level of thick brass. In the form of a folding square with two subsidiary hinged arms which, when fully extended, hold the arms of the square firmly at right angles to one another. A plumb bob was intended for use with the square.

  g Simple microscope. Early 18th century, of a type described by Zahn, in 1702. Eight objects are mounted in a revolving brass wheel which fits between two brass plates. The objective is mounted in a short brass cylinder which is removed for storing the instrument in the box. The set of instruments, which may not all be by

the same maker, appears to date, together with the case, from the early 18th century. (156-1889)
Lit: CLAY/COURT (1928), p33, where another similar, but more elaborately decorated microscope is illustrated.

65 Sector. Brass, length (closed) 6⅞in. Signed: *Butterfield AParis.* (M 146-1925)

66 Gunner's quadrant. Gilt brass, Signed: *C.T.D.E.M. 1617.* From the Bernal Collection. (M 2076-1855) (fig 61)
Lit: ZINNER, p550

67 Refracting telescope, three draw, hand-held. Outer tube covered with red ray-skin, the inner tubes with green vellum. Length 16in when closed. Unsigned, 18th century. (M 1091-1902)

68 Two balance stands. Gilt brass, with elaborate decoration. ? German, 16th century. (M 122 and 121-1953)

69 Lodestone. Silver mounted. Possibly early 17th century. (M 130-1967)

70 Concave mirror. On carved rosewood tripod stand. Ht 5ft. Diameter of mirror and frame 39in. English, *c*1800. (W 17-1961)

71 Book camera obscura. Leather cover, sides covered with marbled paper. The spine is stamped: *OPTIQUES ET CHAMBERE OBSCURE PAR SEANEGATTI.* French, 18th century. (Department of Prints and Drawings)

Lo 26 LONDON
Wallace Collection

Hertford House, Manchester Square, London W1M 6BN

071 935 0687

The collection was assembled by the 3rd and 4th Marquesses of Hertford and the latter's son, Sir Richard Wallace (1818-90). It was presented to the nation by Lady Wallace in 1897. It contains paintings, miniatures, sculpture, furniture, arms and armour, metalwork, ceramics, clocks, etc. There are only a few items which fall within the scope of this survey.
Lit: LONDON (1920) lists on p78 all the items mentioned below. Catalogue numbers are indicated in brackets.

1 Butterfield dial. Brass. Signed: *Butterfield a Paris.* (III J 533)

2 Butterfield dial. Silver. Signed: *Butterfield a Paris.* (III J 512)

3 String-gnomon horizontal dial. Signed: *Christophorus Schissler faciebat Augusta Vindelicorum Anno Domini 1575.* In the form of a diptych dial with hinged lid. Gilt brass decorated with arabesques. Folding gnomon support. (III J/531)
Lit: BOBINGER (1954) pp90, 125, listed p136; BOBINGER (1966) p314; ZINNER (1956) p512

4 Collection of gunner's instruments, probably from various separate sets of the kind carried in a pocket in the sheath of a master gunner's sword. A-H are by members of the Trechsler family of Dresden. All items

are German, early 17th century and gilt brass unless otherwise indicated. (III J 513-530)
Lit: MANN (1962) pp593-5, A 1246

a Quadrant in the form of a right-angled triangle with a pointer attached in one corner which moves over a scale from 1 to 15. Index in form of a cherub's head. Signed: *16 A.T.F.D.20* [the D is reversed] (Abraham Trechsler).

b Gunner's rule and level. Rule calibrated in inches and for weight in pounds of iron, lead and stone shot. Level in quadrant calibrated 12, 8, 0, 8, 12 with index as A. Signed: *C.T.D.E.M.* (Christopher Trechsler).

c Gunner's rule calibrated for lead, iron, brass and pewter shot in pounds and half ounces.

d Dividers. Gilt brass, steel points. Finely engraved. Signed: *C.T.D.E.M. 1617.*

e-h Four rods, three with corkscrew end. Steel with brass handles ornamented with scrolls terminating in a cherub's head.

i Gunner's level for resting on cannon's breech with vertical slide backed by an upright member graduated from 0 to 16. Stamped *A.T.* on the reverse.

j Protractor. Graduated from 0 - 90 - 0°, a pointer pivoted above the 0° point.

k Pair of compasses, gilt brass and steel. Finely decorated. One leg removable to permit use of other attachments.

l Sliding scale for another gunner's level. Graduated into eleven inches.

5 Spyglass. Gold, no hallmark. Signed: *Love Old Bond Street.* Early 19th century. (XX A 57)

6 Spyglass. Enamelled gold. Maker's mark of Pierre Robert Dezarot, Paris, date letter for 1777/8. (XX A 43)

Lo 27 LONDON
Wellcome Museum of the History of Medicine, Science Museum

Science Museum, Exhibition Road, London SW7 2DD

071 938 8000

The immense and comprehensive medical collections of the Wellcome Museum, together with other material not strictly related to the History of Medicine, were assembled by Sir Henry Wellcome (1853-1936). The Museum, first established in 1913, was later financed by the Wellcome Trust and administered by the Trustees, its collections covering not only medical, surgical and pharmaceutical material, but also anthropology, ethnography and archaeology, and including items such as prints, drawings, coins, medals and paintings. The collections were housed in the Wellcome Building, 183 Euston Road, London NW1 2BP. However, in the late 1960s the Trustees decided to concentrate their funds and their efforts on original research in the history of medicine, a decision which caused them to seek a home for the collection. A legal agreement was reached whereby the Trustees were empowered to place the history of medicine collection on permanent loan in the *Science Museum, London,* where the Wellcome Museum of the History of Medicine has, since 1977, formed a Department with a separate advisory committee. Other scientific and technical material, which falls outside the scope of the Wellcome Museum, has been placed on loan to the appropriate department of the Science Museum. Unfortunately the provenance of most items had not been well documented. The medical collections are now, for the first time, fully catalogued. Material which falls outside the scope of the Science Museum has been offered to other institutions.

The fine Wellcome Library (which includes major collections of printed books, western, Islamic and Indian manuscripts, and the recently formed Contemporary Medical Archive) remains at the Wellcome Institute for the History of Medicine at Euston Road. The Academic Unit of the Wellcome Institute is interlinked with the Unit for the History of Medicine at University College, London. Also remaining is the two-dimensional material – paintings, prints, drawings and photographs – as well as five reconstructed pharmacies (though the contents of these have been catalogued with the other medical artefacts). The Wellcome Museum of Medical Science (largely pathological and pharmaceutical specimens) remains at Euston Road.
Lit: TURNER (1980); BRACEGIRDLE (1981); SKINNER (1986)

I   SCOPE OF THE COLLECTION

1 *Medical and surgical equipment.* In numerical terms, there are more than 40,000 items, including surgical, diagnostic and physiological instruments. These range from Graeco-Roman surgical instruments to current medical high technology such as the prototype EMI head scanner (acquired post-1977 by the Wellcome Museum).
Lit: LAWRENCE (1978, 1979A and B) (for the sphygmograph collection)

2 *Dental instruments.* (Approximately 6000 items).

3 *Ophthalmic instruments.* Includes a collection of 3000 spectacles.

4 *Orthopaedic equipment.* (About 500 items).

5 *Pharmacy equipment.* Includes a large collection of about 5000 ceramic and glass drug jars, pill cutters and machines, drug mills, pill tiles and mortars.
Lit: UNDERWOOD (1951A); CRELLIN (1969) for English and Dutch ceramics; CRELLIN/SCOTT (1972) for glassware; CRELLIN/SCOTT (1970); CRELLIN (1972); CRELLIN/HUTTON (1973) for English bell-metal mortars.

6 *Materia medica specimens.* About 10,000, including a large quantity of Indian origin.

7 *Medicine chests.* Approximately 450 examples, including a large Genoese chest of *c*1565.

Lit: BURNETT (1982)

8   Nursing and other domestic items. Includes posset pots, pap boats, veuilleuses, eye baths and bed pans.
Lit: For ceramic items, see CRELLIN (1969)

9   *Marie Stopes collection of contraceptive devices*

10   *Hospital furnishings*

11   *Medical electricity.* A substantial collection of electrical machines and accessories, including the Galvani Collection.

12   *Microscopes, microtomes and slides.* More than 1500 microscopes including numerous important early items, some derived from the Court Collection (Insley (1982)).
Lit: BRACEGIRDLE (1977, 1978)

13   *Scales and weights.* A very large collection, much of which has been transferred to the *Science Museum, London* (Department of Physics).
Lit: CRELLIN/SCOTT (1969)

14   *Chemical furnaces, earthenware and glassware.* Includes many alembics and retorts.

15   *Personalia.* Many groups, including items relating to Edward Jenner and Joseph Lister. The Florey Collection contains microscopical preparations, preserved specimens and laboratory notebooks relating to the work of Howard Florey.

16   *Anatomical models.* Includes a number of 18th-century Italian examples.
Lit: DEER (1977)

17   *Veterinary equipment*

18   *Talismans, charms and amulets.* Some have been retained though most are now transferred to the *Pitt Rivers Museum, Oxford.*

19   *Medals and coins.* Some 6000 examples with medical associations.

20   *Hunteriana.* Personalia, printed books and pictorial material connected with John and William Hunter on loan from the Hunterian Society.

21   *The Hull Grundy Collection.* Largely medical and medicine-related items in precious metals, acquired post-1977.

22   *Eastern medicine.* Items relating to medical systems of China (*Lit:* LONDON – WELLCOME INSTITUTE (1966)), India (both Ayurvedic and Unani), Tibet, Japan and South-East Asia. There is a small representation of Islamic medicine.

23   *Ethnographic medicine.* Although vast quantities of non-medical material has been transferred to other museums, there remains an extensive amount of material from Africa and North America, with smaller collections from Oceania and South America.
Lit: UNDERWOOD (1951B and 1952)

Lo 28 LONDON
Wesley's House and Museum

47 City Road, London EC1Y 1AU

071 253 2262

A museum devoted to the evangelist and reforming preacher John Wesley (1703-1791) in the house in which he lived and died. Wesley was an early user of the electrical machine as a cure for ailments, and the Museum houses an electrical machine he is known to have used. Wesley's first reference to the use of an electrical machine appears in his *Journal* on 9 November 1756. In 1759 he published a book: *The Desideratum: or Electricity Made Plain and Useful.*

1   Cylinder electrical machine, the cylinder measuring 7½ x 4½in diameter. The collector is on a glass insulating column mounted on a board hinged to the main base plate of the machine. The leather pad with black silk attached is mounted on a metal arm whose pressure against the cylinder is controlled by means of a screw. The machine is mounted on four glass insulating legs. As an accessory there is a Leyden jar.
Lit: WOODWARD (1966), sketch on p29 and p93; HACKMANN (1978) pp128-9

M

M 1 MAIDSTONE
Maidstone Museum and Art Gallery

St Faith's Street, Maidstone ME14 1LH

0622 54497

Important collections relating primarily to Kent and containing archaeological, art and natural history material, as well as the Museum of the Queen's Own

Royal West Kent Regiment. In addition there are less specifically local collections of costume, bygones, ceramics, paintings, etc. One of the more unexpected aspects of the Museum is its fine collection of early printed books. All these items are housed in Chillington House, a 16th-century manor house. In the Archbishop's stables is an extensive carriage collection.

The Museum contains a small collection of scientific instruments.

1   Culpeper microscope (2nd form). Ray-skin outer tube, inner tube vellum covered. Ebony eyepiece and base. Brass snout. Dust cover of eyepiece missing.

2   Matthew Loft form of Culpeper microscope. Brass body. Octagonal wooden box-foot. Revolving object holder. Height approx 17½in.

3 Brass Culpeper microscope. Signed: *Shuttleworth London*. Height 10½in. In pyramidal box with accessories.

4 Nuremberg tripod microscope. Wood with cardboard tubes. Branded on the base the initials *IK* (Junker of Magdeburg).

5 Brass solar microscope. Unsigned. 18th/early 19th century.

6 Cary microscope. Mahogany box 4¾in x 4½in. Signed: *Cary London*.

7 Withering folding botanical microscope. Unsigned.

8 Botanical microscope of brass, by IP Cutts, Sheffield & London. With sheet of instructions and box.

9 Compound monocular microscope. Signed: *Powell & Lealand 170 Euston Road London*.

10 Octant. Ebony frame, radius 14in. Signed on brass arm: *Troughton London*. Label in box: *J. & E. Troughton No. 136 Fleet Street London*. Dated *1790*.

11 Oughtred's double horizontal dial. Bronze, octagonal. Signed: *John Allen fecit 1632.* (fig 35)
Lit: TURNER (1981)

12 Cube sundial on stand with compass. Wood, the cube covered with engraved and coloured paper. Signed: *D. Beringer*.

13 Set of drawing instruments in ray-skin case. Sector signed: *Cook & Barton*.

14 Set-square. Signed: *J. Kelton 1739 Oct. 19th*.

15 Camera lucida, together with instructions for use, 'sold by P. Dollond, Newman & W. Cary'.

16 Pocket globe (terrestrial and celestial). Unsigned. 18th century.

17 Pair of globes on stands. Signed: *John Smith, Globe maker to King George IV*.

18 Celestial globe on stand with compass rose in base. Signed: *D. Adams, Charing Cross*.

19 Waywiser. Signed: *G. Adams London*.

20 Small double-barrel air pump on mahogany base. Unsigned. Probably early 19th century. (Fig 2)

M 2 MANCHESTER
Manchester Literary and Philosophical Society

14 Kennedy Street, Manchester M2 4BY

061 228 3638

The Society was founded in 1781, and included among its early members Erasmus Darwin, Joseph Priestley, and James Watt. John Dalton became a member in 1794 and from 1799 onwards was given accommodation for teaching and research in the Society's house at 36 George Street. Dalton's apparatus and instruments were presented to the Society after his death by his friend WC

Henry, but were destroyed in an air raid in 1940, together with the building and the Society's records. Only a few small items associated with Dalton have survived (nos. 2-7) (others can be found in the *Science Museum, London*). Since the War the society has received a bequest (the Garnett Bequest) which consists mainly of optical material of mid to late 19th century date – microscopes, polariscopes, cameras, and lenses – including important material by JB Dancer, a local instrument maker, some of which has now been placed on loan to the *North Western Museum of Science and Industry, Manchester*, and is therefore listed under that heading.
Lit: Dalton's apparatus, including the material destroyed in the war, is described in two articles: JONES (1904) and BARNES (1915-16). Items 2-8 are discussed, together with other surviving pieces of Dalton apparatus, in FARRER (1968).

1 Pocket microscope. Signed: *Cary London*. Garnett Bequest.

2 Barometer signed: *J. Ronchetti Manchester*.

3-4 Two beakers

5 Pottery bottle used with water thermometer

6-7 Two glass vessels

8 Model of a Watt beam engine. Signed by *Clegg, David Street Manchester, c1810*. Believed to have been used by Dalton.
Lit: HATTON/FLOWETT (1963-64)

M 3 MANCHESTER
Museum of Science and Industry

Liverpool Road Station, Liverpool Road, Castlefield, Manchester M3 4JP

061 832 2244

Opened in Grosvenor Street in 1969 as part of the University of Manchester Institute of Science and Technology, the North Western Museum of Science and Industry concentrated on collections of scientific and industrial material of relevance to Manchester and the North West region. In 1983 the Museum was reconstituted and moved to a complex of buildings which includes the world's first passenger railway station, built in 1830.

Collections include aircraft, prime movers, electricity and paper-making, textiles and photography, as well as scientific instruments and apparatus. The latter is chiefly associated with the scientists John Dalton and James Prescott Joule, and with the maker JB Dancer who pioneered microphotography. Material is derived from such sources as the Manchester Microscopical Society (MMS), the *Manchester Literary and Philosophical Society* (MLPS), the Department of Chemistry of the University, and the former Joule Museum, Acton Square, Salford (JMS), where Joule lived *c*1847-1854 and carried out much important work. Items associated with Joule and Dalton are referred to in Section I under the headings Joule and

Dalton. Other instruments are found in the general list (Section II).

In addition to the items listed the collection contains a number of more recent microscopes by makers such as Zeiss and Ross. There are also ophthalmoscopes, optometers and stereoscopes, a series of 19th- and early 20th-century balances, including one balance used by Moseley for the determination of atomic weight. There is an extensive collection of early cameras, in part associated with, or constructed by JB Dancer.

I   ASSOCIATIONS

1   *John Dalton (1766-1844).* Dalton moved to Manchester in 1793 and in the following year became a member of the *Manchester Literary and Philosophical Society* in whose premises he was later given accommodation for teaching and research. A few items, including two mercury thermometers, three glass flasks, two Leyden jars and a walking stick barometer were formerly housed in the Chemistry Department of the University of Manchester. *Lit:* FARRER (1968)

2   *James Prescott Joule (1818-1889).* Joule material is partly derived from the Department of Chemistry of the University, or from the Manchester University Museum, but chiefly from the former Joule Museum, Salford. It includes compressors and force pumps, electromagnets and cores of electromagnets, an electromagnetic engine with rotating armatures, parts of voltaic cells and Daniell batteries, a calorimeter and other pieces of experimental apparatus constructed in connection with Joule's research. A balance, a foot rule and two microscopes, were constructed by JB Dancer (II, nos. 10, 11, 26, 28). Other Joule material can be found in the *Science Museum, London.* *Lit:* LOWERY (1930)

3   *Other associations.* In addition there are quantities of later apparatus from the Department of Chemistry of the University, associated with Sir Henry Roscoe (1833-1915), Sir Ernest Rutherford (1871-1937) and Harold B Dixon (1852-1930).

II   INSTRUMENTS

1   Matthew Loft form of Culpeper microscope. Ray-skin covered outer tube, inner tube covered with green vellum. From Manchester University Museum.

2   Nuremberg tripod microscope. Embossed paper outer tube. Late 18th/early 19th century. From Manchester University Museum. (fig 81)

3   Brass Culpeper microscope. Unsigned. Late 18th century (MMS)

4   Small Cary-type microscope. Marked: *J. Ronchetti, 43 Market Street Manchester.*

5   Cary microscope. Unsigned. Early 19th century.

6   Continental-type drum microscope in box with accessories. Marked: *J. Frank, 114 Deansgate Manchester.* (Loan)

7   Brass drum microscope, *c*1850. (MLPS)

8   Compound microscope signed: *J.B. Dancer Manchester.* Dancer's trade card in box. (1971.66)

9   Compound monocular microscope on tripod foot. Tilting stand. Signed: *Abraham & Dancer Manchester.*

10   Travelling microscope constructed by Abraham & Dancer for Joule. Used for reading as accurately as possible the temperature on a thermometer, *c*1842. Signed: *Abraham & Dancer.* (JMS)

11   Compound microscope made by JB Dancer (JMS)

12   Microscope with binocular and monocular bodies (Regd 27 June 1861) Signed: *J.B. Dancer Optician No. 517 Manchester.*

13   Binocular microscope. Signed: *J.B. Dancer Manchester.* (MLPS)

14   Binocular microscope probably made by Beck but signed by Dancer: *J.B. Dancer Optician Manchester.* (fig 89)

15   Refracting telescope on pillar and tripod stand. Table model. 4in aperture, length 25in. Signed: *Watkins London.* Loan from Manchester Astronomical Society.

16   Refracting telescope on tripod stand. Floor standing model. 4in aperture. Length 63½in. Signed: *Dancer Manchester.*

17   Stereoscopic camera. Signed on ivory label: *J.B. Dancer Manchester Patent No. 12.* (MLPS)

18   Microlantern attachment for projection of microscope slides. Dancer's label in the lid. (MLPS)

19   Optical system for lantern slide projector. Signed: *Dancer Manchester.*

20   Gyroscope. Signed: *J.B. Dancer.* (From MLPS)

21   Cabinet of slides consisting of 31 drawers. Bears nameplate: *J.B. Dancer Optician Manchester.* (Loan)

22   Equinoctial ring dial. 18th century. Diameter approx 6in. (Loan)

23   Octant. Ebony frame. Signed: *Simson Strand St Liverpool.*

24   Surveyor's level. Signed: *W. & S. Jones 30 Holborn London.*

25   Surveyor's compass in rectangular wooden box. Folding sights. Signed: *J.B. Dancer Optician Manchester.* (1968.15)

26   Tapering foot rule. Signed: *Dancer Manchester.* (1970 1-4)

27   Hydrometer for testing silver solutions for photography. Signed: *J.B. Dancer Manchester.* (From MLPS)

28   Balance. Signed: *Dancer Manchester.* Supplied by Dancer for Joule in 1846, probably (JMS). *Lit:* STOCK (1973)

29   Balance, said to have been once the property of Humphry Davy.

30   Double pan balance. Late 18th/early 19th century.

31 Cavendish eudiometer
Lit: FARRER (1963)

M 4 MONMOUTH
Nelson Collection and Local History Centre

Market Hall, Priory Street, Newport NP5 3XA

0600 3519

The Museum, newly housed in the Market Hall, is devoted to relics of Lord Nelson, his contemporaries and Lady Hamilton. It contains a number of instruments:

1 Day and night telescope signed: *Dobson & Baker*

Chiswell St London.

2 Nautical telescope, no draw tubes, leather bound. Inscribed *Admiral Horatio Nelson K.C.B, H.M.S. Theseus, 1797.*

3 Two-draw telescope, brass and wood, signed: *Spencer Browning & Rust, London Day or Night.*

4 Sextant. Radius approx 12in. Ebony frame. Incomplete. Inscribed on ivory plate: *H. Duren.*

5 Richardson's saccharometer by *Joseph Long, 20 Little Tower St London.*

6 Sikes' hydrometer signed: *Loftus 321 Oxford St., London.*

N

N 1 NEWCASTLE UPON TYNE
Museum of the Department of Mining Engineering, University of Newcastle

The University, Queen Victoria Road, Newcastle upon Tyne NE1 7RU

091 232 8511

A College of Physical Science (later Armstrong College) was founded in 1871 but it was not until 1963 that the University of Newcastle upon Tyne became a distinct institution, having formerly been part of the University of Durham. The Museum houses a number of mining and surveying instruments which date mainly from the second and third quarters of the 19th century, as well as one earlier item – the pantograph, which belongs to the late 18th century. In addition to the items listed below there is an extensive collection of miner's lamps.

1 Brass pantograph in mahogany box. Signed on the instrument: *Troughton London.* The box bears the label: *J. & E. Troughton (Successor to Mr Cole) Mathematical, Optical and Philosophical Instrument Maker No. 136 Fleet Street London.*

2 Agricultural drainage level. Telescope for use on a tripod fitted with an adjusting screw working against a spring. Signed on telescope tube: *Cail Newcastle upon Tyne.* Silk-lined mahogany box.

3 Clinometer (? for use with a miner's dial). Semicircle of brass, silver scale. Signed: *Cail Newcastle on Tyne.* Concealed in the box of this instrument (similarly lined to no. 2) is the following inscription in ink: *Made in 1847 by Atkinson Bussey for John Bell.*

4 Miner's dial. Signed: *F. Robson Maker Newcastle upon Tyne.* Mid-19th century.

5 Miner's dial. Signed: *J. Hewitson Newcastle upon Tyne, c*1850-60.

6 Transit theodolite with compass and four screws. Signed: *F. Robson Maker Newcastle on Tyne* amd inscribed: *Durham College.*

7 Transit theodolite. Three screw, no compass. Signed: *F. Robson & Co. Newcastle on Tyne.* Inscribed: *Armstrong College.*

8 Transit theodolite. Three screw, micrometers for reading circles. Signed: *Troughton & Simms.*

9 Plotting protractor. Incomplete. Signed: *Watkins & Hill 5 Charing Cross Rd. London.*

10 Alidade with telescope and vertical semicircle for use with a plane table. Signed: *T. Cooke & Son Ltd. London. York and Cape Town.*

N 2 NEWCASTLE UPON TYNE
Museum of Science and Engineering

Blandford House, Blandford Square, Newcastle upon Tyne NE1 4JA

091 232 6789

The Museum was originally opened in 1934 and was housed in the former Palace of Arts of the 1929 North East Coast Exhibition in Exhibition Park. The extensive collections were transferred in 1981 to their present location. These relate principally to mining, mechanical and electrical engineering, shipbuilding and marine engineering, local trades and industries, and transport. Transport collections, formerly shown at Exhibition Park, are now at Monkwearmouth Station Museum, Sunderland, except for the *Turbinia*, built by Sir Charles Parsons, which is still housed at Exhibition Park (though scheduled to be removed to Newcastle Quayside).

The Museum of Science and Engineering is administered by the Tyne and Wear County Museum Service. Most scientific instruments within the Museum Service are displayed or stored at Blandford House, though examples may move between here and the *Museum and Art Gallery, Sunderland* or the Museum and Art Gallery, South Shields.

Apart from the material listed below there are extensive collections of late 19th- and 20th-century electrical and navigational instruments, ophthalmic optics (260 eyeglasses, spectacles, etc. dating from 1720 to 1960), photographic apparatus, miners lamps (80 examples including five Spedding mills), and metrology, especially engineering examples from Elswick Ordnance Co., NG Armstrong; Armstrong, Whitworth; Vickers Armstrong; and Vickers Ltd, dating from 1850 to 1950. In addition, there are 50 instruments of various kinds on loan from Mr Arthur Frank. The strength of the collection is undoubtedly its representation of makers and retailers from the north-east of England.

Lit: NEWCASTLE UPON TYNE (1950)

MICROSCOPES, ASTRONOMICAL TELESCOPES, ETC.

1   Culpeper microscope. Brass, with rack focusing. Signed: *James Long, Royal Exchange, London. c*1770-1780. (fig 79)

2   Drum microscope. Brass, unsigned. Early 19th century. (C3293)

3   Compound microscope. Signed: *J.B. Young, Newcastle, c*1890. (C10513)

4   Compound microscope. Signed: *Pillischer, London, c*1880.

5   Binocular microscope. Signed: *Crouch, London.*

6   Gregorian reflecting telescope on table tripod, 1½in. aperture. Signed: *Veitch Inchbonny 1828.*

7   Gregorian reflecting telescope on table tripod, 1⅝in aperture. Unsigned. *c*1880.

8   Gregorian reflecting telescope on table tripod, 1¾in aperture. Unsigned. *c*1900.

9   Refracting telescope on stand, signed: *Jas Long, Royal Exchange.* Supplied to Newcastle Literary and Philosophical Society in 1802.

10   Camera lucida. Signed: *Carpenter 21 Regent St London.*

SURVEYING, NAVIGATION

11   Azimuth compass, in 7in square box with lid. Signed: *Harris 50 Holborn London.* Early 19th century. (B2937) (18619)

12   Miner's dial, signed: *F. Robson, Newcastle upon Tyne. c*1840-50. (B2938) (8471)

13   Miner's dial, signed: *Cail, Newcastle upon Tyne* and *135. c*1840. (B2934) (8622)

14   Miner's dial, signed: *Bowie* (of Sunderland). Late 19th century. (B2935) (5132)

15   Miner's dial, signed: *T.B. Winter, Newcastle.* Late 19th century. (C69)

16-18 Miner's levels (3), signed: *Cail Newcastle.* Mid-19th century. (B2933) (8620), (B2936) (8660) and (A38)

19   Theodolite, signed: *T. Heath London, c*1740. (fig 131)

20   Theodolite, signed: *T. Heath London, c*1730 (incomplete)

21   Theodolite, signed: *W & S Jones, c*1800 (C10516)

22   Waywiser, signed: *Dollond.* (C12325)

23   Octant, in case. Ebonised wooden frame, radius 10in. Brass arm, ivory degree scale, vernier and tangent screw. Signed: *P.A. Feathers Dundee.* Mid-19th century.

24   Octant. Mahogany with boxwood degree arc and diagonal scale, radius 18in. Unsigned. Mid-18th century.

25   Octant. Ebony frame, ivory arc, vernier, brass arm, radius 18in. Signed on arm: *Dollond London.* (9176)

26   Octant. Ebony frame, radius 11in. Signed on ivory label: *G. Bradford. 99 Minories London.* (4127)

27   Octant. Ebony with ivory scale and brass index arm. Signed: *S.B.R.* Radius 14in. (2232)

28   Octant. Wood, including index arm. Signed: *H. Carr.*

29   Octant. Ebony with ivory scale and brass index arm. Signed: *L. Simon South Shields.*

30   Octant, in case. Ebony with ivory scale and brass index arm. Signed: *L. Ainsley, South Shields.* (9165)

31,32 Octants. Wood with ivory scale and brass index arm. Signed: *Spencer, Browning & Rust, London.* (2231) and unnumbered.

33   Octant. Wood with ivory scale and brass index arm. Signed: *Braham, Bristol*

34   Octant. Ebony with brass index arm and ivory scale, radius 10in. Signed: *McAll, Tower Hill, London.* Chapman Collection (E1787)

35   Octant. Ebony with brass index arm and ivory scale, radius 10in. Signed: *Spencer & Co. London.* Chapman Collection (E2011)

36   Octant. Ebony with brass index arm and ivory scale, radius 13in. Signed: *T. Smith, Old Gravel Lane, Wapping, London* and *Alex Nicholson. 1806* Chapman Collection (E2012)

37   Sextant, in case. Metal, brass scale. Signed: *Hood, London,* label inscribed: *Frederick Smith, Southampton*

38   Box sextant, diameter 2½in. Brass with silver degree arc. Signed: *Allan London.* In mahogany box with trade card: *John Steele's Nautical Warehouse. Opposite Dukes Dock, Wapping Liverpool.*

39   Box sextant. Signed: *Adie & Son. c*1870. (C10506)

40   Ship's log. Signed: *W. & S. Jones, Holborn, London.* (3291)

41   Telescope. Brass and leather bound. Signed: *Spencer Browning & Co. London.* (E1830)

42   Telescope. Single draw, 1¼in aperture. Signed: *Cail, Newcastle.* With Royal coat of arms (probably an excise officer's instrument).

DRAWING, MEASURING AND CALCULATING INSTRUMENTS

43 Adjustable bow for producing arcs, boxwood. Signed: *Made by Walter Henshaw for Thomas Wrangham 1698.* Length 30in.

44 Scale, boxwood, 12 x 1¾in. Lines of numbers, rhumbs, etc. Unsigned. Early 19th century.

45 Sector, ivory and silver, 6in long. Signed: *F. Watkins, London.*

46 Routledge slide rule, 24in, boxwood. Signed: *Rabone & Sons, Birmingham.*

47 Hawthorne locomotive engineer's rule, 24in. Ivory, with nickel silver mounts. (A486)

48 Cocker patent 2in decimal gauge, signed: *Jos. Whitworth & Co., c*1880.

49 Gunner's calipers, brass. Signed: *R. Whitehead fecit.* and *T. Grey. Mar 25, 1691.* (C3273)

50 Proportional dividers. Signed: *T.B. Winter.* Late 19th century. (C12337)

51 Circular protractor, brass, 2⅜in radius. Unsigned, probably Thomas Heath, 18th century.

52 Circular protractor, signed: *Keyzer & Bendon, High Holborn, London*, in box with trade card signed: *T.B. Winter, Newcastle.*

53 Protractor, brass. Signed: *Robson* (of Newcastle). Late 19th century.

ELECTRICAL INSTRUMENTS

54 Small cylinder electrical machine. Second half of 19th century. (B516)

55 Large multi-plate Wimshurst machine. Signed: *Newton & Co. 3 Fleet St London.* Formerly used by Lord Armstrong, transferred from King's College, Newcastle. (8215)

56 Armstrong hydro-electric machine, with pumping engine for feedwater. Unsigned, though probably by Watson of Newcastle. *c*1843. (3203)

57 Modified form of Armstrong hydro-electric machine, gas-fired. Transferred from King's College, Newcastle. (6478). On loan to Bamburgh Castle, Northumberland.

58 Cooke and Wheatstone double needle telegraph. Signed: *C. Wheatstone Invt.* On loan from *Science Museum, London.*

CHEMICAL EQUIPMENT

59 Chemistry bench, with accompanying glassware, from Newcastle School of Chemistry, *c*1871.

60 Dicas hydrometer. Signed: *Made by S. & M. Dicas, Mathematical Shop, Liverpool.* (3728)

61 Clarke hydrometer. Label, in silk lined box, signed on thermometer: *Dring & Fage No. 248 Tooley St Near London Bridge.*

62 Pair of alcohol slide rules, boxwood. Signed: *Loftus, London.*

MISCELLANEOUS

63 Pyrometer, Musschenbroek type, mounted on red-lacquered wooden base. Signed: *Gallonde a Paris.* (3662)

64 Claude Lorrain glass. Signed: *Carpenter & Westley 21 Regent Street, London*

65 Universal equatorial dial, replaced gnomon. Signed: *Vogler, Augsburg.* (B2923)

66 Dickenson anemometer. Signed: *Casartelli.* (B416)

67 Biram anemometer. Signed: *Watson.* (B2932)

68 Fortin barometer. Signed: *Cail, Newcastle, c*1830. (341)

N 3 NORTHAMPTON
Museum of Leathercraft

Central Museum and Art Gallery, Guildhall Road, Northampton NN1 1DP

0604 34881

The collections of the Museum of Leathercraft, formerly based in London, and partly displayed in the Guildhall Museum, are now housed and displayed in the Central Museum and Art Gallery, Northampton. The Collection contains one or two scientific instruments, assembled not for their intrinsic interest as instruments, but because leather was used in the construction of the instrument itself, or of its case. The most important instrument, an astronomical telescope of 17th-century date, is on loan to the *Museum of London*, and listed under that heading.

1 Triple sandglasses (1, ½ and ¼ hr) in tooled leather carrier, Florentine, *c*1600. (1837-71)

2 Set of drawing instruments in red and green ray-skin covered case. Instruments of silver including a protractor, pen, inkwell, dividers and rule, engraved: *Baradelle Paris No. 1438.* French, 18th century. (617-57)

3 Set of drawing instruments in ray-skin case. Instruments include two pairs of dividers and two slide rules, and are probably of various dates. English, 18th century. (118-49)

4 Drawing instruments in tapering case of silver covered in black morocco leather, and decorated with silver pique work. (653-5)

N 4 NORWICH
Castle Museum

Norwich NR1 3JU

0603 611277

The Museum's collections, housed in the Castle and in several branch museums (part of the Norfolk Museums Service), cover all aspects of art, archaeology, natural history, domestic life and local industry, and contain a substantial number of scientific instruments. At the time of writing these were to be found in the Castle, Strangers' Hall and Bridewell Museum, although not necessarily on exhibition.

As well as the instruments listed below the collection contains a number of medical and surgical instruments, optical toys such as stereoscopes, magic lanterns, Goodchild's trocheidoscope, etc. (see YOUNG (1975) and JONES (1980) for catalogues of the teaching toys), several domestic barometers mainly by local makers (Brook; Beha, Lickert & Co.; Rossi, also by Butler of Somers Town, London), a collection of clockmaking tools and several mid- to late 19th-century microscopes by Watson, Baker and Leitz.

1 Combined spyglass and magnifying glass with needle object holder. Cylindrical with screw cap. Lignum vitae. 18th century. (129.940.2) (fig 73)

2 Screw-barrel microscope fitting into cylindrical ivory tube comprised of several sections containing lenses, slides, etc. Signed: *Ed. Scarlett Soho London.*

3 Botanical microscope with hinged lens holder and ivory handle. Cardboard case, *c*1800. (120.949)

4 Compass microscope with turned ivory handle in velvet-lined shark-skin covered box. Mid-18th century. (104.948)

5 Ellis-type naturalist's microscope (screwing into top of fish-skin covered box). Unsigned.

6 Leeuwenhoek 'aquatic' microscope. Brass. Used by William Arderon FRS (1703-1767). Signed on front plate: *John Yarwell at ye Archimedes & 3 Golden Prospects in Ludgate Street.* Inside marked: *E.C.* (Edmund Culpeper). (65.95). (fig 75)
Lit: CLAY/COURT (1928) p36, fig 12

7 Culpeper microscope (second form) with brass stage and red leather outer tube with gilt tooling, *c*1730.

8 Drum microscope. Red leather outer tube with gilt tooling. Tripod legs separating the cylindrical base (with circular opening to admit light to the mirror) and the microscope proper. Mid-18th century. (312.968)

9 Brass Culpeper-type microscope. Unsigned. Early 19th century.

10 Cary-type microscope, the pillar screwing into the edge of the opened box. Signed: *Cary London.* (102.928)

11 Compound monocular microscope, the limb supported on a pillar mounted on a fixed tripod stand. Signed by *Carpenter and Westley. 24 Regent Street London*, *c*1840. Belonged to the Rev. J Crompton, founder, in 1852, of the Norwich Microscopical Society. (80.938)

12 Binocular microscope. Signed *A. Ross London 615.* Originally the property of the Norwich Microscopical Society. (53.01)

13 Drum microscope. Brass. Signed: *A. Henkel Bonn No. 106.* Presented 1850 (25.50)

14 Handheld telescope. Wood, brass mounted. Signed: *Utzschneider und Fraunhofer* (102.968).

15 Handheld telescope signed: *Spencer Browning & Rust London Day or Night.* Brass, with painted wooden tube.

16 Spyglass with six-draw tube, the outer tube ivory cased. Early 19th century. (111.970)

17 Camera lucida. Early 19th century. (168.949).

18-20 Three brass ring dials. (129.468; 46.94.419; 76.94.17).

21 Horizontal dial. Octagonal. Silver and gilt in leather case. Signed: *Salomon Chesnon ABlois.* (16 4.962)

22 Bronze horizontal garden sundial from Chantry Court, Theatre Street, Norwich. Octagonal. Signed: *B. Martin London.* (37.965)

23 Bronze horizontal garden sundial. Diameter 16½in. Engraved with the arms of the Wyndham family and probably from Earlham Hall, Norfolk. The sundial is unsigned, but the engraving is of the very highest quality. Probably *c*1750-60. (168.970)

24 Diptych dial signed: *David Beringer.*

25 Octagonal ivory diptych dial signed: *Conrad Karner.* (56.B)

26 Equinoctial ring dial. Brass, 18th century. (54.900)

27 Equinoctial ring dial. Diameter 6⅛in. Signed: *Made by Tho. Tuttell of Charing Cross London Mathematical Instrument maker to ye Kings most Excellent Majesty.* (175.959)

28 Magnetic dial signed: *C. Essex & Co.* (on compass card). Label in lid of box: *Simpson jeweller watch & clockmaker Hadleigh.* (53.935)

29 Boxwood Gunter's quadrant with planispheric nocturnal on the back. 17th century. Tudor rose stamped on front. (27.894.2)

30 Brass horary quadrant (Gunter type) incorporating a horizontal sundial on the reverse. The initials VS are engraved on the front of the instrument. (30.952)

31 Boxwood horary quadrant, the Panoraganon of William Leybourn, incorporating trigonometric scales (chords equal parts, etc.). Radius 9 5/16in. 17th century. (27.894.1) (Fig 113)
Lit: LEYBOURN (1672)

32 Octant. Signed: *Henry Gregory London*, third quarter 18th century. (389.971)

33 Sextant. Inscribed *T. Williams Cardiff.* Box with label of Henry Hughes & Son, Marine Opticians. (296.968)

34 Pocket compass in circular brass box. (41.969)

35 Compass in rectangular mahogany box. Paper compass card. Late 18th century. (128.968)

36 Compass in octagonal wooden mount with rectangular projection at one edge; for use with plane table. Late 18th century.

37 Waywiser. Signed: *Moore fecit.* (88.935.122)

38 Waywiser, by George Taylor of Mattishall (nr Dereham). Dated 1866. (186.955)

39 Waywiser. Unsigned. (135.922.367)

40 Circular protractor. Alidade with vernier. Signed: *Wm Collier London.* (130.968)

41 Specific gravity beads (9) in turned wooden box. Handwritten label in lid 'Hydrometers to test liquids'. (135.922)

42 Saccharometer. Signed: *Patent improved saccharometer by Josh. Long, late Gill, Little Tower Street, London.* Thermometer signed: *W. & S. Jones* (158.933)

43 Hydrometer by *J. Long, London.* (557.962)

44 Hydrometer. Signed: *Loftus 321 Oxford Street, London.* (351.962)

45 Hydrometer by *J. Long London.* (551.962.2)

46 Thermometer and pocket compass in circular leather case. Early 19th century. (4.75)

47 Chondrometer. Signed: *Thos. Rubergall, Coventry Strt. London.* (127.968)

48 Chondrometer. Signed: *W. & J. Cooper Lynn: 1826.* (50.952)

49 Plate electrical machine.

50 Small cylinder electrical machine.

51 Large cylinder electrical machine.

52 Clockmaker's lathe (for making tower clocks) by Johnson Jex of Letheringsett, *c*1815. (11.942)

53 Clockmaker's wheel-cutting machine. Index plate mounted above rectangular frame. 18th century. (109.950)

54 Watchmaker's micrometer by Johnson Jex of Letheringsett, Norfolk, 1779-1852. (40.941.3)

55 Watchmaker's lathe by Johnson Jex of Letheringsett, *c*1815. (40.941.2)

56 Clockmaker's turn. Probably early 19th century. (220.946.3)

57 Compass dial plate, copper alloy, marked with the latin initial letters of the points of the compass. Diameter

1⅜in. Early 16th century.
Lit: ATKINS/MARGESON (1985)

N 5 NOTTINGHAM
Industrial Museum

Courtyard Buildings, Wollaston Park, Nottingham NG8 2AE
0602 284602

The Industrial Museum, a branch of Nottingham's City Museum and Art Gallery, contains all the relevant material. Primarily the museum houses material relating to local industry, especially lace making. There are only four relevant items, all of them surveying instruments.

1 Circumferentor. Brass, silvered compass dial. Attachment for tripod. The sights are missing. Signed: *G. Adams Fleet Street, London.* Mid-18th century.

2 Surveyor's compass in rectangular mahogany box with lid. Signed: *Whitehurst & Son Derby*

3 Surveyor's level. Signed: *Troughton & Simms London.* Early 19th century.

4 Surveyor's level. Signed: *Troughton & Simms London.* Late 19th or early 20th century.

N 6 NOTTINGHAM
Department of Physics, University of Nottingham

University Park, Nottingham NG7 2RD
0602 484848

The Department contains, other than a 19th-century locally made sidereal clock by G & F Cope and Co. of Nottingham, one scientific instrument falling within the scope of this study.

1 2½in Gregorian reflecting telescope. Fish-skin covered tube. Simple mounting on pillar and tripod stand. Signed: *Matthew Loft fecit.* Early 18th century

O 1 OLNEY
Cowper and Newton Museum

Orchard Side, Market Place, Olney MK46 4AJ
0234 711516

The Museum contains personal belongings of William

Cowper the poet and Revd John Newton, as well as items of local interest. William Cowper moved to Olney in 1767 after having made the acquaintance of the Rector of Olney, Revd John Newton. The dwelling in which Cowper settled with his friend Mrs Unwin, Orchard Side, is the one in which the Museum is now housed. He remained in Olney until 1786. The only scientific instrument in the museum is an electrical machine said to have been used by Newton on Cowper for therapeutic purposes, and was obtained from members of Cowper's family. Cowper's use of an electrical machine is documented in a letter of 24 May, 1792, to Lady Hesketh.
Lit: SPILLER (1968)

1 Cylinder electrical machine on rectangular base 20 x 12in. Cylinder 11in long held on two wooden supports. Friction cushion and pillar are missing. The prime conductor on a separate oval base is supported on a glass pillar. The collecting comb is mounted in one end of the prime conductor. Similar to Cavallo's electrical machine. Accessories include a Leyden jar filled with brass filings and a long narrow jar with a slight shoulder.

O 2 OXFORD
Ashmolean Museum, University of Oxford

University of Oxford, Beaumont Street, Oxford OX1 2PH

0865 278000

The nucleus of the Ashmolean's collections is the 'cabinet of curiosities' assembled by John Tradescant, enlarged by his son, and presented to Elias Ashmole (*Lit:* McGREGOR (1983)). Ashmole offered it in 1675 to Oxford University with the proviso that a suitable repository be found. The collection was originally housed in a specially erected building in Broad Street, now the *Museum of the History of Science*. It was the first public museum to be opened in the British Isles (1685). The museum has greatly expanded since its foundation and certain sections of the early collections (e.g. natural history and ethnography) have been transferred to other, more recently founded, university museums. The present-day collections cover painting, the decorative arts, both European and oriental, prints, drawings and watercolours, coins and medals, as well as British, Mediterranean, Egyptian and Near Eastern archaeology. Scientific instruments are, however, few, and are in the care of the Department of Antiquities. As well as the items listed below there are two guinea balances, two small balances from Tartus and Smyrna, and an extensive collection of watches.

1 Universal equinoctial dial. Brass, 2 x 2in. Signed: *Johan Schretteger in Augsburg.* (1913.916)
Lit: BOBINGER (1966) p356; ZINNER (1956) p53; GUNTHER (1920-23) vol.2, p131

2 Diptych dial incorporating a windvane. Ivory, 3½ x 2¼in. Signed: *Lienhart Miller 1614* (1836 Cat. p5 no. 68)
Lit: GUNTHER (1920-23) vol.2, p129, Illd.; ZINNER (1956) p447.

3 Horizontal dial. Bronze, 6 x 6in. Signed: *H. Sutton fecit 1657* (1888.488)
Lit: GUNTHER (1920-23) vol.2, p110

4 Gunter's quadrant. Brass, radius 3½in, constructed for 51°40' and incorporating an almanac for 1660-84. On the back of the quadrant is a planispheric nocturnal. Inscribed: *I.W. 1665.* (1892.45A)
Lit: GUNTHER (1920-23) vol.2, p178

5 Ring dial. Bronze, *c*1700. (1886.5828)

6 Ring dial. Bronze, *c*1700. (1886.5829)

7 Diptych dial. Ivory, with red and black inlay. Underside with incised 4. German, Nuremberg, second half 17th century. Probably by Melcher Karner. (1889.80)

Lit: ZINNER (1956) p404

8 Bottom leaf of diptych dial of bone. Probably S. German, 17th century. Found in New Bodleian Libraryexcavations. (1937.252)

9 Horizontal plate dial. Brass, octagonal, 8 x 8in. ? 18th century. (1931.554)

10 Equinoctial ring dial, brass, unsigned, *c*1730. Presented by Lewis Evans. (1894.19)

11-14 Clog almanacs. Wood, Danish, 7th century. Presented by J Heysig in 1683 (1836 Cat. p133, nos. 343, 345-7). Another, no. 344, transferred to *Museum of the History of Science, Oxford*
Lit: GUNTHER (1920-23) Oxford vol. 2, p285, no. 158

15 Compass. Brass, probably 18th century. From St Aldates, Oxford. (1880.257)

16 Abacus. Pinewood, Russian, early 17th century. Tradescant Collection.
Lit: GUNTHER (1920-23) vol. 1, p126; RYAN (1972); McGREGOR (1983) (frontispiece)

17 Alembic, greenish glass, from Iran. (Reitlinger gift)

18 Alembic, pottery, and other fragments of distillation apparatus from Oxford sites, 16th/17th century.
Lit: MOORHOUSE (1972)

O 3 OXFORD
Merton College, University of Oxford

Merton Street, Oxford OX1 4JD

0865 276380

Founded in 1264. During the 14th century, Oxford, and Merton College in particular, was a major centre for the study of mathematical sciences, and at least some of the instruments listed may well have been made for, or used by, members of the Merton School of Astronomy. Geoffrey Chaucer's *Treatise on the Astrolabe* was written *c*1400 for his son, Lewis, who was an undergraduate at Merton.

1 New quadrant of Profatius. Bronze, radius 12in. French, first half 14th century.
Lit: GUNTHER (1920-23) vol. 2, pp165-9; WARD (1965-8)

2 Astrolabe and equatorium. Brass, diameter 14¼in. No interchangeable plates, since the astrolabe was designed only for use in Oxford, and is inscribed: *Lat. 52°6' Oxonia.* Rete for 32 stars. Two lugs carrying vanes with pinhole sights are on either side of the shackle. Near the foresight is a fixing point for a plumb line for use with a quadrant scale engraved on the border. English, *c*1350. Possibly this is the *astrolabium magnus* left to Merton in 1372 by Simon Bredon.
Lit: GUNTHER (1920-23) vol. 2, pp208-10; GUNTHER (1932) no. 297; PRICE (1959B); TURNER (1973A) no. 13, illd. p23.

3 Astrolabe. Brass, diameter 9¾in. Rete and rule missing, lunar dial and index added later. Three plates engraved for Climates III-VII (latitudes 30, 36, 41, 45 and

48°. In place of Climate II is a plate for London, 52°. Climate I is in the base of the mater (16°). English, *c*1390. *Lit:* GUNTHER (1920-23) vol. 2, pp210-13; GUNTHER (1932) no. 303, plate 133.

4 Physician's quadrant. Circular, diameter 6⅛in, with 3¾in quadrant engraved on one face, below which is an engraved human figure where each part of the body is marked with its appropriate sign of the zodiac. The instrument has been subject to various alterations. On the reverse, a volvelle for lunar and solar calculations.

5 Astrolabe. Brass, diameter 13½in: Signed: *Gualterus Arscenius nepos Gemme Frisius Louanni fecit anno 1571*. Rete for 51 stars, plates for 48, 50, 52, 54 and 56° and table of horizons. Quadratum – nauticum in base of mater. Universal stereographic projection of Gemma Frisius on the reverse. The astrolabe was left to the College in 1598 by Dr John Woodward (fellow in 1547). *Lit:* GUNTHER (1920-23) vol. 2, pp216-8; GUNTHER (1932) no. 232, plate 96.

O 4 OXFORD
Museum of the History of Science

Old Ashmolean Building, Broad Street, Oxford OX1 3AZ

0865 277280

The Old Ashmolean Building was erected between 1679 and 1683 to house the Ashmolean Museum, the School of Natural History of the University and the Chemical Laboratory.

The Ashmolean Museum moved to its present site in Beaumont Street in 1895/6. In 1925 when the Lewis Evans Collection of scientific instruments was presented to the University, it was housed on the first floor of the Old Ashmolean Building. To the Lewis Evans Collections were soon added items on loan from Colleges of the University. The Museum of the History of Science was established by statute in 1935, and now occupies the whole building. The moving spirit behind this venture was RT Gunther (1869-1940), Fellow of Magdalen College and first Curator. Its collections include scientific instruments and apparatus, clocks and watches, and an extensive library of early printed books on dialling and scientific instruments.

Of particular importance are the early astronomical, mathematical and surveying instruments, as well as the microscopes, the latter augmented by the loan of the important Royal Microscopical Society Collection. Of great significance too, are the astronomical instruments from the Radcliffe Observatory, and the less numerous but much earlier instruments of the Savilian professors of astronomy.

No attempt can be made here to list individually the instruments housed in the Museum of the History of Science. Astrolabes alone number more than one hundred, the largest collection in the world; there are several thousand sundials. Instead, in common with other large collections, an attempt has been made merely to indicate its scope. Section I indicates the provenance of major groups of instruments, Section II the main types

of instruments represented, but it must be recognised that this survey does not seek to be exhaustive, and the absence of reference to a particular type of instrument or maker does not imply that they are not represented in the collection.

Lit: For general information, see GUNTHER (1935 and 1939). GUNTHER (1967) includes biographical details of the Museum's founder, RT Gunther, and material dealing with the foundation and development of the Museum, and SIMCOCK (1985) and GUNTHER (1920-23) list many instruments now in the Museum, under their original Oxford location. Collections donated by JA Billmeir are catalogued by JOSTEN (1955) and MADDISON (1957). Catalogues exist of the chemistry apparatus, HILL (1971), watches, MADDISON/TURNER (1973) and drug jars, HILL/DREY (1980). Details of some of the important early instruments are found in MADDISON (1969). Some of the French instruments are catalogued in BEDARIDA/MADDISON (1970). Additional literature referring to specific aspects of the collection is listed under the section to which it relates.

I   SOURCES OF THE COLLECTION

1   *Lewis Evans Collection*. An extensive collection of early scientific instruments presented to the University of Oxford in 1924 by Dr Lewis Evans. It includes astrolabes, perpetual calendars, a fine and extensive collection of portable sundials and nocturnals, as well as other scientific instruments.

2   *Daubeny Collection*. Chemical apparatus associated with CGB Daubeny (1795-1867), Aldrichian Professor of Chemistry from 1822-1854. Daubeny bequeathed the collection, which includes some material which may have belonged to his predecessor, John Kidd (1775-1851), to Magdalen College; it is now displayed in the basement of the Old Ashmolean Building, in the *Officina Chimica* where Daubeny's early laboratory was situated. *Lit:* HILL (1971); GUNTHER (1920-23) vol. 1, pp81-86

3   *Orrery Collection*. Assembled by Charles Boyle, 4th Earl of Orrery (d. 1731), and bequeathed, with a collection of books, to Christ Church. The collection has been displayed in the Old Ashmolean Building since 1925. The instruments, primarily dating from the end of the 17th or the early years of the 18th century, include microscopes, telescopes, a scioptric ball, mathematical, drawing and surveying instruments, portable quadrants, and some philosophical instruments. Makers such as James Wilson, John Marshall, John Worgan and Henry Sutton are represented, but a large proportion of the instruments are signed by John Rowley. One item from the Orrery Collection – the 1713 orrery by J Rowley – was not presented to Christ Church and remained until 1974 the property of the family. It was then purchased by the *Science Museum, London* to whom it had been on loan for many years. *Lit:* GUNTHER (1920-23) vol. 1, pp378-82 (reprints Thomas Wright's inventory of the Orrery collection, compiled 1731). TURNER (1973B) illustrates a number of items from the Orrery Collection.

4   *Oriel College*. Instruments on loan to the Museum include an important medieval astrolabe and quadrant of

14th-century date as well as several philosophical items from the late 18th century.

5 *St John's College.* Instruments from this source include Oughtred's Circles of Proportion (the earliest dated example of a logarithmic slide rule) and an astrolabe by George Hartmann.

6 *Other Colleges.* Instruments from Magdalen, All Souls, The Queen's College and Balliol are deposited in the Museum. Of the Colleges, only *Merton* (see above) retains instruments.

7 *Savilian Professors of Astronomy.* In 1936 a number of early instruments came to light which were used by the Savilian Professors of Astronomy in the 17th century. As well as an astrolabe constructed by Thomas Gemini for Queen Elizabeth I, dated 1559, three observatory instruments were found: a large quadrant, a small two-foot quadrant and a sextant, all three with scales graduated by Elias Allen in 1637.
Lit: GUNTHER (1936); GUNTHER (1937B)

8 *Radcliffe Observatory.* Erected as the result of a petition to the Radcliffe Trustees by Thomas Hornsby, Savilian Professor of Astronomy, for astronomical instruments and a building to house them. Hornsby's proposals were accepted in 1771 and a number of astronomical instruments were obtained from John Bird between 1772 and 1774. The building was in part use in 1777, but was not finally completed until 1793. Further instruments were acquired in the 19th century by Stephen Riguard and WF Donkin (appointed 1849). Since 1930, when the Radcliffe Trustees sold the grounds and the Observatory (now part of Green College), instruments have been housed in the *Museum of the History of Science.* Predating the Radcliffe Observatory is one important item, a 12ft focus Gregorian reflecting telescope by James Short, constructed in 1742 for the Duke of Marlborough, and presented in 1812 to the Radcliffe Trustees.
Lit: GUNTHER (1920-23) vol. 2, pp88-91, 394-6; ROBBINS (1929-30)

9 *Thompson collection.* A collection of surveying instruments owned and used by John Thompson of Witherby, Leicestershire (1722-83) and his two sons, Ralph and Samuel, all three professional surveyors. The collection includes instruments by George Adams, G Search and T Wright.
Lit: GUNTHER (1967), pp439, 440

10 *Royal Astronomical Society.* Part of the Society's collection of instruments was presented in 1930. The gift, consisting primarily of instruments of early 19th-century date, included transit instruments, theodolites, altitude, reflecting and repeating circles, sextants, a diffraction grating, polarimeters, an armillary sphere and a cometarium. Among the few 18th-century items is a reflecting telescope on an equatorial mount by James Short.
Lit: LECKY (1875-76) for a brief list of the more important RAS items; GUNTHER (1935) pp29-30, 118-23; GUNTHER (1967)

11 *Royal Microscopical Society.* An important collection of microscopes and accessories, together with a few other optical instruments, which was built up chiefly by presentations from members of the Society, but also includes a number of instruments purchased by Fellows of the Society at the sale of Professor JT Queckett's effects in 1861. The collection of instruments was placed on loan to the Museum in 1970.
Lit: DISNEY HILL/WATSON-BAKER (1928); TURNER (1989)

12 *Clay Collection.* A large collection purchased from Dr RS Clay (d. 1954) in 1944. The collection consists mainly of microscopes, but also includes some telescopes and navigational instruments.
Lit: CLAY/COURT (1928) (illustrating many microscopes from the collection); TURNER (1972)

13 *Billmeir Collection.* An outstanding collection of early scientific instruments, derived in part from the collection of Henri Michel in Brussels. After being on loan for some years the instruments were presented to the Museum of the History of Science by Mr JH Billmeir CBE in 1957. The collection, of nearly three hundred early scientific instruments, consists primarily of astrolabes, quadrants and sundials, but includes also important examples of surveying and mining instruments of 16th- and 17th-century date. Outstanding makers such Philipp Danfrie, Erasmus Habermel, Christopher Trechsler, Jobst Burgi, Thomas Gemini and Gemma Frisius are represented by important items.
Lit: JOSTEN (1955); MADDISON (1957)

14 *Barnett Collection.* This consists of sundials, clocks and watches, bequeathed to the Musem in 1935.

15 *Beeson Collection.* A collection of clocks by Oxfordshire makers, presented to the Museum of the History of Science by Dr CFC Beeson in 1966.
Lit: BEESON (1962)

II *SCOPE OF THE COLLECTION*

1 *Air pumps and accessories.* The collection includes an 18th-century double-barrel pump of Hauksbee type, and a double-barrel table pump by Nairne, as well as a pump by Watkins and Hill. There are a number of accessories, some of them for use with the Nairne pump.

2 *Armillary spheres.* A collection ranging from a small Ptolemaic armillary sphere of *c*1500 to 18th-century instruments by John Rowley and Richard Glynne. Both the Copernican and the Heracleidian systems are illustrated. Phillipp Danfrie, Carolus Platus and Cornelius Vinchx (fig 4) are represented among makers of fine 16th- and early 17th-century armillary spheres.

3 *Astrolabes.* The largest single collection of astrolabes of all periods and types, ranging from an instrument of the 9th century AD and including a number of universal astrolabes (figs. 9 and 10) and the only surviving complete example of a spherical astrolabe. (Fig 11)
Lit: MADDISON (1962). Many are listed in GUNTHER (1932); JOSTEN (1955); MAYER (1956); MADDISON (1957); PRICE (1955) and GIBBS/HENDERSON/PRICE (1973).

4 *Astronomical compendia.* A large collection of instruments by German makers, with a few examples of compendia by makers of French, Spanish, Dutch or Italian origin. The earliest example, dated 1481, may be by Hans Dorn of Vienna.

5 *Astronomical observation instruments.* The Museum houses an important body of instruments, some of them from the Radcliffe Observatory. Among other items there are equatorials, equatorial and zenith sectors, mural quadrants, a meridian circle, transit instruments, and several large telescopes, including a 7ft focus Newtonian reflector by William Herschel, a 12ft focus Gregorian reflector by James Short, and an 8ft long refracting telescope. Many of the 18th-century instruments are by John Bird. There are two astronomical regulators one by John Shelton of London, the other by John Hawting of Oxford. Three 17th-century observatory instruments were used by the Savilian Professors of Astronomy. Portable astronomical instruments are to a large extent derived from the collection of the Royal Astronomical Society, and from the Clay collection.

6 *Barometers and meteorology.* A relatively small collection of barometers, but including several interesting and important items. There are a number of meteorological instruments.

7 *Calculating instruments.* As well as numerous early slide rules (including an example of Oughtred's Circles of Proportion), sectors, abaci and Napierian rods (fig 98), the collection contains a range of calculating machines including Morland's adding machine, 1666, and a calculating machine invented by Earl Stanhope.

8 *Chemical apparatus.* To a large extent derived from the Daubeny collection and of early 19th-century date. It includes furnaces, bunsen burners, blowpipes, a wide range of chemical glassware (alembics, retorts, receivers, glassware used in the preparation of gases, etc.), stills, air pumps and accessories, eudiometers, chemical balances (including a Hauksbee-type hydrostatic balance on loan from St John's College), crucibles, hydrometers, and apparatus associated with HGJ Moseley's work on X-rays of 1913-14.
Lit: GUNTHER (1920-23) vol. 1; HILL (1971)

9 *Drawing instruments.* The collection includes beam compasses, a curve bow (fig 45), plotting scales, a sectograph (for dividing lines into equal parts), geometric curves and solids, parallel rulers, sets of drawing instruments and ellipsographs.

10 *Electrical apparatus.* Range of electrical machines and accessories, including globe machines by Nairne and Dollond, c1780, Read's electrical machine of c1770 signed by G Adams, a globe machine with table clamp by the same maker, purchased in 1790 by John Birch (the original bill still survives), a large twin plate electrical machine by Watkins & Hill, and an electrostatic cabinet by W & S Jones, c1830, with an extensive range of accessories. Later items include Wimshurst machines, electric motors, electrometers, galvanometers, electric batteries and arc lamps.

11 *Globes.* Include both oriental (Persian and Indo-Persian) (*Lit:* SAVAGE-SMITH (1985) nos. 7, 28, 32, 60, 80, 81, 86, 88, 93, 94, 100, 103, 105, 115) and European celestial globes, as well as an Indian Bughola (earth ball)

dated 1571 (*Lit:* DIGBY (1973); ANDERSON (1982) no. 122, plate 8) and a range of 18th- and 19th-century English terrestrial globes.

12 *Gunnery.* Gunner's instruments, including callipers, levels and sights, sectors, and a military architect's rules and protractor are represented, dating mainly from the 17th but in part also from the 18th century.

13 *Horology.* A wide range of clocks and watches, mainly from the Beeson and Barnett Collections, falls outside the scopes of this survey. In addition, however, there are a number of clockmaking tools, including a wheel-cutting engine of c1700 signed *J. Hook London* and a late 18th-century engine signed: *Daniel Fenn.*
Lit: MADDISON/TURNER (1973)

14 *Magnetism.* Lodestones are chiefly 18th century in date, although the earliest may be 17th century; they include two terrellae, one mounted in oval end-pieces, as well as a very large lodestone (supporting 160lb), pre-1756 in date (from the *Ashmolean Museum, Oxford* to whom it was given by the Countess of Westmorland), and a laminated magnet of early 18th-century date. A variety of compasses are represented, both as independent instruments and forming part of surveying, navigation and mining instruments, as sundials. There are several examples of Chinese geomantic compasses.

15 *Medical and surgical instruments.* In addition to *electrical machines* used for medical purposes, and magneto-electric machines, there is a collection of surgical instruments (in part on loan from the *Royal College of Surgeons, London*) and dental instruments. Among more recent items are culture cells associated with early research on penicillin in Oxford (1939-43).

16 *Microscopes.* The extensive collection of microscopes derived in the main from the Clay collection has been enhanced by the deposit on permanent loan of the collection of the *Royal Microscopical Society*, with the result that a very complete range of English microscopes is represented, including such outstanding items as two Marshall microscopes and a silver compound microscope constructed c1770 by George Adams for King George III. The extent of the collection, and the variety of instruments of any one type, make it invaluable for study. An important modern item is Burck's prototype ultra-violet reflecting microscope dating from 1946/7. Continental microscopes are less numerous, but there are instruments of German, French, Dutch, American and Australian origin.
Lit: DISNEY/HILL/WATSON-BAKER (1928); CLAY/COURT (1928); TURNER (1966)., Microscopes from the Museum of the History of Science are illustrated and discussed in BRADBURY/TURNER (1967), TURNER (1972) and TURNER (1989).

17 *Mining.* Represented by early Tyrolean mining compasses, 16th-18th century in date. These are simple hanging compasses, usually accompanied by ruler and sector, and the miner's theodolite occurs in examples of various dates.

18 *Navigation instruments.* As well as reflecting and repeating circles, nocturnals, sandglasses and ring dials, navigation instruments include a traverse board (on loan from the National Trust, *Snowshill Manor, Broadway*) a mariner's astrolabe, cross staves, tide tables, compasses (including a fine azimuth compass by Richard Glynne), artificial horizons, octants, sextants, etc.

19 *Nocturnals*. A range of instruments including several gilt brass nocturnals of 16th/17th-century date, English pearwood nocturnals, and German brass, or wood and paper, instruments of the 18th century. An unusual instrument is a peripole, a form of nocturnal invented by Loysel in 1671.

20 *Optics*. Included are eyeglasses, prisms (including salt and liquid prisms) polarimeters, saccharimeters and a crystal goniometer by Troughton and Simms of *c*1900. There are examples of camera obscuras, including one in book form, as well as zograscopes, zoetropes, praxinoscopes, a phenakistoscope and numerous stereoscopes.

21 *Oriental instruments*. In addition to Chinese and Japanese sundials and compasses, the Museum houses three important Chinese astronomical items – an 18th-century celestial map painted on silk, a Hsuan Chi and a Pi, ?600-900 AD.

22 *Orreries and astronomical demonstration apparatus*. Orreries include examples on loan from Christ Church (Thomas Wright) and All Souls (by Heath and Wing). There is an orrery by Tompion and Graham, as well as several examples, with tellurium and lunarium, by W & S Jones. Other, unsigned, orreries are of French or Italian origin of late 18th-century or early 19th-century date. An uncommon later item is a tellurian, *c*1790, constructed by Richer of Paris to the designs of the Abbe Grenet. An 18th-century lunarium of unknown maker was used to demonstrate eclipses.

23 *Perpetual calendars and almanacs*. Including clog almanacs of Scandinavian, German and English origin, perpetual calendar token discs, and silver and ivory perpetual calendars in the form of notebooks.

24 *Photographic apparatus*. The collection includes a number of early cameras, including daguerrotype cameras from *c*1841, and a panoramic camera with a fisheye lens of *c*1870. Dating from *c*1850 is CL Dodgson's (Lewis Carroll) portable developing apparatus, and a camera used by TE Lawrence during his travels in Arabia. There are a number of early prints (including some by WH Fox Talbot, *c*1841), Daguerreotypes, ambrotypes and a number of early colour photographs.

25 *Physical instruments*. See *electrical machines, air pumps, barometers and meteorology* and *chemical apparatus*. In addition there are a series of thermometers, hygrometers and hydrometers, and a few mechanical models from the Orrery Collection and Oriel College.

26 *Quadrants (portable)*. Both medieval quadrants, Near Eastern instruments (Persian and Indian) (fig 114A & B) and a wide range of European quadrants of 15th-17th century date are included. Among English quadrants there are examples of Gunter's and Sutton's quadrants, as well as others of less usual type, including examples in brass and printed on paper by John Prujean of Oxford.

27 *Sundials*. The sundial collection covers the full range of English and European instruments. The earliest item is a Roman dial of *c*150 AD. Oriental and Islamic instruments (quadrants) are also represented. A large part of the collection derives from the Lewis Evans, RT Gunther and Billmeir collections, although many important instruments have been acquired from other sources. No other collection of sundials is comparable (figs 37 and 38). The collection includes several equinoctial sundials, probably of 15th-century date, and some very unusual early dials, such as a navicula venetiis (or ship dial) of mid-15th century date, an unusual form of equinoctial dial (*c*1480), an apparently unique spoon dial and an astronomical ring by Gemma Frisius of Louvain.
Lit: MADDISON (1969)

28 *Surveying*. The collection ranges from an important series of Renaissance instruments many with a military bias, by such makers as Erasmus Habermel, Jobst Burgi, Philipp Danfrie, and Christopher Trechsler, and including Burgi's triangular instrument, a radio latino and an early surveyor's cross, to two contrasting collections of 18th-century instruments – the *Orrery* and the *Thompson* collections – which cover the entire range of instruments in use in the 18th century, including theodolites, plane tables, circumferentors, pantographs, levels etc. Several waywisers are included in the collection.

29 *Telescopes (portable)*. (For observation instruments see Section II/5.) These include refracting and reflecting telescopes of late 17th-19th century date, of both English and continental make, numerous examples of achromatic telescopes, and an example of a Japanese telescope of *c*1800. Table-mounted tripod telescopes include instruments by John Bird, James Short and Benjamin Martin. Floor-standing tripod telescopes include achromatic telescopes by Dollond.

O 5 OXFORD
Pitt Rivers Museum, University of Oxford

University of Oxford, Parks Road, Oxford OX1 3PP

0865 270927

The name of the collection and its manner of arrangement derives from its founder, Lieutenant-General Augustus Henry Lane Fox Pitt Rivers (1829-1900). The collection combines archaeological and ethnological material. It was housed originally in the Bethnal Green Museum, and then later at the South Kensington Museum. In 1883 General Pitt Rivers offered the collection in its entirety to the University of Oxford, which housed it in the present building, constructed alongside the University Museum. The collections cover firearms and armour, fire-making and lighting equipment, tools and implements, toys, musical instruments, metalwork, ceramics, textiles, ornaments and religious and magical objects. A section illustrating means of navigation includes a few scientific instruments, while a number of time-measurement instruments are also included in the collection. Some items have been transferred to the *Museum of the History of Science, Oxford*.

In addition to the instruments listed below, the collection contains four Chinese compasses, and a geomantic compass, weights and scales, including steelyards and nested weights. Lighting devices include Argand lamps. As well as sundials proper there are sandglasses and a number of primitive time-measurement instruments.
Lit: BLACKWOOD (1970)

1 Indian astrolabe. Brass, diameter 4½in. Dated 1673 and inscribed: *Jos'i Indraji-Kasya Yantram*. Rete for 23 stars, four plates for 8 latitudes from 18 – 32°, and table of horizons.
Lit: GUNTHER (1920-3) vol. 1, p198, no. 80; GUNTHER (1932) no. 79

2 Backstaff. Lignum vitae with boxwood arcs. Name plate missing, but decorated with fleur de lys and Tudor rose stamps.

Lit: GUNTHER (1920-23) vol. 1, no. 257, Illd. p356. (Another backstaff, signed *Edm. Culpeper fecit* is on loan to the *Museum of the History of Science, Oxford* (GUNTHER *ibid.* no. 258).)

3 Octant. Ebony frame, brass arm, ivory arc. Signed: *C.W. Dixie Optician to Her Majesty New Bond Sᵗ London*, 1840 or later.

4 Horary quadrant. Wood. Dated AD 1625.

5 Universal equinoctial ring dial. Brass, unsigned, 18th century.
Lit: GUNTHER (1920-23) vol. 1, no. 64

6 Universal equinoctial ring dial. Brass, unsigned, 18th century.

7-9 Three ring dials. Brass.

10-11 Two Chinese diptych dials.

12 Chondrometer. Signed: *Fraser & Son Bond St London 135*.

P

P 1 PETERHEAD
Peterhead Arbuthnot Museum

St Peter Street, Peterhead, Aberdeen AB4 6QD

0779 77778

The Museum, now part of the North East of Scotland Library Service, is housed in the same building as the public library. The collection illustrates local history, and as such is chiefly connected with the sea, and particularly with whaling. The few scientific instruments are also primarily navigational.

1 Backstaff. Probably pearwood. Radius 23in. Sights renewed. Inscribed: *John Gordan 1736*.

2 Octant. Ebony frame. Radius 16in. Vernier, but no tangent screw. Brass and wood arm. On ivory label: *William Sloman Gregory London*.

3 Octant. Mahogany frame. Radius 17¾in. Brass and wood arm. Vernier but no tangent screw. Signed on ivory label. *Made by Jno Gilbert Tower Hill London for Capt. Alexr. Milne Mar 12 1763*.

4 Octant. Ebony frame. Radius 13½in. On ivory label: *Spencer Browning & Rust London Lieut. James Green 1803*. *SBR* engraved on the degree arc.

5 Octant. Ebony frame. Radius 11in. On ivory label: *Spencer Browning & Rust London*.

6 Traverse board. Wood with bone pegs. Length 16½in. Eight sets of 12 holes and 4 sets of 4 holes.

7 Refracting telescope on folding tripod stand. Brass. Signed: *Cary London*.

P 2 PETWORTH
Petworth House

Petworth, West Sussex GU28 0AE

0798 42207

Petworth House, as it currently stands, is largely a rebuilding of the ancient house of the Percy family by Charles Seymour, 6th Duke of Somerset, from 1688 to 1696. A famous sculpture gallery was added by George Wyndham, 3rd Earl of Egremont, in *c*1780. The house contains rich collections of paintings, sculpture and furniture. It is administered by the National Trust.

The scientific instruments and apparatus at Petworth were accumulated by Elizabeth Ilive (1770-1822), mistress of the 3rd Earl of Egremont (1751-1837), who established a well equipped laboratory between 1790 and 1800. A number of supplier's bills survive. Apart from this there is an important, large terrestrial globe, signed by Emery Molyneux and dated 1592; it is traditionally held that it was given to Henry Percy, 9th Earl of Northumberland (1564-1632), by Sir Walter Raleigh, when both were held prisoner in the Tower of London. There is also a smaller globe by Nathaniel Hill, *c* 1750.

Besides material listed below, there are large quantities of chemical glassware and ceramics. A complete listing is provided by McCann (1983); numbers in brackets refer to her published Catalogue on pp646-649. Only one object (no. 15) is currently displayed.
Lit: PETWORTH (1982) for a general guide; McCANN (1983)

1 Air pump, double-barrelled. Signed: *T Harris, Hyde*

Street, Bloomsbury, London. (1, Illd.)

2 Air pump, double-barrelled. Probably by W & S Jones. (2)

3 Two vacuum fountains. (5, 6, Illd.)

4 Intermittent fountain. (7)

5 Guinea and feather apparatus. (10)

6 Turned mahogany base wih a brass pillar with a tap, receiver plate with a leather washer and fountain jet. Probably the single transferer supplied by W & S Jones on 22 November 1797. (12)

7 Two pairs of Magdeburg hemispheres. (14, 15)

8 Eighteen bell jars. (3, 4, 11, 16-29)

9 Three imploding bottles (30-32)

10 Weight-lifting bladder demonstration. (34)

11 Cylinder electrical machine, in mahogany chest, with accessories. Signed: *J. Simons, London.* (37, Illd.)

12 Leyden jar. (38)

13 Nooth's apparatus, to Parker's design. Numbered: *N6432.* (39, Illd.)

14 Two Wedgwood pyrometer pieces and five clay containers. Probably supplied with pyrometer (missing) on 26 January 1798. (55-59)

15 Alembic, coarse ceramic, with knob. (92, Illd.)

16 Five graduated glass gas jars. (93-97)

17 Orrery fragments. Probably part of the instrument supplied by W & S Jones on 12 October 1797. (110)

P 3 PLYMOUTH
Plymouth City Museum and Art Gallery

Drake Circus, Plymouth PL4 8AJ

0752 668000 ext 4878

The City Museum and Art Gallery houses important collections of English paintings porcelain and silver, as well as furniture, archaeological, historical, natural history and maritime material, especially ship models. There is an important collection of English and Italian drawings and early printed books. The scientific instruments, though not numerous, include several important and interesting items, housed either in the Museum at Drake Circus, or at Buckland Abbey, once the home of Sir Francis Drake and his descendants, and now the property of the National Trust, administered and maintained by the City of Plymouth. It is at Buckland Abbey that the greater part of the folk life collection, the ship models and Drake relics are housed. In addition to the instruments listed below, the Museum owns clocks, watches and chronometers and a number of weights, measures and scales. Two of the instruments form part of the Cottonian Collection (consisting mainly of fine art, manuscripts and books).

Lit: PLYMOUTH (n.d.)

Instruments displayed at Buckland Abbey are indicated (BA) in the following list:

1 Drake cup. A goblet of silver, partially parcel gilt. The globe is constructed in two halves which join at the equator. The upper half of the globe is surmounted by a miniature armillary sphere. Constructed by Abraham Gessner, Zurich, 1571, the goblet was presented to Sir Francis Drake by Queen Elizabeth I in 1582. (fig 55)

2 Celestial globe on wooden tripod table stand with compass mounted between the legs. Compass signed: D. Adams London and globe: *Dudley Adams......60 Fleet Street London 1798.* Diameter 12in. (BA)

3 Calendar roll. Drawn in ink on papyrus strip 80in long. The calendar is preceded by an illustration of a group of European astronomical and surveying instruments including an astronomical quadrant with telescopic sights on a stand, and a graphometer. Persian, 18th century. (Cottonian Coll.)

4 Universal equinoctial dial. Silver, rectangular. Case with hinged cover engraved on the inside with a list of latitudes of major cities, and on the outside with the latitudes of harbours on the Spanish main. A notched scale at the side of the compass, together with a hinged arm, permits the equinoctial plate to be adjusted to allow for latitude. Spanish, late 16th century. The dial has been tentatively associated with Drake since the harbour bearings would have been familiar to few other than Drake at the end of the 16th century.

5 Napierian rods. Boxwood in wooden slip-in case, 3½ x 2 9/16in. *c*1700. (Cottonian Coll.)

6 Lodestone with scalloped silver mounts, maximum dimension 1in. In black ray-skin case. From the Drake family. Probably 18th century. (BA)

7 Refracting telescope. Three-draw with wood-cased outer tube. Closed length 7⅝in. Signed: *Berge London late Ramsden*, early 19th century.

8 Refracting telescope. Three-draw with wood-cased outer tube. Closed length 9¼in. Signed: *Ramsden London.* Late 18th century.

9 Refracting telescope. Single-draw, wooden outer tube, closed length 27in. Signed: *Dollond London.* Late 18th century

10 Refracting telescope. Brass and wood, single-draw tube. Closed length 20⅝in. Signed: *Dollond London Day or Night.* Leather cover for outer tube. Early 19th century.

11 Sextant. Ebony frame, ivory scale and vernier, brass arm, arc graduated to 125°. Four shades for index glass, two for horizon glass. Maker's label missing. First half 19th century. (53-64)

12 Sextant. Brass, silver scale. Inscribed on sextant: *Alfred William 1866.* The maker's name is illegible, but the box bears the label of Whyte Thomson & Co. Glasgow.

13 Microscope, the 'Star', by *R. & J. Beck 68 Cornhill London*, Serial No. *19504, c*1890. (Judge Bequest 1954)

14 Apparatus for inflating lungs. A mahogany box 17 x 9 3½in contains bellows, pipes, tubes, etc. constructed of wood, ivory and brass. The bellows bears the impressed mark: *Evans.* Accompanied by a leaflet illustrating and describing the apparatus, and including the Royal Humane Society's Instructions. The leaflet records that the apparatus can be obtained from *Evans & Co. Surgical Instrument Makers No. 10 Old Change London.* Early 19th century. (Judge Bequest 1954). (fig 72)

P 4 PORTSMOUTH
The Mary Rose Exhibition

HM Naval Base, Portsmouth PO1 3LX

0705 750521

The ship *Mary Rose* was built for King Henry VIII in Portsmouth in 1509-10 in the earliest dry dock in the world. In 1545 *Mary Rose* keeled over and sank during an engagement with a French invasion fleet a mile and a quarter from the entrance to Portsmouth Harbour. Underwater excavations started in 1971 and between 1979 and 1982 the ship was totally excavated, recorded and prepared for recovery. By September 1982, over 16,000 registered finds had been brought ashore. The exhibition opened in 1984.

Amongst these are a number of navigational instruments, most of them being the earliest examples known of their type from an English source. In addition to the instruments listed below, there are the remains of eleven large balances (? to weigh rations), nine simple shot gauges, nine rulers, four sandglasses and two tally sticks. The barber-surgeon's box was recovered, and fragmentary remains of surgical instruments have been identified, as well as drug jars, a mortar, and three well-preserved metal urethral syringes.
Lit: RULE (1982 and 1989).

1-3 Mariner's compasses in boxes, mounted on gimbals. (79A0766/7, 81A0118, 81A0071, 81A0802)

4 Compass rose (?) oak. (81A111)

5-12 Pocket sundials (8) circular, wooden cases, with folding brass gnomons. Diam 1½ in. (80A0942, 1669; 81A0420, 0730/1-3, 1992, 2026; 82A5076/1-2, 5077)

13 Pocket sundial, as above, with spherical wooden case. (81A5681/102)

14 Pocket sundial (as above) in leather and wood folding book-like case. (81A4123)

15 Calibre gauge, in the form of a hinged rule, copper alloy, one scale marked *ISERN* (iron), the other *BLI* (lead) for different types of shot. (81A0249)

16 Mortise gauge for timber joints. (82A2286)

17 Plank gauge, simple stepped gauge of wood. (81A1201)

18 Carpenter's rule incorporating line of timber measure, of a type first described by Leonard Digges in 1556. Constructed from oak, length 24 in, width 2 in, thickness ½ in. (B1A4477).
Lit: KNIGHT (1990)

19 Plotting board (?) or possibly gaming board, wooden board with engraved grid, similar to an example of ca 1665 from a wreck in Red Bay, Labrador. (78A0279)

20,21 Pairs of dividers (2) brass. (81A0084, 0085, 0512, 0201)

22 Log reel, all-wood construction, handles missing. (81A1851)

23-25 Sounding weights (3) lead. (79A476, 894, 81A2847)

25 Apothecary or money-changer's balance, brass pans (both stamped with a star, and a crown), beam and hanger. (MR81A2014)

P 5 PORTSMOUTH
Portsmouth City Museum and Art Gallery

Museum Road, Old Portsmouth PO1 2LJ

0705 827261

The City's collections, housed in a main museum building and several branch museums, include furniture, paintings, glass and ceramics, local history especially the military history of Portsmouth, natural history, geology, and archaeology. There are a few scientific instruments, chiefly nautical in character.

1 Refracting telescope, three-draw wooden outer tube, silvered brass draw tubes. Signed: *Saml. J. Browning late Stebbing 66 High St Portsmouth.*

2 Refracting telescope, single-draw leather-covered outer tube. Signed: *E. & E. Emanuel by Appointment to Her Majesty, 3 The Hard Portsea and 101 High St Portsmouth.*

3 Refracting telescope, four-draw, brass, with altazimuth mounting on tripod stand. Signed: *Russell London.* In box with label of *George Lee & Son Manufacturer of Optical and Nautical Instruments 33 The Hard Portsmouth.*

4 Stuart's Marine Distance Meter 477. Signed: *G. Lee & Son The Hard Portsmouth*

5 Barograph. Signed: *A. Barratt & Son, Guildhall Square Portsmouth.*

6 Chondrometer. Signed: *Bleuler London.*

P 6 PORTSMOUTH
Royal Naval Museum

HM Naval Base, Portsmouth PO1 3LR

0705 733060

The Museum houses a collection of ship models and marine paintings, as well as relics associated with Admiral Lord Nelson and with naval life in general. Instruments are included in the collection as far as they were used for navigation or associated with naval figures (e.g. Nelson's horizontal garden sundial). A number of nautical instruments were on loan from the *National Maritime Museum* at the time of writing.

1 Octant. Ebony frame. Radius 10¾in. Signed on arm: *SBR* and inscribed: *HN 1798*. In box with label of Richard Hornsby 36, South Castle Street, Liverpool.

2 Octant. Ebony frame. Radius 9½in. Unsigned. Early 19th century.

3 Octant. Ebony frame, radius 13¾in. Signed: *Wm. Harris London*, *c*1810. (loan)

4 Sextant. Double brass frame and index arm, silver scale. Vernier with magnifier and tangent screw. One telescope. Signed: *G. Bradford 99 Minories London*. (Reputed to have belonged to Thomas Hardy, Captain of *HMS Victory*.)

5 Sextant. Brass frame, silver scale, radius 8¾in.

6 Nautical telescope signed: *Thomas Jones 62 Charing Cross*. (Belonged to John Pasco, Signal Lieutenant on *HMS Victory*.)

7 Telescope with decagonal wooden tube. Signed: *Dollond London*.

8 Brass telescope with wooden casing to outer tube. Signed: *Ramsden London*. (Said to have been used at the battle of Navarino, 1827.)

9 Brass telescope, wood-cased outer tube. Signed: *J. & W. Watkins*.

10 Brass telescope, wood-cased outer tube. Signed: *Watkins Charing Cross London*.

11 Brass telescope, leather-covered outer tube. Signed: *Cox Devonport*. (Reputed to have been used at the Battle of Camperdown, 1797.)

12 Horizontal garden sundial, signed: *Sterrop London Lat. D 17 M 18*. (Said to have belonged to Nelson, *c*1787.)

13 Augsburg-type equinoctial sundial. Brass, octagonal in leather case. Signed: *Butterfield AParis*.

14 Two day chronometer by Charles Cope, dated: *1820*.

P 7 PORT SUNLIGHT
The Lady Lever Art Gallery

Port Sunlight Village, Wirral L62 5EQ

051 645 3623

Now part of the National Museums and Galleries on Merseyside, the Museum was established by the first Viscount Leverhulme in memory of his wife. The foundation stone was laid in 1914 and the gallery opened officially 1922. Almost all the collection was assembled by the donor. It consists of paintings and watercolours and sculpture, engravings and miniatures, mainly of the British School. In addition there is furniture, tapestries, Chinese ceramics and Wedgwood ware.

1 Horizontal sundial, brass, rectangular, with lid. Dated *MDCC XXVIII*. (M.1532)

2 Horizontal dial. Square on ebony base. Printed paper compass card, folding brass gnomon. Unsigned, but *B* at bottom right, ? D. Beringer, Nuremberg. Probably late 18th century. From the Rosenheim collection.

3 Diptych dial. Ivory, unsigned. Probably German (Nuremberg), 17th century. From the Rosenheim collection. (X 4643 (39))

4 Diptych dial. Ivory. Signed: *Alberrecht Karner* and dated *1667*. From the Rosenheim collection. (K 4642(38))

5 Diptych dial. Wood covered with engraved and coloured paper. 4 x 2½in. Unsigned. German, late 18th century. (M 1416)

6 Universal equinoctial dial. Brass, gilt and silvered, octagonal. Signed: *Lorenz Grasl Augspur*.

7 Universal equinoctial dial. Silver and gilt brass circular base. Signed: *Cole fecit*. English, 18th century.

8 Crescent dial. Hour ring in the form of two semicircles set back to back. Square base, silvered and gilt brass. Perpetual calendar mounted beneath the base. Signed: *Christoff Schener in Augspurg*. Late 17th or early 18th century.

R

R 1 READING
Reading Museum and Art Gallery

Blagrave Street, Reading RG1 1QH

0734 399809

The collections cover natural history and archaeology, including Roman finds from Silchester, and the Thames Conservancy collection of prehistoric and medieval

metalwork. There are displays of historical material and a fine collection of Delftware.

In addition to clocks and watches, chiefly by local makers, weights and scales, dental and medical equipment (including magneto-electric machines, tooth extractors, etc.) there are a number of scientific toys (zoetrope, magic lantern stereoscope) as well as several scientific instruments.

1 Set of drawing instruments signed: *R.B. Bate No.21 Poultry* including ivory sector signed: *Dollond*. (1534:64)

2 Universal equinoctial ring dial. Brass. Signed: *Walter Hayes in Moorfields Londini fecit*. (441:47)

3 Two brass ring dials. Brass, unsigned. 18th century. (440:47, 439:47)

4   Chinese geomantic compass. (16:54)

5   Aquatic microscope in fish-skin covered box. Signed: *W. & S. Jones 30 Holborn London.* (132:6912)

6   Terrestial telescope. Handheld. 19th century. (1538:64)

7   Waywiser. Diameter of wheel 63in. There is no dial but a bell rings after each revolution of the wheel.

8   Sikes' hydrometer. Signed: *Loftus 6 Beaufoy Terrace London.* (1265:64)

9   Saccharometer. Signed: *Ash & Sons, 4 Bull Street, Birmingham.* (1264:64)

10   Wooden apothecary's chest. 19th century (1101:64)

11   Medicine chest made by Edward Hume and Co., Walworth Road, London. With accompanying booklet written by a Reading doctor, 1868. (708:73)

R 2 ROCHESTER
Guildhall Museum

High Street, Rochester, Kent ME1 1QU

0634 48717

The Museum is chiefly devoted to local and natural history, archaeology, geography, bygones, etc. It has a collection of Dickensiana, ship models, clocks and several scientific instruments.

1   Butterfield dial. Silver, octagonal in leather case. Signed: *Butterfield AParis.*

2   Part of a mechanical equinoctial sundial inscribed: *S.C. 1787.*

3   Crow's patent seaman's octant. Brass, radius 11in. *Inscribed: Crow's patent seaman's quadrant London.* This instrument (c1832) permitted the solution of problems concerning right-angled triangles without calculation. (fig 103)

4   Battenburg's course indicator. Signed: *Elliott Bros London No. 527.*

5   Sextant. Brass, radius 10in in fitted mahogany box. Signed: *Horne & Thornthwaite, 121, 122, and 123 Newgate Street, London.* Mid-19th century.

6   Cylinder electrical machine, unsigned, but one of the accessories, a Henley quadrant electrometer, bears the maker's name, *W. & S. Jones London.* The equipment includes a conductor mounted on a glass pillar, a thunder house, 3 Leyden jars, an aurora flask, electric cannon, a supply of lead foil, a Lane electrometer, straight and curved medical directors, and a joined discharger, c1800.

7   Binocular microscope. Signed: *R. & J. Beck London.*

8   Spectroscope by *W. Ladd, Beck St Regent St W.*

9   Fuller's spiral slide rule, constructed by WF Stanley & Co., London.

R 3 ROTHBURY
Cragside

Rothbury, Morpeth, Northumberland NE65 7PX

0669 20333

Cragside was built in a remote spot by William George Armstrong (1810-1900) (created 1st Baron Armstrong of Cragside in 1887) between 1864 and 1885. From 1869 its architect was Norman Shaw. The house was the first in the world to be lit by electricity derived from water power.

Armstrong was a Newcastle engineer who made a large fortune from armaments and warships. His earliest interest was hydraulics, but following the Crimean War he produced improved rifled ordnance, later developing naval artillery and ironclad warships.

In his later years, when established at Cragside, Armstrong took up again an interest in electricity (he had designed the hydroelectric machine which generated static charges from effluent steam – see fig 50). Working with Joseph Swan, Armstrong installed incandescent electric light in 1880. He established a laboratory and observatory in the house and spent his last years experimenting on high-tension currents. Some of the surviving instruments in the house doubtless were used in this work. It is noteworthy that most of the instruments were supplied by the London firm of Harvey and Peak.

Apart from the material listed, there are collections of electric lamps, early electric switches (Browett's patent), Crookes tubes, chemical glassware and a cabinet of minerals supplied by W Edwards. One item, no. 45, has no connection with Cragside and Armstrong, being recently acquired.

Cragside is now owned by the National Trust.
Lit: TAYLOR (1980)

1   Air pump Sprengel-type (?), glass tubing partly missing. Mounted on wooden stand, 72in high, signed: *F. Robson Optician 45 Dean Street Newcastle.*

2   Double-barrel air pump, table variety, base 19 x 19in, signed on top: *Harvey & Peak London.* With bell jar. (CRA/MISC/18)

3   Frame of pulleys, wood, also incorporating windlass, capstan and pile driver. Frame size 26in high x 27in wide. Signed: *Harvey & Peak late W Ladd & Co 6 Charing Cross Rd London WC.*

4   Inclined plane and trolley, wood, base 12in x 3¼in. Signed: *Harvey & Peak London.* (CRA/MISC/1258)

5   Frame of levers, wood, 10½in high. Signed *Harvey & Peak London.* (CRA/MISC/1258)

6   Series of 17 weights for use with nos. 3-5.

7   Worm drive on stand (gearwheel diameter 3in).

8   Dissected cone, wood, height 6in, diameter of base 5¼in. (CRA/MISC/1230)

9   Hollow prism (for liquids), glass in wooden frame 4½ x 3¾ x 3¾in. Signed: *Harvey & Peak London.* (CRA/MISC/1257)

10 Wheatstone photometer, brass, diameter 2¼in. Signed: *Elliott Bros 30 Strand London* (CRA/M/63)

11 Dewar flasks (2), spherical, diameter 8in. One with neck, the other closed with a brass cap. (CRA/C/686)

12 Planimeter in box, signed: *Stanley, Great Turnstile, Holborn, London.* (CRA/MISC/1274)

13 Wimshurst machine, diameter of plate 17in. Signed: *Harvey & Peak Late W Ladd & Co 6, Charing Cross Rd London WC.*

14 Glass discs (2) for plate electrical machine, diameter 24in. (CRA/C/685)

15 Leyden jar, lid missing, diameter 6in x 12¼in high. (CRA/MISC/1259)

16 Leyden jar, lid missing, diameter 3½in x 7½in high. (CRA/MISC/1244)

17 Leyden jar with diamond patterned metal coating ('diamond spotter jar'), 9¾in high. (CRA/MISC/1246)

18 Leyden jar with movable coating. (CRA/MISC/1260)

19 Luminous tube (hand spiral).

20 Electrical collector.

21 Discharging tube (aurora borealis tube), 42½in long x 2½in diameter. Key signed: *W. Ladd* and *London*. The tube screws into a base bearing a label signed: *Apps, 433 Strand London.* (CRA/MISC/18)

22 Gold leaf electroscope (base missing).

23 Induction coil, with box, 39½ x 19½in (base). Signed: *Apps 433 Strand London* and *Patent 967.* (CRA/MISC/17(b))

24 Induction coil, 26 x 17in (base). Signed: *Apps 433 Strand London* and *Patd 1881-264/No 1102.*

25 Primary condensers (2) for induction coils, signed: *Apps 433 Strand London.*

26 Electrolytic interruptor (? Wehnelt type) for use with induction coils, 11in (high) x 6in (diameter). Consisting of a glass jar with two insulated wires piercing the lid.

27 Insulators (17) of green glass, some signed: *Folembray Depose.*

28 Leclanche cells (14), some embossed: *Carporous* (referring to the construction, not a maker).

29 Oersted's apparatus (Sturgeon's modification), compass needle missing, signed: *Harvey & Peak London.* (CRA/M/673)

30 Voltameter, with single graduated collection tube. (CRA/C/677)

31 Voltameter, 'V' shaped, on wood stand. Signed: *Harvey & Peak London.* (CRA/MISC/1253)

32 Voltameter tubes (2) 'U' shaped. (CRA/C/675)

33 Apparatus for experiments in strong magnetic fields (e.g. diamagnetic properties), coils missing.

34 Wire coil on hollow brass former, 10in (high) x 3½in (diameter). (CRA/MISC/1726(E))

35 Recording voltmeter with case, signed (on case): *Jules Richard Impasse Fessart 8, Paris* and (? patent) number on instrument: *31341.*

36 Electric dynamo, consisting of armature rotating between two massive cylindrical iron magnets, wood base. Signed: *Harvey & Peak late W. Ladd & Co 6 Charing Cross Rd London WC.* Post-1891. (CRA/MISC/1726(C))

37 Grooved disc with handle for rotating, mounted on wood frame, all on wood base 38in long. Possibly used to turn no. 35. (CRA/MISC/1726(A))

38 Carbon filament lamps (2) made by Joseph Swan, formerly used by Andrew Richardson, Armstrong's technician.

39 Glass tubes (3), ground glass ends, 4¾in diameter, 24, 12½ and 8½in long. One fitted with iron flange. (CRA/MISC/18)

40 Gas jar, 19in (high) x 6in (diameter).

41 Mortar and pestle, glass, rim diameter 9in. (CRA/C/678)

42 Jugs (2) for pouring mercury, one wood, the other ceramic. (CRA/C/641 and CRA/MISC/1227)

43 Binocular microscope, signed: *Smith, Beck and Beck* and *4265.*

44 Manometer, glass capillary tube bent twice and fitted to brass screw-threaded mount. (CRA/MISC/1271)

45 Nooth's apparatus, top stopper missing, height 25in, number scratched on all three parts: *N 150.*

R 4 ROTHERHAM
Rotherham Museum

Clifton Park, Rotherham S65 2AA

0709 382121

The museum collections are locally orientated, containing material relating to the botany, zoology, prehistory and archaeology (including material from the Roman fort at Templeborough) of the region and local history of Rotherham itself. There are significant collections of local ceramics, including a range of Rockingham porcelain, geology, silver, glass, and weights and measures which are housed in an extension completed in 1974. A new Art Gallery in the Library and Arts Centre was opened in 1976. Less specifically local are collections of gemstones and jewellery. The collection contains only one scientific instrument.

1 Cylinder electrical machine, Cavallo type, the friction pad now missing. Unsigned. First half 19th century.

S

S 1 ST ANDREWS
Department of Chemistry, University of St Andrews

The Purdie Building, University of St Andrews, St Andrews KY16 9ST

0334 76161

The University of St Andrews was founded in 1411. Dr John Grey, one-time student at the University, who died in 1811, left a sum of money to pay the salary of a Professor of Chemistry. However, an appointment to the Chair of Chemistry was not made until 1840, although a course in chemistry was held from 1811 onwards. A collection of chemical glassware and crucibles was found in the tower of St Salvator's Chapel in about 1925, and the suggestion has been made that it may be that bought from Thomas Thomson (1773-1852), who abandoned his private practical chemistry course in Edinburgh in 1811. This is on the basis of an inventory of the apparatus acquired in 1811 by Robert Briggs (Professor of Medicine and Anatomy) from the Edinburgh dealer Alexander Allan, who had obtained much of it from Thomson.
Lit: ST ANDREWS.   READ (1953) for an account of chemistry at St Andrews.

1   Long beam balance. Signed: Made by Geo. Adams Instrut Maker to his Majesty's Office of Ordnance at Tycho Brahe's Head in Fleet Street London. Beam length 37in. Constructed of brass, on turned circular stand, *c*1750. Pans are later replacements.
Lit: READ (1935)

2   Collection of chemical glassware, probably all late 18th century in date. All items are of clear white glass.

   a   Receiver. Length 28in.

   b   Receiver. Length 29½in.

   c   Receivers (2) with single tubulure, quilled.

   d   Quilled receiver.

   e   Quilled receiver with two tubulures.

   f   Retort, tubulated and quilled, the quill bent at right angles.

   g   Retort, quilled.

   h   Alembic heads (2). Ground joints for connecting with body, and ground stopper at top. (fig 19)

   i   Alembic head, plain. (fig 19)

   j   Bottle on circular foot with four necks, two at the side and two at the top. The upper two receive cylindrical tubes, each furnished with glass stoppers.

   k   Straight-sided glass beaker on circular foot.

   l   Spherical flask with two ground necks opposite one another.

   m   Glass bottle with two ground necks, one centrally placed, the other offset. Into the centrally placed neck fits the narrow end of an elongated vessel whose upper end is equipped with a ground neck and a subsidiary neck below the shoulder.

   n   Funnel-shaped tube. (fig 19)

   o   Two straight-sided glass domes (or beakers). (fig 20)

   p   Three flasks.

   q   Three ovoidal bottles.

3   Numerous graphite crucibles. Stamped: Patent Plumbago Battersea Work London Crucible Company.

4   Retorts. Glazed earthenware. Stamped: *Wedgwood*.

S 2 ST ANDREWS
Department of Physics, School of Physical Sciences, University of St Andrews

North Haugh, St Andrews KY16 9SS

0334 76161

The Department of Physics houses an important collection of scientific instruments which ranges from 16th-, 17th- and 18th-century items of outstanding importance, to a wide collection of 19th-century apparatus. The earliest instruments were constructed by Humphrey Cole during the last quarter of the 16th century. A mariner's astrolabe (by Elias Allen) and a Dutch circumferentor belong to the early years of the 17th century. From the latter part of the 17th century come an horary quadrant and a dip circle, as well as a tympanum for the 16th-century astrolabe constructed for the latitude of St Andrews. Some of these instruments may have been among those purchased in 1673 by James Gregory, Professor of Mathematics and Astronomy at St Andrews, 1668-1674, to form part of the equipment of an observatory.

Other instruments – an orrery by Benjamin Cole (*c*1750) and a Gregorian reflecting telescope dated 1736 – were acquired by David Young, Professor of Natural Philosophy, 1747-59, who also took over Gregory's astronomical apparatus. It is known that Young ordered a reflecting telescope from Short and it is probable that he was also responsible for the acquisition of the orrery.

The early 19th century was apparently a period of active acquisition for the department, to judge from the signed instruments by makers such as Troughton, Robinson, Cary and Thomas Jones.

The date of purchase of much of the later equipment can be determined from an accessions book, commenced in 1859 by Professor Swan, with apparatus being purchsed from Fuess, and from Gruet, Berlin, Ilgmann in Breslau, Hartman and Braun, Harvey and Peak, Baird and Tatlock, and Kemp and Co., Edinburgh.

Lit: WRAY (1983A), WRAY (1983B)

1 Gregorian reflecting telescope of brass. 4½in aperture, 24in focus, altazimuth mounting fitted in 1748. Signed: James Short Edinburgh 1736 1/82.
Lit: BRYDEN (1968) p8; BRYDEN (1972B) p14; WRAY (1983B) pp[3-4]

2 Horary quadrant, 'the Panorganon, or a universal instrument', described by William Leybourn in 1672. Brass, unsigned, late 17th century.
Lit: LEYBOURN (1672)

3 Universal equinoctial ring dial. Brass, signed: Dollond London. 18th century.

4 Armillary instrument consisting of meridian and equinoctial circles fixed at right angles to one another, and mounted on a circular base incorporating a compass. Inside the circles is mounted a circular disc which can be rotated so that it lies in the plane of the meridian or of the equator. It is equipped with an alidade with sights which rotates around the central point of the disc. The plate is engraved with an orthographic projection of the celestial sphere on the plane of the colura of the solstices (Roias projection). Brass, signed: *Humphrey Cole fecit 1582.* Height 18in.
Lit: GUNTHER (1927) p300, plates 75 and 76; WATERS (1958) plate 566, WRAY (1983B) pp[11-12]

5 Grand orrery. The mechanism is contained in a twelve-sided glazed case. Mounted above the ecliptic circle, the five principal circles of the sphere. Signed: Cole Maker at the Orrery in Fleet Street London. Mid-18th century. (fig 104)
Lit: WRAY (1983B) pp[2-3]

6 Astrolabe. Brass, diameter 24in. Combines in the normal stereographic projection, Gemma Frisius's universal projection, his quadratum nauticum and a table of horizons. The instrument was designed to take three tympana, one of which (engraved for 52°) survives, while another has been replaced at a later stage by a plate signed: Iohn Marke fecit and constructed for latitude 56°25'. This plate must date from the 1660s or 1670s. The instrument is signed on the quadratum nauticum: *Humfridus Cole Londinensis hoc instrumentum fabricavit 21 die Maii Ao Dni 1575.*
Lit: GUNTHER (1927) pp274ff, plates 63-66; GUNTHER (1932); PRICE (1955); EDINBURGH (1959); WRAY (1983B), pp[7-8]

7 Mariner's astrolabe. Brass, diameter 15⅝in. Signed: Elias Allen fecit 1616. Probably one of the instruments acquired by James Gregory for the University Observatory in 1673.
Lit: GUNTHER (1932) no. 321; PRICE (1955) no. 321; EDINBURGH – Scottish Museum (1959); WATERS (1966) p32; ANDERSON (1972) no. 10; STIMPSON (1988)

8 Lodestone, brass-mounted. 17/18th century.

9 Compass mounted in gimbals. 18th/19th century.

10 Prismatic compass. Signed: Cary London.

11 Dip circle. Brass, 12in diameter. Inscribed: *Dono Archibaldi Areskini Armigeri Londini.* 17th century.

Lit: WATERS (1966), plate 63

12 Dip circle, 10in diameter. Brass, on tripod foot, the circle enclosed in a glazed mahogany case. Horizontal graduated circle. Signed: Robinson Devonshire St Portland Place, London.

13 Dip circle. Signed: Robinson, 38 Devonshire St Portland Place, London.

14 Quintant. Double brass frame with silver scale and vernier, radius 10¼in. One telescope, tangent screw and magnifier. Signed: *Troughton London* and engraved with the serial number 793, *c*1800-1810. (fig 123)

15 Quintant. Brass T-section frame, silver scale, radius 8½in. Telescope, vernier and tangent screw, magnifier. Signed: *Cary London.* Serial number *334.*

16 Circumferentor. Gilt brass, with inset compass, double shadow square, 32 point windrose, fixed sights on the main radii and sights on alidade. Dutch, early 17th century. (fig 22)

17 Theodolite. Brass. Signed: *Cary London.* Early 19th century.

18 Pantograph. Brass, signed: *Cary London.* Early 19th century.

19 Micrometer. Signed: *Thomas Jones Charing Cross London,* dated 1849.

20 Atwood machine. Clock mechanism signed: *Troughton London.*
Lit: GUNTHER (1920-23), plate opposite p80.

21 Atwood machine. Clock mechanism signed: W. Wilson, 1 Belmont St. London NW.

22 Wheatstone wave machine. Signed: *C. Wheatstone Invt.*
Lit: WRAY (1983B), pp[13-14]

23 Balance. Wooden beam and frame. Designed by Professor W Swan, c1865, to demonstrate the theory of the balance.

24 Gold-leaf electroscope with Zamboni piles. Mid-19th century.

25 Astatic needle galvanometer. Signed: *E. Ilgmann a Breslau fct.*

26 Reflecting mirror galvanometer.

27 Wimshurst machine. Signed: *Harvey & Peak, late W. Ladd & Co. Charing Cross Rd. London W.C.*

28 Wimshurst machine. 24in diameter plates. Signed: *Ruhmkorpf a Paris.*

29 Width gauge of brass, dated 1804. Signed: *A Allan Edinr 1804.*

30 Regulator, with mercury pendulum. Signed: *J.& A. McNab Perth.*

31 Compound microscope, 'most improved' type, signed: *Dollond London.*

32 Standard resistance, numbered '33' and inscribed *British Association Unit.*

33 Bloxam's patent dipleidoscope, signed: *Dent's patent meridian instrument 82, Strand & 33 Cockspur St London.*

34 Pulley frame. Unsigned, 1803.
Lit: WRAY (1983B), p[14]

35 Patent Medical Electrical Machine, Nairne pattern. Signed: Gilbert, Leadenhall Street, London.
Lit: WRAY (1983B), pp[16-17]; MORRISON-LOW (1984), p88

36 Achromatic microscope. Signed: Andw. Ross & Co. Opticians 33 Regent St Piccadilly. Purchased by the University in 1840 and used by Sir David Brewster.
Lit: WRAY (1983B), pp[18-19]

37 Small altazimuth instrument, Kater's pattern. Signed: Adie & Son, Edinburgh.

38 Holtz Electrical Machine. Signed: *RUHMKORFF A PARIS.* Purchased 1875.
Lit: WRAY (1983B), p[22]

39 Quadrant Electrometer, Thomson pattern. Signed: No 52 J. White Glasgow.
Lit: WRAY (1983B), pp[24-25]

S 3 ST ANDREWS
University Library

University of St Andrews, St Andrews KY16 9TR

0334 76161

The Library houses three clocks by Joseph Knibb, including one split-second clock, all of which were purchased by James Gregory (Professor of Astronomy at St Andrews, 1668-1674). Gregory built up a collection of instruments by London makers for his observatory. Makers included Hilkiah Bedford, John Marke, Christopher Cock and John Yarwell. They are referred to in inventories of 1697 and 1714 still preserved in the Library. The clocks and an incomplete telescope are both referred to in the 1697 inventory. Other instruments acquired by Gregory are in the *Department of Physics.*

1 24 hour split-second bracket timepiece by Joseph Knibb, adapted later as a longcase clock by the addition of a narrow waisted trunk. Probably purchased in 1673.
Lit: BROOK (1900-01); TURNBULL (1939) p514, plated; LLOYD (1944) p341; LLOYD (1958) p75; WRAY (1983B) pp[5-6]

2-3 Pair of longcase clocks by Joseph Knibb, London.
Lit: BROOK (1900-01); LLOYD (1958) p75

4 Refracting telescope, incomplete. 12in sectional tube of parchment, about 3in diameter. The outer tube and all mountings and lenses are missing. Presumably one of the Galilean refracting telescopes listed in the 1697 inventory.

5 Oughtred's double horizontal dial. Bronze. Signed: Hilkiah Bedford Londini fecit, *c*1660-80.

6 'The Mercantile Chronometer' by *S. Saunders engraved by Hewitt Queen Str Bloomsby Entered at Stationer's Hall.* Printed paper mounted on card. Mid-18th century.

7 'A new astronomical instrument'. Paper volvelle for showing day of month, change and age of moon, places of the sun, etc. from 1817-1864. Inscribed: *Published by Thomas Jones, 62 Charing Cross, London, November, 1817.*

8 Volvelle mounted on circular wooden plate. Inscribed: *I. Shackloch, September, 1814.*

9 Terrestial globe by Cary, London. Dated 1806.

10 Two mathematical solids. Boxwood.

11 Meridian line constructed across the floor of the upper library hall. Used for meridional observations in conjuction with an iron trident with three small holes in the prongs upon a cairn of masonry at Scooniehill. Both still in existence.

S 4 ST FAGANS
Welsh Folk Museum

St Fagans, Cardiff CF5 6XB

0222 569441

St Fagans Castle, its gardens and grounds were presented to the National Museum of Wales in 1946 by the Earl of Plymouth for the creation of a national Welsh Folk Museum. The Museum is now housed in St Fagans Castle, a museum block (recently constructed) and a number of reconstructed Welsh buildings. The collections cover all aspects of life in Wales through the centuries. They contain an extensive collection of Welsh clocks and watches, as well as medical, surgical and obstetric equipment (e.g. cupping glasses, amputation instruments, scarifiers, bleeding lancets, stomach pumps, etc.) in addition to a number of scientific instruments.
Lit: Most of the scientific instruments are unpublished, but nos. 6,27,28,29 and a number of garden sundials with Welsh connections, or constructed by Welsh makers are listed by PEATE (1975).

1 Islamic astrolabe. Brass with coloured inlay. Three double-sided plates for six different latitudes. Persian or Ottoman Turkish, 17th century. (09.7)

2 Gunter's quadrant. Brass, in wooden case. The back of the instrument with a planispheric nocturnal and the owner's name: Nathaniell Jeynes 1678. The quality of the engraving, particularly of the star map, is rather coarse. Engraved on the box: *N.I. 1703.* (23.34)

3-4 Pillar dials. Boxwood. Probably 19th century. (15.70/7 and 8)

5 Butterfield dial. Silver, octagonal, signed: *Butterfield A Paris.* (15.70/2)

6 Horizontal sundial. Inscribed: *Ellis Wynne 1711.* (59.478)
Lit: PEATE (1975) no. 2

7 Magnetic dial in circular wooden box with lid. 18th century. (58.2/2)

8 Horizontal compass dial in round brass box with lid. Printed paper compass card in base. 18th century. (61.512)

9 Horizontal compass dial in round brass box with lid. Inscribed: I.H. Penryn 1757. (03.199)

10 Ring dial. Bronze, 18th century. (P.81)

11 Universal equinoctial dial. Brass, signed: *L. Grassl.* Augsburg, late 18th century. (15.70/1)

12 Diptych dial. Chinese. (15/107)

13 Diptych dial. Wood with engraved and coloured paper. German (? Nuremberg) late 18th century. (15.70/3)

14 Vertical plate dial. Boxwood, 18th century. (15.70/9)

15 Nocturnal and vertical disc dial, lunar volvelle and aspectarium. Dated 1598. Apparently French, but probably a fake. (15.70/9)

16 Nocturnal. Boxwood. Inscribed: *Richard Goddard 1727.* (53.239)

17 Nocturnal. Boxwood. Inscribed: *Fisher Combes Londini Fecit for Nathaniel Thomas 1730.* (53.484)

18 Compass in circular brass case with lid. Paper compass card. 18th century. (34.349/3)

19 Compass in wooden box with paper compass card. Probably late 18th century. (60.247)

20 Sector. Brass, signed: *Butterfield AParis.* The quality of the engraving is quite unlike Butterfield. (49.131)

21 Chondrometer, unsigned. (52.490/1)

22 Chondrometer. Signed: *Dring & Fage 22 Tooley St London.* (52.93)

23 Camera lucida, Wollaston type. (89.91)

24 Waywiser. Circular dial on arm. Small metal wheel. Signed: *Fra. Watkins London.* Mid-18th century. (42.342/1). (fig 137)

25 Waywiser. Large metal wheel. Rectangular wooden box, with dials at top and side, fixed at centre of wheel. Mid-19th century. (40.319)

26 Cylinder electrical machine with accessories including discharging tongs, Leyden jar and insulating stool. The instrument, which is unsigned, was used by John Rees (1813-1895). (44.200)

27 Clockmaker's wheel-cutting engine of late 17th-century type with index plate mounted below rectangular frame. (18.68)
Lit: PEATE (1975) fig 11, no. 123

28 Clockmaker's wheel-cutting engine, late 18th century (57.106)
Lit: PEATE (1975) no. 124

29 Clockmaker's lathe. Early 19th century. (45.68)
Lit: PEATE (1975) no. 125

30 A collection of clock- and watch-making and repairing tools (including lathes), used by the Peel family of Tonypandy, c1850-1920. (F.72.393/1-520)
Lit: PEATE (1975) no. 178

31 A collection of clock and watch repairing tools used by the firm of Abbot, Cardiff Docks, Cardiff. Late 19th

and early 20th century. (F.72.397)
Lit: PEATE (1975) no. 179

S 5 ST HELENS
St Helens Museum and Art Gallery

College Street, St Helens WA10 1TW

0744 24061 ext 2959

The Museum collection contains items relating to the natural history, local history and industry of the area. It includes three items of relevance to this survey, listed below, as well as dental instruments of 19th-century date, gas and X-ray tubes, an early projector and an Edison phonograph.

1 Model beam engine. Brass mechanical parts, mahogany box. Signed: *W. & S. Jones 30 Holborn London.* Second quarter 19th century. (fig 94)

2 Model of a pit winding engine. 19th century.

3 Brass Culpeper microscope on rectangular base, ht 17in. In pyramidal box. Unsigned. Late 18th or early 19th century. (loan)

S 6 ST HELENS
Pilkington Glass Museum

Prescot Road, St Helens WA10 3TT

0744 692499

The museum houses collections and displays relating to the evolution of glass-making techniques. It contains a fine and expanding collection of glass of all periods. There are a few scientific instruments, illustrating the use of glass in this context, as well as a series of spectacles on loan from the British Optical Association.

1 Matthew Loft form of Culpeper microscope. Lignum vitae(?) eyepiece. Black ray-skin outer tube with unusual broad wooden band around its lower end. Octagonal box foot of ebonised wood. Unsigned.

2 Refracting telescope (3in aperture) on pillar and tripod stand. Signed: *Dollond London.* Early 19th century.

3 Altimeter. Signed: *Negretti & Zambra Inst. Makers to Her Majesty London.* Mid-19th century.

4 Cistern barometer. Signed: *Wm. Cox Devizes.* First half 19th century.

S 7 ST MARY'S
Isles of Scilly Museum

Church Street, St Mary's, Isles of Scilly TR21 0JT

0720 22337

The Museum, founded in 1967, is run by the Isles of Scilly Museum Association. Its collections illustrate all aspects of past and present life in the Scilly Isles, covering the flora, fauna, geology and archaeology of the islands, as well as containing historical material from more recent periods. There is a section on shipbuilding, and a display of material retrieved from shipwrecks, including the wreck, in 1707, of the *Association*.

1 Octant. Mahogany frame, radius 14in. Brass arm, vernier and tangent screw. Two shades. The ivory label and degree arc are missing. Early 19th century.

2 Sextant, brass, with silver scale, radius 8½in. Four coloured shades for the index glass, and three for the horizon glass. One telescope, vernier with tangent screw and magnifier. Signed: *Elliott Bros. London*. Late 19th century.

3 Lid of a Pieter Holm tobacco box with Dutchman's log.

S 8 SALFORD
Salford Mining Museum

Buile Hill Park, Eccles Old Road, Salford M6 8GL

061 736 1832

The Salford Museum of Mining contains several scientific instruments, two of which are of considerable importance. These are two microscopes by Benjamin Martin, both of which belonged to the Reverend William Cowherd (1762-1816). Cowherd was the founder and pastor of the Bible Christian Church at Salford and erected an observatory on the roof of his church. None of his astronomical instruments are known to survive, but the microscopes came into the hands of his trustees, and were presented to the Salford Museum.

1 Benjamin Martin's Grand Universal microscope. Signed: *B. Martin invenit et fecit Londini*. Another example of this rare model, belonging to the Royal Microscopical Society (now in *Museum of the History of Science, Oxford*) is illustrated in CLAY/COURT (1928).

2 Opaque solar microscope as described by Benjamin Martin in his pamphlet of 1774. Signed: *B. Martin Fecit London No. 15.*

3 Compound microscope in mahogany box. Signed: *G. Adams No. 60 Fleet Street London*. Ht 19in. A fixed vertical rectangular pillar is mounted on a folding tripod foot. The snout of the tube is screwed into the upper side of a sliding transverse arm.

4 Brass Culpeper microscope with rack focusing. Numerous accessories but lacks pyramidal case. Signed: *F. Day 37 Poultry London*. Ht approx 16in.

5 Calculating machine. Stamped: *Wertheimber patentee*. A calculating machine was patented by David Isaac Wertheimber in 1843, British Patent 9616.

6 Wimshurst machine, *c*1900.

S 9 SALISBURY
Salisbury and South Wiltshire Museum

The King's House, 65 The Close, Salisbury SP1 2EN

0722 332151

The Museum was founded in 1860 and houses extensive and important collections which cover archaeology, natural history, geology and the history of Salisbury and the surrounding country, as well as English pottery and porcelain. There are a number of clocks and watches, primarily by local makers, some weights and measures, and a few scientific instruments.

1 Lodestone, brass mounted. Possibly 16th/17th century.
Lit: STEVENS (1870) p78

2 Ring dial. Brass, unsigned. 18th century.

3 Universal equinoctial ring dial. Silvered brass, in a fish-skin case. Signed on dial: *Culpeper fecit*. Reputed to have belonged to Sir Isaac Newton.

4-5 Two horizontal garden sundials. Brass, unsigned, 18th century.

6 Inclining dial. Brass, unsigned. 18th/19th century.

7 Spyglass. Single-draw, brass, with wood-cased outer tube. Signed: *Salom & Co. Edinburgh*.

8 Spyglass. Single-draw, outer tube ray-skin covered, the inner tube leather covered with gold tooling and the maker's name: *G. Adams London*. Late 18th century.

9 Theodolite. Brass, signed: *Watkins London*. Early 19th century. Said to have been used to survey the railway between Salisbury and Westbury in 1854. (93/1963)

10 Waywiser. By William Lander of Mere. Early 19th century. (32/1928)

11 Chondrometer. Unsigned. Early 19th century.

12 Microscope. Brass, under a glass dome. Constructed by Henry Brooks FGS of Salisbury. Together with a range of microscope accessories and slides. 19th century. (65/1973)

S 10 SALTCOATS
North Ayrshire Museum

Manse Street, Kirkgate, Saltcoats KA21 5AA

0294 64174

The North Ayrshire Museum contains collections relating to the life, local history and industry of the region. The Kirkgate museum contains only one scientific instrument (no. 1 below) whilst the Maritime Museum in the Old Custom House has a number of navigation instruments, notably two backstaves.

1   Theodolite. Brass with silver vernier. Radius of base 4½in. Incorporating bubble levels in base and on telescope. Signed: *Miller & Adie Edinburgh*. Early 19th century.

2   Backstaff. Lignum vitae and boxwood, length 24½in. Vanes missing. Inscribed: *Bernard Hurley Jany 26 1779*.

3   Backstaff. Pearwood, length 25in. Vanes missing. Unsigned, probably early 18th century.

4   Octant. Ebony frame, radius 11in. Signed: *Crichton London*. 1840s.

5   Octant. Brass frame, ivory arc. Radius 7¾in. Arc graduated in 130°. Marked: *Patent composite arc*. ? Mid-19th century.

6   Quintant. Brass, silver scale, radius 7½in. Signed: *G.H. & C. Gowland, Sunderland*.

7   Quintant. Brass, partly lacquered. Radius 7in. Signed: *J. Coombes Devonport 11252*.

S 11 SCUNTHORPE
Scunthorpe Museum and Art Gallery

Oswald Road, Scunthorpe DN15 7BD

0724 843533

The Museum acted as the regional museum for what was North Lincolnshire. It contains important collections of prehistoric and Roman material, and also covers geology, natural history and local industry. There are collections of bygones and period rooms. A branch museum, Normanby Hall, a country mansion of early 19th century date, is appropriately furnished and houses some of the costume collection. Other than a few domestic barometers in the Museum and Art Gallery and at Normanby Hall, there is only one scientific instrument.

1   Octant. Ebony frame, brass arm, radius 15¾in. Signed: *J. Gilbert Tower Hill London*.

S 12 SHAFTESBURY
Shaftesbury Local History Museum

Gold Hill, Shaftesbury SP7 8JW

0747 52157

The Museum is owned and run by the Shaftesbury and District Historical Society, founded in 1946. The present building, formerly a doss-house attached to the Sun and Moon Inn, was opened in 1957. The collections are devoted chiefly to local history.

In addition to a magic lantern, a zoetrope, a fine early stereoscope, a biograph (early cinema projector), domestic medical equipment and a mortar dated 1675, the collection contains four items within the scope of this survey.

1   Tellurium. Wooden base on three turned legs, brass zodiac ring. The mechanism is mounted above the base

and encased in a brass box. The central position, normally occupied by the Sun, carries a candle holder. The terrestrial globe is by Cary and dated 1791. The instrument is signed: *John Handsford Maker Birmingham*. Probably *c*1800.

2   Clockmaker's wheel-cutting engine with fine rococo engraving in the centre of the dividing plate. The machine belongs to the type described by TR Crom as the 'Lancashire engine' and dating from *c*1770-1780 (see CROM (1970), fig 97). Inscribed: *Tho. Tonkin Penzance* (possibly the owner rather than the maker). Tonkin, clockmaker of Penzance, is recorded as active in 1768. *Lit*: BROWN (1970) p81

3   Guinea balance, signed: A. Wilkinson (late of Kirby) Ormskirk, with a printed label inside case lid with name: *Joseph Miles ...Shaston*. *c*1780.

4   Kiln pyrometer, consisting of 'Seger cones' set in ceramic base, for estimating the temperature of a pottery kiln. Late 19th century.

S 13 SHEFFIELD
Sheffield City Museum

Weston Park, Sheffield S10 2TP

0742 768588

The Museum contains general collections covering archaeology and local history, natural history, geology and bygones. It owns important specialised collections of cutlery and Sheffield plate. There are few scientific instruments, though Sheffield was a centre of instrument-making during the 19th century.

In addition to the microscopes listed below are others by Ross; Watson and Son; Powell and Lealand; J Swift and Son; Smith, Beck and Beck; and R and J Beck Ltd. They are either incomplete or belong to the later decades of the 19th or even the early 20th century.

1   Solar microscope in oak box. Mahogany plate, fish-skin covered tube, gut drive to mirror. A small fish-skin covered box contains a screw barrel microscope and accessories. Second half 18th century. (X1975.500)

2   Brass Culpeper microscope. Unsigned. Late 19th century. (L.1909.16)

3   Drum microscope of brass. Unsigned. Contained in the box is an engraving of a similar microscope and a pamphlet. *Description of an Improved Compound Microscope*. First half 19th century. (X1969.274)

4   Compound monocular microscope. Signed: *Ross 5381 London* and inscribed: *Metallurgical microscope. Specification of Thos. Andrews F.R.S. c1850*.

5   Four-draw telescope. Sheffield plate with japanned outer tube. Signed: *J.P. Cutts. Sheffield, c*1840. (1970.624)

6   Orrery. Designed and made by Benjamin Gorrill, Sheffield, 1897. (1921.43)

7   Miner's dial. Signed: *Davis & Son Derby. c*1860-70.

8   Massey's log. Brass. (1967.479)

9   Sikes' hydrometers (3) by Buss, Hatton Garden; Dring and Fage; and L. Lumley and Co. (X1976:671; X1976. 670; X1976.672)

S 14 SHUGBOROUGH
Staffordshire County Museum

Shugborough, Stafford, ST17 0XB

0889 881388

A fine 18th-century mansion designed by James Stuart and Samuel Wyatt, the home of the Earls of Lichfield. House and grounds are now the property of the National Trust. In the estate buildings (brewhouse, laundry, stables, coachhouses) is housed the Museum with farm equipment, crafts, domestic life, costume, geology and natural history. Other than a garden sundial, there is a single portable dial.

1   Universal equinoctial dial, brass, the underside of the compass box engraved with the maker's initials *LTM* (Ludwig Theodor Miller, Augsburg). Late 18th century.

S 15 SIDMOUTH
Sidmouth Museum

Church Street, Sidmouth EX10 9AH

03955 6139

The museum, established by the Sid Vale Association, has collections relating to the locality, including costumes and bygones. It houses one item of scientific interest.

1   Seven-prism spectroscope believed to have been constructed by Sir Norman Lockyer and used by him in the discovery of helium in the spectrum of the Sun, 1868. *Lit:* LOCKYER/FRANKLAND (1869, 1870)

S 16 SKIPTON
Craven Museum

Town Hall, High Street, Skipton BD23 1AH

0756 4079

The Museum contains local collections, including Iron Age and Roman archaeological material, bygones, geological and natural history specimens.

1   Miner's dial. Unsigned, 19th century.

2   Clockmaker's wheel-cutting engine. Incomplete. Signed: *Thos. Green Liverpool, No. 153.* Early 19th century.

S 17 SOUTHAMPTON
City Museums and Art Gallery

Civic Centre, Southampton SO9 4XF

0703 223855

The City's collections, covering fine art, archaeology, history, shipping, etc. are housed in the Art Gallery and four separate branch Museums. In one of these, the Maritime Museum, Wool House, Bugle Street, are number of navigation and other instruments which are not on exhibition.

1   Octant. Ebony frame, brass arm, radius 11¾in. Vernier and fixing screw. Signed on arm: *Dollond London.* Late 18th century.

2   Octant. Ebony frame, radius 11¾in. Vernier and fixing screw. Signed on ivory label: *Spencer Browning & Rust* and on ivory degree arc: *SBR.*

3   Octant. Ebony frame, brass arm, radius 9½in. Vernier, tangent screw. One telescope. Unsigned. First half of 19th century.

4   Sextant. Brass frame, silver arc, radius 7in. Signed: *Troughton London.*

5   Sextant. Double brass frame, silver arc, radius 8in. 5 telescopes. Signed: *Troughton & Simms, c*1830.

6   Sextant. Solid cast brass frame, silver arc radius 7in. Signed: *Walker Liverpool.* 19th century.

7   Universal standing equinoctial ring dial incorporating a compass and a table of the equation of time. The back of the meridian ring is engraved with a list of Scottish and Irish latitudes. Brass, unsigned, second quarter 19th century.

8   Brass Culpeper microscope in pyramidal box. Unsigned: Late 18th/early 19th century.

S 18 SOUTH SHIELDS
Marine and Technical College

St George's Avenue, South Shields NE34 6ET

08943 60403

The College, one of the main centres for the teaching of nautical studies in the country, has assembled, by gift and on loan, a collection of navigation instruments, most of which are permanently on display. A number of items have been loaned to the College by the Tyne and Wear Museum Service (marked TW in the ensuing list). Not listed is a number of instruments – compasses and sextants – belonging to the latter part of the 19th and the early 20th century.
Lit: There is no published catalogue, but a valuable source of information for these and many other navigation instruments in the North East is the manuscript catalogue prepared by Captain DM Robinson of the Marine and Technical College (ROBINSON n.d.). References to this catalogue are indicated by the letter R

followed by the entry number.

1 Backstaff. Lignum vitae frame with boxwood arcs. Length 24in. Unsigned but inscribed: John Durkie. Possesses original horizon vane. First half 18th century. (R1)

2 Octant. Mahogany frame and index arm, radius 18in. Boxwood arc with diagonal scale. Unsigned. Label reads: *Mr Robert Bland Junr 1770*. (R2)

3 Octant. Ebony frame, radius 15in. Signed: *Spencer Browning & Rust London*. c1790. (R3)

4 Octant. Ebony frame, radius 14in. Unsigned but with marking behind arm: and slip card marked: *Lower Shadwell*. (R4)

5 Octant. Ebony frame, radius 15in. Unsigned, but with slip card marked: *Bristol*. Late 18th century. (R9) TW

6 Octant. Ebony frame, radius 11in. Signed: *Spencer Browning & Rust London. Sold by W. Heaton Newcastle*, c1800. (R10) TW

7 Octant. Ebony frame, radius 11in. Inscribed: *H. Helmsley 138 Ratcliffe Highway London*. Second quarter 19th century. Inscribed behind index glass: *July 2 – 55 Meldrun*. (R8)

8 Octant. Ebony frame, radius 10in. Signed: *Crichton London*. Mid-19th century. (R6)

9 Octant. Ebony frame, radius 10in. Inscribed: *M. Moncrieff S. Shields* and marked behind index horizon glasses: *T.C. Sargent*. Mid-19th century. (R7)

10 Octant. Ebony frame, radius 10in. Unsigned, mid-19th century. (R12) TW

11 Octant. Ebony frame. 9½in radius. Inscribed; *H. Gibson North Shields*. Mid-19th century. (R13) TW

12 Octant. Ebony frame, 9½in radius. Unsigned, mid-19th century. (R15) TW

13 Octant. Ebony frame, radius 11½in. Unsigned but inscribed: *W.H. Barling*. (R14) TW

14 Octant. Ebony frame, radius 9in. Signed: *Thos. L. Ainsley S. Shields*. (R11) TW

15 Sextant. Ebony frame, radius 8in. Signed: *Jones Gray & Keen Strand Liverpool*. c1840. (R5)

16 Sextant. Brass frame, radius 8in. Signed: *Cary London*. (R18)

17 Sextant. Brass frame, radius 6¼in. Sold by Reid & Son Newcastle. Mid-19th century. (R16)

18 Azimuth compass. 7in brass bowl, dry card compass. Signed: *Crichton 112 Leadenhall St London*. Label of John Crichton in box. 1840-50. (R25)

19 Artificial horizon (glass hood, mercury trough) signed: *John Crichton 112 Leadenhall St., London*. 1840-50. (R28)

20 Celestial globe on floor-standing tripod stand. 22in globe signed: *Newton and Son, 66 Chancery Lane, London, 1848*. (R27)

21 Orrery. Signed: *Bleuler 27 Ludgate St London*. Late 18th/early 19th century. (R26)

S 19 SPALDING
Spalding Gentlemen's Society Museum

Broad Street, Spalding PE11 1TB

0775 4658

The Society grew up in 1710 among a group of men who used to hold informal meetings in a coffee house in Spalding. About two years later it was organised on a more formal basis on the lines of the Royal Society. It is described in its rules as a 'Society for Gentlemen for the supporting mutual benevolence and their improvement in the Liberal Sciences and Polite Learning'.

Over the centuries the Society has acquired, mostly by gift but sometimes also by purchase, an extensive library and collections of glass, ceramics, coins and medals, bygones, local historical and archaeological material, as well as a number of scientific instruments. These are housed in premises erected in 1910. In addition to the instruments listed below the Museum contains clocks and watches and one or two medical items – a set of surgical instruments and a case containing trepanning instruments of late 18th-century date.

The most important instrument in the Museum is probably one of the Society's earliest acquisitions, the large Arsenius astrolabe (no. 1) which was presented in 1729 by Mr William Gilby, the Recorder of Lincoln. Subsequently, in 1761, the Society purchased four instruments from the maker Benjamin Martin, one of which survives (no. 6).

The Museum is open to the public by appointment with the Curator.

1 Astrolabe. Brass, diameter 11½in. Includes both the universal projection of Gemma Frisius (on the reverse of the mater), as well as the normal stereographic astrolabe projection with 5 plates for latitudes 33, 35⸴ 37, 39, 48, 51, 46 and 56°. Signed: *Regnerus Arsenius nepos Gemmae Frisii fecit Lovanii 1565*. William Gilby of Gray's Inn, who presented the astrolabe, was Recorder of Lincoln and Hull, and became a member of the Society in 1724. The Society's Minute Book records that the instrument had previously belonged to the Lord High Chancellor Hatton. *Lit:* GUNTHER (1932), no. 230; ZINNER (1956) p238

2 Universal equinoctial ring dial. Brass. Unsigned, 18th century.

3 Universal equinoctial ring dial. Brass, unsigned, but dated 1715.

4 Horizontal dial. Brass with folding gnomon for one latitude, mounted on rectangular ebony base containing the compass box. Late 18th century.

5 Orrery. Wooden base covered with engraved paper, mechanism of brass. With attachments for planetarium. Signed: *A new portable Orrery invented and made by W. Jones Sold by P. & J. Dollond St Paul's Ch.Yard London*, c1800.

6 Gregorian reflecting telescope on pillar and tripod stand. Signed: *B. Martin Fleet Street London*. Acquired by the Society in 1761 and, according to the Minute Book, purchased together with an air pump and a compound and a solar microscope from Benjamin Martin who received a payment of £28.18.0. The last three instruments no longer survive. The telescope was repaired by George Adams in 1789, and a payment of £0.12.0 to the instrument maker is recorded in the Minutes.

7 Culpeper microscope. Brass, in pyramidal box with trade card of *William Harris & Co. 50 High Holborn Corner of Brownlow Street London and at Hamburgh*. Probably second decade 19th century.

8 Compound microscope in mahogany box with accessories. Microscope signed: *T. Cooke & Son York*. Mid-19th century.

9 Backstaff. Lignum vitae with boxwood arcs. Lacks vanes. Tudor rose stamped on arc. Early 18th century.

10 Sextant. Ebony frame, ivory arc, brass arm. Graduation 0 – 125°. Vernier, tangent and fixing screw 4 and 3 shades. Signed: *D. Filby Hamburg*. Early 19th century.

11 Octant. Ebony frame, brass arm, ivory arc graduated 0 – 105°. Vernier, tangent and fixing screw, three shades. Signed on ivory label: *J.W. Norie & Co. London*. First half 19th century.

12 Octant. Ebony frame. Unsigned, early 19th century.

13 Sextant. Brass frame, silver scale graduated 0 – 120°. Vernier with magnifier, tangent and fixing screw, telescope. Signed: *Wm Ford 102 High Street Shadwell London*. Early 19th century.

14 Surveyor's level. Brass. Bubble level mounted above telescope which is mounted on a plate incorporating a compass. The base of the compass has been cut at a later stage to insert a further bubble level. Signed: Made by *GE...Fleet Street London*. Presumably George Adams. Third quarter 18th century.

15 Pantograph. Brass, signed: *Dollond London*. Early 19th century.

16 Dutch barometer, 'barometre liegeois'. 18th/19th century.

17 Chondrometer. Signed: *Corcoran & Grigg London*.

18-20 Sandglasses. Two mounted in wood, the third, mounted in iron, appears to be early in date, possibly 16th or 17th century.

21 Sikes' hydrometer. In box inscribed: *Made by Dring & Fage No. 20 Tooley St London and Established by Act of Parliament throughout the United Kingdom*.

S 20 STIRLING
Smith Art Gallery and Museum

Dumbarton Road, Stirling FK8 2RQ

0786 71918

The Museum was founded by Thomas Stuart Smith of Glassingall, and opened in 1874. It houses collections of oil paintings and watercolours, natural history, domestic material, archaeology and ethnology. In addition there are several interesting scientific instruments, and two mid-19th century stereoscopes, one a table model, the other of pedestal type.

1 Circumferentor/theodolite signed: *E. Nairne London*. The circular base plate (12½in diameter, graduated in degrees) is equipped to take either slit and thread sights or a vertical semicircle with a telescope carried on arm with a bubble level (length of telescope 13in, ht of semicircle 7in). (B 4997) (figs. 23 and 132)

2 Mariner's compass (dry card) in gimbals in oak box (11⅛ sq in). The paper compass card is signed around the centre *JEAN FRANCOIS HERVOUET A VANNES 1764*. Diameter of compass rose 7⅝in. (B 1287) (fig 25)

3 Octant. Probably mahogany frame, ivory arc brass arm. Radius 16in. Label in box: *J. Mann at the Hadley's Quadrant & Hourglass Water Street Liverpool Makes and sells all sort of Hadley's Quadrants and other Instruments compleat of reasonable Terms*. This instrument, together with the mariner's compass belonged to a Captain Forrest (d. 1818).

4 Sextant. Ebony and ivory. Radius 12in. Vernier. Label in box *P.A. Feathers Nautical instrument maker Dock St Dundee*. Inscribed: *Moncur Williams Dundee 1857*.

5 Orrery and tellurium by W Jones. Tellurium 12¾in diameter, brass mechanism, wooden base. *Designed for the new portable orreries by W. Jones and made and sold by W. & S. Jones 30 Holborn London*. Orrery diameter 7¾in. *A new portable orrery invented and made by W. Jones and sold by him in Holborn London*. Accompanied by booklet W Jones, *The Description and Use of a New Portable Orrery* (London 1799).

6 Pocket globe in black shagreen case. Signed: *Lane's Pocket globe, London 1818*. Diameter 3in.

7 Terrestrial globe, signed: *Alexr. Donaldson South Niddry Street Edinburgh 1822*. On wooden stand with four legs. Globe diameter about 5in.

8 Large terrestrial globe, signed: *Bardin, 16 Salisbury Square, London, 1802*. Badly damaged.

9 Terrestrial globe, signed: *J. and W. Cary, Strand, London 1st January 1812*. Diameter 12in. (B 10973)

10 Celestial globe, diameter 12in, on a four-legged wooden stand. Inscribed: *The New Twelve Inch British Celestial Globe 1800*. (B 3694)

11 Aneroid barometer in a case resembling a mantel clock. Dated 1862. (B 5220)

S 21 STOCKPORT
Stockport Museum

Vernon Park, Turncroft Lane, Stockport SK1 4AR

061 474 4460

The Museum houses collections relating to local history, geology, natural history and Victoriana. At the time of writing it houses, on loan from the *Lady Lever Art Gallery, Port Sunlight*, a number of instruments, 17 all told, all but one of which are sundials. These items have not been examined, and therefore only those which are signed have been listed below. However, unsigned instruments include diptych or string-gnomon dials, stone garden dials, and a sandglass. The majority of the instruments would seem to be of 18th-century date.

1   Universal equinoctial dial. Octagonal, 2½ x 2½in. Engraved on underside: *I.G.V.* (Johann Georg Vogler) Augsburg, mid-18th century. (XX15)

2   Universal equinoctial dial. Brass, octagonal, 2¼ x 2¼in. Signed on underside: *T.N. Holderich Augsburg*. Late 18th century. (XX14)

3   Universal equinoctial dial. Brass, octagonal, 2¼ x 2¼in. Signed: *And. Vogler* on underside. Augsburg, second half 18th century. (XX13)

4   Universal equinoctial dial, 3 x 3in. Signed: *J. Schrettegger Augsburg*. Late 18th or early 19th century. (XX10)

S 22 STOKE-ON-TRENT
The Wedgwood Museum

Josiah Wedgwood and Sons Ltd, Barlaston, Stoke-on-Trent ST12 9ES

0782 204141

The Museum contains an extensive collection of Wedgwood ware from earliest times until the present. Josiah Wedgwood (1730-1795) constructed ceramic apparatus for a number of his scientific friends including mortars, retorts and crucibles.
Lit: CHALDECOTT (1981)

1   Wedgwood pyrometer for gauging very high temperatures in a kiln. The description of this 'thermometer' was communicated by Josiah Wedgwood in a paper to the Royal Society in 1782. It is based on the measurement of the contraction of small pieces of clay which slide down a groove with an empirically constructed scale.
Lit: WEDGWOOD (1978) pp35-39, plate 3

2-3   Two clocks by Whitehurst, Derby, one a timing-clock supplied to Wedgwood in 1806.

4   Engine-turning lathe used for fluting wares. Possibly supplied by Boulton and Watt in the 1780s.

S 23 STRANRAER
Wigtown District Museum

The Old Town Hall, George Street, Stranraer DG9 7JP

0776 5088

The Library building houses a small collection of items of local interest, including three scientific instruments.

1   Octant. Ebony frame, radius 12in. Signed: *J. Holm Amsterdam 1765*. Label in lid of box: *Culmer & Tennant, Mathematical Instrument Makers and Ships Chandlers, 126., opposite Wapping New Stairs, Wapping*.

2   Octant. Ebony frame, radius 16in. Engraved on the arm: *Thos. Burnett, 61, Gt. Tower Street, London*.

3   Mercury artificial horizon – brass hood, glass panes, trough. Used by Sir John Ross on his voyages of exploration for the North-West Passage (1829-33).

S 24 STROMNESS
Stromness Museum

52 Alfred Street, Stromness, Orkney KW16 3DF

0856 850025

The Museum, run by the Orkney History Society, houses collections which in the main are related to the life and history of Orkney. In addition to collections of zoology, geology and botany, there are ship models, whaling relics and Eskimo material (Stromness was once a port of call for arctic whalers). Agricultural items and bygones relating to the islands are also on display.

    In addition to two rare traverse boards, and one or two other navigation instruments, the Museum houses a collection of telegraphic equipment of the type used in Orkney in the early days of telegraphy, i.e. a Wheatstone bridge, receiver, sounder, double current key, Wheatstone transmitter and perforator; most of this equipment is lent by the *Science Museum, London*.

1   Octant. Rosewood, ivory arc, brass arc, radius 18½in. Central zero vernier. Unsigned, *c*1760-80. Was the property of the Arctic explorer John Rae (1813-1893).

2   Traverse board, 15 x 11in.

3   Traverse board, painted wood, 16 x 9in.

4   Stuart's marine distance meter. Signed: *G. Lee & Son The Hard Portsmouth*, late 19th century.

S 25 SUNDERLAND
Sunderland Museum and Art Gallery

Borough Road, Sunderland SR1 1PP

091 514 1235

The Museum, now part of the Tyne and Wear Museum Service, houses collections covering geology, botany, zoology, archaeology, shipping, ceramics and English silver. There are a number of scientific instruments which are predominantly navigational; not listed are pelorus compasses and ships' logs of early 20th-century date. In addtion there is also apparatus associated with Sir Joseph Wilson Swan's invention of the incandescent electric lamp in 1878. As well as the air pump mentioned below, the Swan material includes early electric lamps, Geissler tubes, Leyden jars, a gold-leaf electroscope, Crookes' cathode ray tube and a radiometer.

1 Brass drum microscope. Unsigned. First half 19th century.

2 Gregorian reflecting telescope, 4¾in. aperture. Lacks mounting. Late 18th/early 19th century.

3 Backstaff. Lignum vitae with boxwood scale. Horizon vane marked *W.W* (? William Wright of Bristol). *c*1730. On loan from the *National Maritime Museum, London.*
Lit: HILL/PAGET-TOMLINSON (1958), p12; LONDON – NATIONAL MARITIME MUSEUM (1971) Section 5-3 (Ref. S99)

4 Backstaff. Lignum vitae with boxwood arcs, lacks vanes. Inscribed: Mr Charles Burns 1737. On loan form the River Wear Commissioners.

5 Backstaff. Lignum vitae with boxwood arcs, lacks vanes. Apparently incomplete – arcs are not graduated. Marked: *CRW*. On loan from the River Wear Commissioners.

6 Octant, mahogany frame, radius 20in. A mid-18th century instrument which has been adapted to an unknown purpose. The horizon glasses, index arm, etc. have been removed, and the frame planed flush. The zero end of the frame is secured at right angles to a mohogany board fitted with castors. A hinged mahogany board is so mounted as to sweep out to any angle.

7 Octant, ebony frame and index arm with brass sleeve at end, radius 17¾in. Signed: *Made by Jno Gilbert on Tower Hill London for Jnº Fenton April 4th 1769.*

8 Octant, ebony, with brass arm, radius 16in. Signed: *Nairne & Blunt London.*

9 Octant, ebony frame, radius 10in. Signed: *Norie & Co. London.*

10 Octant, ebony frame, radius 10in. Signed: *Harris & Son London.*

11 Octant, ebony frame, radius 10in. Signed: *Critchton London Made for A. Baharie Sunderland.*

12 Octant, radius 10in. Signed: *C. Wilson & Wilson late Norie London.*

13 Octant, ebony frame, radius 10in. Signed: *P.A. Feathers Dundee.*

14 Octant, radius 10in. Ebony frame. Signed: *Rennison Sunderland.*

15 Octant, radius 10in. Ebony frame. Signed: *I.W. Carew Wapping Wall London.*

16 Sextant. Brass frame. Signed: *W.F. Cannon 177 Shadwell High St., London.*

17 Compass in turned and red painted wooden box with lid. Signed on compass card: *A. van Santen fecit Rotterdam.* Around the outside of the box is inscribed: *Ricd Dobson Pilot No. 10 Sunderland.* 18th century.

18 Compass in rectangular wooden box with lid. Late 18th century.

19 Chrondometer, signed: *Charles Frodsham, 84 Strand London No. 3491.*

20 Surveyor's level. Telescopic sight, bubble level, compass. Signed: *Adams London.*

21 Two circular protractors. Signed: *T.B. Winter Newcastle-on-Tyne.*

22 Sail-maker's sector. Signed: *I.B. Paton 1792.* Mahogany and boxwood. Length of arms 20in.

23 Hydrometer. Signed: *Buss Hatton Garden London.*

24 Double-barrel air pump. Mid-19th century. Used by Sir Joseph Wilson Swan.

25 Universal equatorial dial. Signed: *Vogler, Augsburg.*

S 26 SWANSEA
University College of Swansea and Royal Institution of South Wales Museum

Victoria Road, Swansea SA1 1SN

0792 53763

The Royal Institution was founded in 1835 as a literary and philosophical society on the lines of the Royal Institution in London. The Museum, which formed part of the Institution, has displays relating to Swansea and district – Welsh pottery and porcelain, paintings engraving and watercolours, archaeological, geological, ornithological and botanical collections. (The former Industrial Museum of South Wales (now The Swansea Maritime and Industrial Museum) is no longer housed in the Royal Institution building, but at South Dock.) The Museum's collections relate primarily to the industries of South Wales – steel, oil, copper, etc. In 1973 the University College of Swansea came into partnership with the Royal Institution of South Wales in the administration of the Museum. The Museum possesses a small collection of scientific instruments.

1 Quintant: Signed: *Troughton and Simms London.* Double brass frame, arc graduated in 150°.

2 Quintant. Brass frame with silver scale. Vernier, magnifier. Unsigned, second half 19th century.

3 Quintant signed: *Critchton London.* Brass frame, silver arc.

4 Sextant. Brass with silver scale. Vernier. Label in box: *Cosons & Son, 20 Vind Street, Swansea.*

5 Navigation compass/direction finder signed: *J.A. Webber & Son, 104 Oxford Street Swansea.*

6 Air pump, early 19th century. This instrument has been attributed to Wood of Colchester.

7 Cylinder electrical machine, probably early 19th century.

8 Cylinder electrical machine signed: *Elliott Bros. 449 Strand London.*

S 27 SWINDON
Swindon Museum and Art Gallery

Bath Road, Swindon SN1 4BA

0793 26161 ext 3129

In addition to the Museum and Art Gallery in Bath Road,

four branch museums come under the care of the Borough of Thamesdown. The Bath Road Museum contains archaeology, natural history, geology, bygones, coins and tokens, as well as the Art Gallery, and houses only one instrument of relevance to this survey, while the Great Western Railway Museum in Faringdon Road contains an instrument of some historical importance.

1 Waywiser. Signed: *Louch St Albans, c1790.* (Swindon Museum and Art Gallery)

2 Tide gauge made by J Newman, Regent Street, London. Used by IK Brunel during the construction of the Saltash Bridge. (Great Western Railway Museum)

T

T 1 TALGARTH
Howell Harris Museum

Coleg Trefeca, Talgarth, Powys LD3 0PP

0874 711423

The Museum houses a collection of personalia relating to the Methodist reformer Howell Harris, founder of a religio-industrial settlement at Trevecca in the 18th century.

Howell Harris's brother Joseph, an astronomer, writer of mathematical works and Assay Master of the Royal Mint, spent some time at Trevecca and observed the Transit of Venus there in 1761. The Museum contains the telescope which he apparently used for this observation. In addition, it houses an electrical machine recorded by Howell Harris as having been purchased in 1763.
Lit: BENYON (1958) pp210-13

1 Newtonian reflecting telescope with painted wooden tube. The secondary mirror is missing, but the primary speculum survives, and so does the arc which comprised part of the alt-azimuth mounting of the telescope. Aperture 5½in, length of tube 73in.

2 Cylinder electrical machine, 1763. Table model. Unsigned, but of good quality and workmanship. Dimensions of base 35½ x 23¾in. Total length of cylinder 13½in.

T 2 TAUNTON
Somerset County Museum

Taunton Castle, Taunton TA1 4AA

0823 255504

Local and natural history collections. Insects, extinct

animals from Mendip caves, geology, archaeology, bygones, ceramics, metalwork and glass. The collection also includes a small number of scientific instruments. In addition to the items listed below there are two microscopes of the latter half of the 19th century, one of them by Watson, as well as some early X-ray tubes.

1 Culpeper microscope of brass in box with the trade label of: *Wm Harris & Co., 50 High Holborn London and at Hamburgh.*

2 Day and night telescope signed: *Jones London.*

3 Day and night telescope signed: *H. Bracher Impd. Day or Night.*

4 Cylinder electrical machine. Signed: *Nairne's Patent Medical electrical machine.*

5 Plate electrical machine. Incomplete and unsigned. Probably early 19th century.

6 Electrometer made by James A Pring *c*1850, dial by *E.M. Clarke, Magnetician 428 Strand London.* Also home-made electric coil by JA Pring.

7 Pantograph. Unsigned, ebony and ivory, probably late 18th or early 19th century.

8 Horizontal dial of silver signed: *Butterfield Paris, c*1700. The dial was possibly for use in Scotland, since the latitudes of several Scottish towns are engraved on the underside of the dial. Lent by the National Trust from Tintinhull House.

9 Diptych dial. Ivory, French, early 17th century. The gnomon is missing and there is no compass.

T 3 TEDDINGTON
National Physical Laboratory Museum

National Physical Laboratory, Teddington, Middlesex TW11 0LW

081 977 3222

The National Physical Laboratory, the national institution for the establishment and testing of standards, was set up in 1899, shortly afterwards occupying its present site

based at Bushy House. Many of its functions, such as testing of thermometers, barometers, watches, chronometers and sextants, had previously been carried out at the Kew Observatory. The role of the NPL rapidly developed in the early years of the 20th century especially in engineering, electrical and metallurgical work. A museum was set up in 1977 (previously, some items had been donated to the *Science Museum, London*). Most of the objects lie, chronologically, outside the scope of this inventory. Admission to the museum is by prior arrangement only.
Lit: NATIONAL PHYSICAL LABORATORY (1967) and (1983); PYATT (1983)

1 Standard yard, brass, signed: *Kew Standard 1852* and *Troughton & Simms London* and *No 49.*

2 Imperial standard yard, in cast iron. 1845. Copy no. 63.

3 Metre/yard, painted on ceramic, signed: *L. Casella No. 31.* In wooden frame (numbered 2602), painted with label 'Measures of length published by the Metric Committee of the British Association 1869'.

4 Large bullion-type balance said to have been used by JH Poynting *c*1886. Possibly constructed by L Oertling. Still in use and not displayed.

5 Balance, 'Board of Trade Standard', signed: *L. Oertling*, dated 1914. Still in use and not displayed.

6 Set of rock crystal metric weights, in box signed: *Jos Nemetz Wien V/1.*

7 Two Fahrenheit thermometers made by Casella, dated 1854, for Kew Observatory.

8 One Fahrenheit and three Celsius thermometers, dated 1878 and 1879, made for Kew Observatory.

9 Two mercury in verre dur thermometers made by Tonnelet, Paris, dated 1893 and 1897, for Kew Observatory.

10 Standard 10 Candle Power pentane lamp, designed by Vernon Harcourt in 1877 (certified by the NPL in March 1903).

11 Arithmometer (No 3149) signed: *L. Payen, Paris.*

12 Clock, signed: *J. Morrison & Sons. 27 Packington St London.No 8702*, with invar pendulum, used for frequency experiments.

13 Collection of electrometers, displayed within semicircular wooden housing which itself acted as a semicircular scale.

14 Kelvin's multicellular, electrostatic voltmeter, Board of Trade volt standard, pre-1894, signed: *James White Glasgow No. 256.*

15 Standard barometer constructed in 1855 by John Welsh at Kew Observatory.

16 Cathetometer, used to read no. 15, signed: *R.W. Munro London 1876 No 3.*

17 Metallurgical microscope, designed by Walter Rosenhain, constructed by R. & J. Beck, *c*1910.

18 Malby's terrestrial globe, published by Edward Stanford, *c*1887.

T 4 TORQUAY
Torquay Museum

529 Babbacombe Road, Torquay TQ1 1HG

0803 23975

The Torquay Natural History Society was founded in 1844. In 1845 it obtained permission to explore Kent's Cavern, and its Museum contains material from its own excavations, as well as from other sources. There is prehistoric material from other Devon caves, as well as collections illustrating Devon natural history, local history and folk life, ceramics, etc. The collection contains a small group of scientific instruments, the most interesting of which is the 'Graphic telescope' patented by Cornelius Varley in 1811.

1 Equinoctial ring dial. Brass, late 18th century. Unsigned.

2 Universal equinoctial dial. Brass in wooden case. First half 19th century.

3 Horizontal dial with mid-day cannon. Circular marble base, brass cannon. 19th century.

4 'Graphic telescope' patented by Cornelius Varley in 1811. Inscribed: *Cornelius Varley's Patent Graphic Telescope Tottenham Court Road, London.*

5 Culpeper microscope. Brass, unsigned. Late 18th/early 19th century.

6-7 Drum microscopes. Brass, in wooden box. Unsigned. Mid-19th century.

8 Compound monocular microscope on circular brass base. Pillar inclinable by means of compass joint. 19th century.

9 Simple microscope. The pillar fits into the lid of its wooden box. Lenses can be used singly or combined. Late 18th century.

10 Simple microscope. Circular brass base with pillar supporting mirror, stage and objectives.

11 Pocket botanical microscope. Brass with ivory handle. Cardboard and leather case. Early 19th century.

12 Octant. Ebony frame, ivory scale, brass arm. Fixing screw, vernier. Graduation to 95°. Early 19th century.

13 Octant. Ebony frame, ivory scale, brass arm. Vernier, tangent and fixing screw. Signed: *Dollond London.*

14 Banker's slide rule, 12in. Wood and engraved paper. *Published by W. Cary 182 Strand April 1 1813.*

T 5 TRURO
Royal Institution of Cornwall, Royal Cornwall Museum

25 River Street, Truro TR1 2SJ

0872 72205

The Royal Institution of Cornwall was founded in 1818 'for diffusion of science and promotion of literature' and from the beginning sponsored a museum which has received considerable bequests over the years. The Museum is still largely the responsibility of the Royal Institution of Cornwall, receiving grants from the City and County Councils. It houses important collections relating to Cornish archaeology, life, industry and art, including a number of interesting instruments, as well as model beam and mill engines.
Lit: TRURO (1966) (includes an illustration of one of the miner's theodolites)

1   Surveyor's level with telescope signed: *E.T. Newton (late Wilton), Camborne.* Second half 19th century.

2-3 Lean's dials (miner's theodolites) (2). One signed: *W. Wilton St Day Cornwall,* the other: *Willm. Wilton St Day and Camborne.*

4   Circular protractor with arm and vernier. Signed: *Willm. Wilton St Day Camborne.* In box with William Wilton's trade card.

5   Clockmaker's wheel-cutting engine, made and used by Caleb Boney Padstow, early 19th century.
Lit: BROWN (1970) p40, plate 4

6   Clockmaker's wheel-cutting engine. Used by the Ham family of Liskeard. Late 18th/early 19th century.
Lit: BROWN (1970) p40, plate 3

7   Milemeter invented by Thomas Hickes of Truro. Signed on dial: *Thomas Hickes Truro 1843.* In the form of a tricycle with large measuring wheel supported by two smaller wheels. Short shafts for pulling.

8   Astronomical telescope, 4in aperture refractor signed: *T. Cooke and Sons, York.*

9   Brass Culpeper microscope. Unsigned, second half 18th century. In pyramidal oak case.

10   Jones 'Improved' microscope. Fixed square vertical pillar, stage focused by means of pinion acting on a rack in side of pillar. On rectangular base with drawer for accessories. Signed: *W. & S. Jones 135 Holborn London.* Late 18th century.

11   Compound microscope on folding tripod foot – a development of Benjamin Martin's microscope, focused by adjustment of position of stage through rack and pinion. English, late 18th century.

12   Compound microscope signed: *J.B. Dancer.* Second half 19th century.

13   Compound microscope signed: *Millikin and Lawley, Strand.* 19th century.

14   Botanical microscope. Brass, folding into flat case, late 18th century.

15-16 Two blow-pipe sets for mineral testing. Signed: *J.T. Letcher, Truro.*

W

W 1 WADDESDON
Waddesdon Manor

Waddesdon, nr Aylesbury HP18 0JH

0296 651211/651282

The manor, set in extensive wooded parkland, was built in the French Renaissance style for Baron Ferdinand de Rothschild between 1880 and 1889. Together with its outstanding collections it was bequeathed, with an endowment, to the National Trust in 1957 by Mr James de Rothschild. The core of the art treasures at Waddesdon were acquired by Baron Ferdinand, a lover of French 18th-century art, of English-18th century portraits, and of earlier Dutch and Flemish masters. Most of his collection, with the exception of his late Medieval and Renaissance bequest to the *British Museum, London,* can be seen at Waddesdon. The manor also contains collections assembled by Baron Ferdinand's sister and successor at Waddesdon, Miss Alice de Rothschild, who

assembled the collections of arms and armour and Sevres porcelain. A third source of the collection was Mr James de Rothschild, who inherited the Manor on the death of Miss Alice and who bequeathed with the house his father's important collection of china and pictures.

Among the outstanding and wide-ranging collections at Waddesdon, as well as some early clocks, there is a small number of important and unusual scientific instruments, some of which have been included in catalogues of the collection, published by the Office du Livre, Freibourg.
Lit: BLAIR (1974) and BELLAIGUE (1974) for catalogues for the brass instruments and standard lengths respectively.

1   Watch in book form. Gilt brass, the lid incorporating a horizontal sundial. Movement replaced in 19th century. South German, late 16th century. (W 1/143/1)
Lit: BLAIR (1974) p471, no. 219

2   Equinoctial ring dial. Silver. Inscribed: *Johannes Fischer Gen. Superintenden per Livan* and signed: *Daniel von Berthold feci. in Liv. A° 1584.* This dial differs from the usual universal equinoctial dial since the inner equatorial ring consists of a flat disc with a cut-out slit for the pin hole slide. The disc is decorated with engraved ornament. The inscribed date has been modified from

1684 to 1584.
Lit: TURNER (1979)

3 Butterfield type dial. Octagonal, silver. Signed: *Cadot gendre Macquart A Paris 1734.*

4 Drawing instruments and travelling inkpot. Signed: *Baradelle A Paris 1765.*

5 Proportional compass. Brass, steel points, length 16½in. Ornamented with finely engraved floral decoration. German, first half 17th century. (W 1/110/3)
Lit: BLAIR (1974) p473-6, no. 220

6 Set of gunner's instruments. Steel and brass in leather case with gilt copper locket and chape. Consists of gunner's rule with scales for STEIN, BLEY, EISEN and ZOLL, terminating in baluster-shaped finial, four steel rods, and a pair of dividers (maker's mark a stork). German, mid-18th century. (W1/89/1)
Lit: BLAIR (1974) p480, no. 222

7 Set of gunner's instruments. Steel and brass, in leather-covered wooden case with gilt brass chape. Contents: gunner's rule (scales for LOTT, BLEI, EISEN, STAIN and ZOLL, inscribed: *MINCHNER GWICHTS* and signed and dated *NK 1693*) and three tapering steel rods with corkscrew, point and spear point terminations. South German, ? Augsburg, 1693. (W1/88/3)
Lit: BLAIR (1974) p476, no. 221

8 Ell rule. Wood veneered with ivory and tortoiseshell. Length of measure 57.1cm, i.e. Leipzig ell. Decoration of foliage and engraved figures. Scratched on the handle: *Aus dem Nachlass der am 6/6 1795 verst. Grosmutter Aheder Herrn Baron v. Beaufort Jabitz d. 9/7 1861.* German, c1630. (W 1/95/2)
Lit: BELLAIGUE (9174) p664, no. 155

9 Ell rule. Square section, walnut inlaid with staghorn and mother of pearl, in part stained and engraved. Measure either Flemish ell or Venetian *braccio*. South German, dated *1592.* (W1/88/5)
Lit: BELLAIGUE (1974) p665, no. 156

10 Refracting telescope. Multi-draw, leather-covered outer tube with gold pressings. Signed: *Paulus de Beletis Fecit Anno 1682.*

W 2 WARLEY
Avery Historical Museum

W & T Avery Limited, Foundry Lane, Smethwick, Warley B66 2LP

021 558 1112

The origins of the firm can be traced back to a blacksmith and steelyard maker, James Ford, who established a workshop at No. 11, Well Street, Birmingham c1730. The Workshop passed into the hands of William Barton, and subsequently Thomas Beach, maker of scales, steelyards and weights. The Avery family became associated with the Beaches during the latter half of the 18th century, and also, through marriage, with Joseph Balden who took over Beach's

workshop. In 1818 Joseph Balden Junior made the business over by deed of trust to William Avery and three others. In 1841 Joseph Balden made William Avery his sole executor.

The firm moved to Soho Foundry in 1895. Over the years a museum has been established which, while owning many products of the firm of Avery, has accumulated items relating to the history of weighing in all periods and countries. The earliest items are ancient Egyptian weights, Roman beam scales and weights and a Romano-British steelyard. There is an extensive collection of coin scales from the 17th century onwards, of steelyards of various forms, agricultural scales, platform scales, person and counter scales, spring balances, English wool weights, etc. Strictly, although in its own field this is a collection of great importance, most of the material in the Museum falls outside the scope of this survey, but there are a number of items which deserve particular mention, and are listed below.
Lit: BROADBENT (1949) for an account of the history of the firm; SANDERS (1960) discusses and illustrates many items in the Museum; BIRMINGHAM (1940) is a typescript and photographic catalogue of the Museum; a copy is to be found in the Science Museum Library, London

1 Precision balance in glazed mahogany case. Two hollow brass cones joined at their bases form the beam which is of the type used by Ramsden and Troughton. Unsigned, but attributed to Troughton, early 19th century.
Lit: STOCK/BRYDEN (1972) p46, fig 2

2 Precision balance by De Grave, Short and Fanner. 10in beam. In glazed wooden case. Late 19th century.

3 Assay balance, 1907. Signed: *L. Oertling London.*

4 Chondrometer. Unsigned, early 19th century.

5 Chondrometer. Box with label of De Grave, Short and Fanner, 59, St Martin-le-Grand London.

6 Sikes' hydrometer in wooden box. Unsigned, first half 19th century.

W 3 WHITBY
Museum of the Whitby Literary and Philosophical Society

Pannett Park, Whitby YO21 1RE

0947 602908

The Whitby Literary and Philosophical Society was founded in 1823 by the Revd George Young DD, supported by a group of enthusiasts who wished to benefit the town of Whitby by 'supporting a Museum and promoting the interests of Science and Literature by such other means as may be found practicable'. The Museum moved to its present location in 1931. The maintenance of the building is undertaken by Whitby Town Council while the Society remains responsible for the control and management of the Museum exhibits. The collections are extensive and contain much interesting and important material. They cover the fields of natural history, geology, archaeology, bygones, local history and shipping, as well as items of less specifically local interest such as coins and medals, arms and armour, etc. From the point of view of this survey the museum houses an interesting collection of instruments, connected in part with the arctic explorer William Scoresby, and in part generally with navigation.

William Scoresby (1789-1857), the son of a whaling captain of the same name, carried out important observations in the Arctic, especially on magnetism, before turning to the ministry. On his death he bequeathed his instruments, books and papers to the Whitby Literary and Philosophical Society, where they still remain. Items marked (S) below are associated with Scoresby.

In addition to the items listed below, the Museum houses coin balances, weights and measures, some surgical instruments and four ivory syringes.
Lit: WHITBY (1972) is a guide which summarises the exhibits. McCONNELL (1986) lists the Scoresby relics and apparatus, and relates the latter to published accounts of Scoresby's work.

1   Two cases of bar magnets, the larger as described and illustrated in Scoresby (1831, 1832). (S)

2   Magnetic battery consisting of numerous thin 14in bars bound together. Suspended in a 21in wooden frame. The magnet supports a series of steel balls. (S)

3   Two further 14in magnetic batteries as item 2.(S)

4   Magnetic battery consisting of 9 horseshoe magnets. Fitted with a suspension ring. Length, excluding ring, 11½in. Constructed by Scoresby in 1822 when locked in the ice off the coast of Greenland. (S)

5   Three magnetic batteries suspended from 28in lightwood frame with turned columns:

  a   Battery of five horseshoe magnets, length 12in.

  b   Battery of fifteen horseshoe magnets, length 6in.

  c   Battery of six horseshoe magnets, length 6½in.

All three have keepers equipped with hooks for supporting weights. (S)

6   Magnetic battery. Two sets of thin bar magnets have been screwed together. One set is connected with the other by means of three turned metal supports. (S)

7   Three compound compass needles, length 7½in, 7½in, and 24in. (S)

8   Nine compass needles with agate caps. Varying types. (S)

9   Dip circle. Signed: *W. Wilton St Day Cornwall.* *c*1840-50. (S)

10   Declination compass for measuring diurnal changes. Consists of a compound magnet needle mounted with graduated arcs (15 – 0 – 15°) in a glazed oblong wooden box. Turned brass columns support a horizontal member equipped to relieve weight on the swivel point of the compass needle by means of a thread. (S)

11   Compass in heavy circular brass case. Diameter 7in. Magnet needle with broad ends and agate cap. Brass strip on one arm of the needle. (S)

12   Prismatic compass. Card signed: *Thomas Jones....62 Charing Cross London.* 2nd quarter 19th century. (S)

13   Mariner's compass in gimbals in wooden box. Compass diameter 5¼in. Signed: *Spencer Browning & Co. London* on compass card. (S)

14   Compass cards:

  a   Spencer Browning & Rust, London (S)

  b and c   Spencer Browning & Co. (S)

15   Compass card with agate cap. Signed: *W. & T. Gilbert 148 Leadenhall St London.* The card is mounted on a compound magnet needle. The reverse of the card bears the number 9 and records, in ink, measurements dated between 1839 and 1855. (S)

16   Compass needle mounted on brass stand equipped with graduated arc for determining the inclination of the earth's magnetic field to the horizontal. (S)

17   Pivoted bent magnet on a stand. Magnet and stand can be connected to a battery in order to demonstrate the fact and direction of a magnetic field due to a current in the neighbourhood of a straight conductor. (S)

18   Magnetometer. Constructed by Scoresby, incorporating a circular protractor signed: *Cary London.* *Lit:* SCORESBY (1827) (S)

19   Compass card. Inscribed: *H 15* in ink and signed: *J. Dickman Leith.* Note on back: *Leith 23 Febr. 1836 Mr. Dickman Nautical Instrument Maker to his Majesty in Scotland No. III(S).* Said to have been used in 1822 to support a ship's chronometer so that it was always in the same position relative to the earth's magnetic field. (S)

20   Mariner's compass in gimbals in a wooden box 9½in x 8½in. ? First half 19th century. (S)

21   Mariner's compass in gimbals. Circular brass bowl, diameter 10½in. Signed: *Mangavel A Bordeaux.* ? Late 18th/early 19th century. (S)

22   Mariner's compass. Sir W Thomson patent no. 7818. Made by James White Glasgow, *c*1900.

23   Gold leaf electroscope. Second half 19th century. (S)

24   Cylinder electrical machine on base 17¾ x 5in. Probably a home-made machine. Cylinder diameter 8in. *c*1760-70.

25   Magnetic generator on the lines of Saxton's (1834). Constructed by William Scoresby and James Prescott Joule and described by them in Scoresby/Joule (1846). (S)

26   Astatic needle galvanometer with glass dome. Signed: *Watkins & Hill Charing Cross.* (S)

27   Ampere's apparatus to illustrate electromagnetic rotation. Horseshoe magnet mounted with poles upwards on wooden stand with 'buckets' placed on each pole which revolve when currents pass through. (S)

28   Backstaff. Traditionally used by James Cook. ? Pear wood. Signed: *Made by Geo McEvoy Temple Bar Dublin.*

29 Octant. Radius 18in. Ebony frame. Index arm ebony, with lower part of brass. Centre zero vernier. Signed: *Made by Jno Gilbert on Tower Hill for Charles Richardson May 19 1767.*

30 Octant. Radius 18in. Ebony frame, brass arm. Centre zero vernier. Lacks index mirror. Inscribed: *Corney [illegible] 1788.*

31 Octant. Ebony frame, radius 12in. Two sockets, one set of shades. Signed: *Spencer Browning & Rust London.* Late 18th/early 19th century.

32 Octant. Ebony frame, brass arm, radius approximately 12in. Signed: *Spencer Browning & Rust.* First quarter 19th century.

33 Octant. Ebony frame, brass arm, radius approximately 12in. Marked: *SBR* on arc. Said to have been made by William Morley, Church St, Whitby.

34 Octant. Radius 12in. *SBR* on arc. Signed on ivory label: *Browne Bristol. c1825.*

35 Octant. Ebony frame, brass arm, radius 10in. Signed: *Spencer Browning & Co. London* (on ivory label), *SBR* on arc. Box with label of *D. van Ketwick Kalkmarkt U 196, Amsterdam, c1825.*

36 Octant. Brass frame with telescope, ivory degree arc, radius 12in. First half 19th century

37 Octant. Ebony frame, brass arm, radius 10in. Signed: *Mrs. Janet Taylor 104 Minories London.* Box with label of W Collyer, 3 Bickley Row, Commercial Docks, London SE, *c1850.*

38 Octant. Ebony frame, brass arm, radius 10in. Signed: *Galley London.* Mid-19th century.

39 Octant, ebony frame, radius 9⅞in. Unsigned. Mid-19th century.

40 Octant. Ebony frame, radius 9⅞in. Signed on ivory label: *D.W. Laird Leith.*

41 Sextant. Radius 9in. Signed: *Dalton Hartlepool.* Late 19th century.

42 Mechanical log, 'New Yacht Log'. Signed: *Edwd Massey Patentee London,* c1840.

43 Mechanical log. *T. Walker's patent harpoon ship's log London,* c1860-1870.

44 Mechanical log. *The pendant log, Reynolds patentee London,* c1870.

45 Dutch barometer, 'barometre liegois', glass, 18th century.

46 Screw barrel microscope. Ivory, bone and brass, in fish-skin case, lined with silk and velvet. Mid-18th century.

47 Nautical telescope. Signed: *Troughton & Simms London.* Inscription with presentation date of 1863.

48 Nautical telescope. Signed: *R. Imme Berlin 1873.* Box contains document recording presentation from Emperor of Germany to Capt. Dobson of Whitby.

49 Orrery and tellurium. Wood, covered with engraved paper. Brass mechanism. Diameter 14in. Signed: *W. Harris & Co. 50 Holborn London.* Second quarter 19th century. (S)

50 Clarke's hydrometer. In mahogany box. Complete set of weights. Metal hydrometer inscribed: *Clarke c 5105.*

51 Hydrometer. Glass, paper scale. With engraved paper explanatory diagram signed: *J. Croce fecit* and engraved by *J. Battles York.* Probably second quarter 19th century.

52 Pedometer. Gilt metal and shagreen case. Signed: *Spencer & Perkins.*

53 Diptych dial. Ivory with red and black inlay. Signed: *Michael Lesel.* First quarter 17th century.

54 Marine barometer in gimbals. Signed: *M.O. 641 Adie London.*

55 Pocket globe (terrestrial globe with celestial gores inside fish-skin case). Signed: *R. Cushee.* First half 18th century.

W 4 WINCHESTER
Hampshire County Museum Service

Chilcomb House, Chilcomb Lane, Winchester SO23 8RD

0962 846304

The administration, workrooms and stores of the County Museum Service are housed at this address, but there is no display (see Alton and Basingstoke). The instruments listed below can therefore only be inspected by prior arrangement.

1 Tellurium. Wooden base, 7¾in. diameter, covered with engraved and coloured paper, brass mechanism. Signed: *A new portable orrery invented & made by W. Jones London.* (WOC 4495)

2 Waywiser. Mahogany frame and iron-bound wheel. Heart-shaped handle with counting mechanism and dial mounted below. Signed: *Watkins Charing Cross London.* Early 19th century. (ACM 1952.180)

3 Waywiser. Light metal wheel and frame, counting mechanism in box mounted at centre of wheel. Signed: *Adie London.* Mid-19th century. (WOC 4646)

4 Waywiser. Iron wheel. Signed: *Elliott London.* Mid-19th century.

5 Chondrometer. Signed: *Loftus Oxford St London.* (1969.75/1-4)

6 Sikes' hydrometer. Signed: *Loftus 146 Oxford St London.* (ACM 1939.169)

7 Sikes' hydrometer. Signed: *Loftus 321 Oxford St London.*

8 Bate's saccharometer. In box inscribed: *Revenue Saccharimeter Loftus 116 Oxford Street, London.* Serial no. 4313.

9 Bate's saccharometer. Signed: *Loftus 321 Oxford St London.*

10 Saccharometer. Signed: *Dring and Fage*, serial no. 16089.

11 Saccharometer. Signed: *J. Long*, serial no. 6020.

12 Pedometer. Signed: *Spencer and Perkins.* (WOC 2089)

13 Compound monocular microscope. Unsigned. Ht 6in.

14 Microscope of brass, signed: *Keith 24 St Georges Street, Plymouth.*

15 Microscope of brass, signed: *David Marr, 27 Little Queens Street, Holborn, London.*

16 Chondrometer, signed: *Payne London.* (1970.434)

W 5 WISBECH
Wisbech and Fenland Museum

Museum Square, Wisbech PE13 1ES

0945 583817

The Museum was founded in 1835 and is housed in a building of considerable architectural interest designed for the purpose by James Buckler and erected in 1845-7. The collections relate chiefly to the surrounding Fenland area, covering archaeology and natural history, as well as bygones and ceramics.

1 Equinoctial ring dial. Brass. Signed: *Ralph Greatorex fecit Londini.* Late 17th century. (Acq. 1910)

2 Inclining dial. Brass with silvered compass plate and two inset levels. Signed: *W. & S. Jones 30 Holborn London.*

3 Part of a pocket sundial found in the River Ouse at Earith. Consists of the hour plate for a horizontal sundial for which the circular compass box and gnomon are missing. Roman numerals. ? *c*1500. (Acq. 1847)

4 Octant. Ebony frame. Tangent screw and vernier. Signed: *W.E. Elliott Jun. Maker London.* In box with label: *J.M. Wimple, Nautical and mathematical instrument maker, 7 Anne's Place, New Church St., Bermondsey.* (Acq. 1931)

5 Octant. Ebony frame. Tangent screw and vernier. Signed: *Parnell London.* Label in box: *Reynolds & Wiggins.* (Acq. 1921)

6 Quintant. Second half 19th century. Unsigned. Brass, silver scale, one telescope, magnifier. Vernier. Labels on box of: *Van der Voodt-Cornet, 52, First Dock Antwerp* and *Newton Brothers, 11 Princess Dock Walls, Hull.* (Acq. 1921)

7 Patent Course Corrector by *H.R. Ainsley, Cardiff.* (Acq. 1921)

8 Two-draw telescope. Wooden tubes, the outer in four sections, brass bound. Closed length approximately 4 ft. 18th century. (Acq. 1850)

9 Set square. Brass, engraved with diagonal scale. Inscribed: *W.C. 1756.* (Acq. 1908)

10 Perpetual calendar. Paper in glazed and turned wooden box. Signed: *Satchy Fec.* Inscribed on back: *Perpetual regulator of time.* 18th century. (Acq. 1851)

11 Spy glass. Six-draw. Lacquered brass with ivory outer tube. Signed: *Bate London.* (Acq. 1935)

12 Glass prism. ? 18th century.

13 Brass compass in gimbals. Lacks box. Signed: *D. McGregor & Co Liverpool.*

W 6 WORTHING
Worthing Museum and Art Gallery

Chapel Road, Worthing BN11 1HD

0903 39999 ext 121

The Museum contains collections of archaeological material, geology, costume, Sussex bygones, early English watercolours, dolls and jewellery. There are a number of watches, some photographic apparatus (*c*1890 onwards), a number of apothecaries' scales, coin scales, and a diamond merchant's balance, of early 19th-century date, as well as the instruments listed below.

1 Napierian rods. Boxwood in flat boxwood case, 3¼ x 2½in. Unsigned, probably 18th century. (54/256)

2 Inclining dial. Brass, circular, in circular leather case. Signed: *L.Casella Instrument Maker to the Admiralty London, c*1856-60. (1820)

3 Magnetic dial. Engraved paper compass card mounted on needle in turned wooden box with domed screw-top lid. Diameter of base 1¼in. Probably early 19th century. (2986)

4 Magnetic dial. Engraved paper compass in turned wooden box. Signed: *C. Essex & Co. London* and marked: *Entered at Stationer's Hall.* Scratched on bottom: *Stockdale 1836.* Second quarter 19th century. (768)

5 Compass in brass case. Signed: *W. Brown Leith.* (769)

6 Protractor. Brass in wooden box. Signed: *Fraser London.* Second half 18th century. (58/323)

7 Octant in wooden box. Marked on box: *J.T. Needs, 100 New Bond Street, late J. Bramah, 124 Piccadilly* and *E.G. Amphett Worthing.* (714)

8 Telescope, hand-held. Brass inner tube, outer tube wood-covered. Signed: *Watkins & Hill Charing Cross London.* Early 19th century.

9-11 Three telescopes, hand-held. Single-draw tubes of brass, outer tubes covered in black-painted linen cloth. Signed: *Ramsden London.* Late 18th century. (55/34-36)

Y

Y 1 YORK
Castle Museum

York YO1 1RY

0904 653611

The Castle Museum houses what is probably the most extensive and important collection of folk life and bygones in the country. The basis of this collection is the material amassed by Dr JL Kirk, the founder of the Museum. It includes costumes, toys, musical instruments, crafts, domestic and agricultural equipment, Yorkshire militaria, as well as numerous clocks, watches and scientific instruments. A large part of the Museum is organised in the form of reconstructed streets of shops or workshops, including the frontage of T Cooke, the York optician and instrument maker.

The collection of scientific instruments is one of the more extensive outside the major scientific museums. There is an emphasis on microscopes and telescopes, but there are many other interesting items besides. Not listed is the extensive collection of weights, scales, rules, etc.

MICROSCOPES, TELESCOPES

1 Magnifying glass. Horn-mounted, with leather case. Probably early 18th century.

2 Combined microscope and magnifying glass with needle object holder. Turned ivory. 18th century.

3 Screw-barrel microscope. Ivory, with accessories in velvet-lined fish-skin covered box.

4 Screw-barrel microscope. Ivory handle, brass body. 18th century.

5 Aquatic microscopes (2), signed: *R. Field and Son, Birmingham.*

6 Brass Culpeper microscope on rectangular base containing drawer for accessories. Signed: *Silberrad Aldgate London.*

7 Nuremberg tripod microscope. *IM* (Junker of Magdeburg) branded into base.

8 Ayscough 'single and compound' microscope. Signed: *Ayscough London No. 24.*

9 Jones's 'most improved' microscope. Signed: *Dollond London.*

10 Combined compound and simple microscope in a box with accessories on the lines of the Cuff microscope. Accompanied by an engraving of a box mounted microscope entitled 'Improved Compound and Simple Microscope'. The instrument itself however, is mounted on a folding tripod foot with compass joint. Late 18th century. Unsigned.

11 Cary-type microscope. Unsigned. Early 19th century.

12 Cary-type microscope in velvet-lined wooden box. Unsigned.

13 Cary-type microscope. Plain wooden box. Unsigned.

14 Drum microscopes (2). In boxes with accessories. Unsigned. *c*1840.

15 Compound microscope. Signed: *R. & J. Beck London.*

16 Single-draw refracting telescope. Signed: *Dollond London.*

17 'Day or Night' refracting telescopes (2). Signed: *Dollond London.*

18 Nautical telescope. Decagonal wooden tube, brass-bound at centre. Length 37½in.

19 Nautical telescope. Single-draw, decagonal wooden tube. Length 21in. Signed: *Dollond London.*

20 Single-draw refracting telescope with wooden tube. Length 19in. Signed: *Made by Jas Chapman St Catherine's London for day or night.*

21 Two-draw refracting telescope. Length 7in. Wood-cased outer tube. Signed: *T. Harris & Son London.*

22 Gregorian reflecting telescope, aperture 1½in, box mounted. Green ray-skin covered tube. Box covered in black fish-skin. First half 18th century.

23 Gregorian reflecting telescope, aperture 2¾in. Brass on pillar and claw stand. Signed: *G. Adams Mathematical Instrument Maker to his Majesty, Fleet Street, London.*

24 Gregorian reflecting telescope, aperture 3in, on pillar and claw stand. Signed: *G. Adams No. 60 Fleet St London.*

25 Gregorian reflecting telescope, aperture 4¼in, on pillar and claw stand. Signed: *James Short London No. 266/1770 = 18. c*1770.
Lit: BRYDEN (1968) no.38

26 Gregorian reflecting telescope, aperture 3in, on pillar and claw stand. Signed: *James Short London 1741 12/288 = 12.*

27 Gregorian reflecting telescope. Pillar and claw stand. Brass, lacks eyepiece. Signed: *Fra. Watkins Charing Cross London.*

28 Spy glass. Outer tube ray-skin covered; inner, vellum. Fish-skin covered case. Unsigned. 18th century.

ORRERY

29 Orrery. Wooden base plate on turned stand. Mechanical parts of brass. Surmounted by glass cover. Unsigned, late 18th century.

SUNDIALS AND NOCTURNALS

33 Equinoctial ring dial. Brass, diameter approximately 6in. Unsigned. 18th century.

34 Equinoctial ring dial. Brass, diameter approximately 10in. Unsigned. 18th century.

35 Butterfield dial. Oval, silver. Signed: *Butterfield Paris.* In black leather velvet-lined case.

36 Inclining dial. Brass, circular, on three screw feet, spirit levels mounted in base of compass. Signed: *J. Gilbert Ludgate Street London*.

37 Universal equinoctial sundial. Brass. Signed: *Johann Schretteger in Augsburg*.

38 Horizontal sundial. Signed: *Abraham Bath*.

39 Horizontal garden sundial. Bronze. Signed: *Hindley York*.

40 Garden sundial. Bronze. Unsigned, c1800. (436/38)

41 Nocturnal. Boxwood. Unsigned. English, c1700.

NAVIGATION AND SURVEYING

42 Backstaff. Ivory label missing. Tudor rose and star stamps. Lignum vitae frame with boxwood arcs, 18th century.

43 Octant. Ebony frame, brass arm engraved with trophies, radius 16in. Vernier and fixing screw. Unsigned, probably late 18th century.

44 Octant. Ebony frame, radius 14in. Signed: *F. Watkins Charing Cross London*.

45 Octant. Ebony frame, radius 9in. Signed on arm: *W. & S. Jones 30 Holborn London*.

46 Sextant. Brass frame, radius 12in. One telescope. Signed: *Ramsden London*.

47 Sextant. Brass frame, radius 8½in. Signed: *Cary London No. 309*.

48 Sextant. Brass frame. Signed: *Thos L. Ainsley, South Shields*. 19th century.

49 Dent's patent meridian instrument.

50 Course corrector. HR Ainsley's patent.

51 Quadrant of wood, the interior cut away leaving the rim and one central support in its centre with a screw hold. The arc is graduated into 90° and the radii into twelve equal divisions.

52 Surveyor's quadrant to Vernier's 1631 design. Brass. Fixed sights on the radius. Signed: *C. Caccia F.Ra. Ao 1688*.
Lit: KIELY (1947) 176-7

53 Surveyor's levels. One signed: *Ronchetti Manchester* (second quarter 19th century), the other *T. Cooke & Sons York*.

54 Spirit level. Signed *T. Cooke York*.

55 Miner's dial. Folding sights. Signed: *Abraham Liverpool*.

56 Compass, turned wooden box attached to a wooden arm with holes for nuts and bolts. Signed *J. & W. Cary Strand London*.

57 Compass in rectangular wooden box with lid. Folding sights. Unsigned, c1800.

58 Spirit level. Ebony and brass. 18th century.

59 Compass in mahogany box with lid. Signed: *Dollond London*.

DRAWING AND CALCULATING INSTRUMENTS

60 Draughtsman's rule (part of parallel rule). Brass, engraved with diagonal and other scales. Signed: *H. Sutton fecit 1655*. Length 18in.

61 Sector. Signed: *Renoldson London*.

62 Pantograph. Brass. Signed: *Newman, Successor to Heath & Wing Exeter Change London*. Trade card of Newman in mahogany box.

63 Pantograph. Brass. Unsigned.

64 Pantograph. Brass. Signed: *T.B. Winter, 21 Grey Street, Newcastle on Tyne*. In wooden box with Winter's trade card. c1850. (187/60)

65 Engineer's slide rule. Signed: *J. Cail Newcastle on Tyne*.

66 Farnley's Universal Calculator 'by Royal Letters Patent'.

MISCELLANEOUS

67 Thermometer. Signed: *Hardy York*. In red leather case.

68 Thermometer. Signed: *Nairne & Blunt London*. Black fish-skin case.

69 Set of specific gravity beads in a silk- and velvet-lined box. A total of 210 beads. A slide rule is signed *Lovi Edinburgh* and the box is signed: *Lovi Edinr Patentee No. 1*. Lovi took out a patent in 1805 for 'Apparatus for determining the specific gravity of fluid bodies' (British Patent No. 2826). In ink on box: £7.17.6.

70 Field's patent alcoholometer. Signed: *Joseph Long Maker Little Tower Street London* (Patent no. 10,729, 1845).

71 Sutcliffe's specific gravity apparatus.

72 Double-barrel air pump and accessories. Unsigned, probably late 18th century. Accessories include Magdeburg hemispheres and several receivers.

73 Cylinder electrical machine. Separate prime conductor now missing. Signed on ivory label: *Tuther 221 Holborn London*. Second decade 19th century.

74 Cylinder electrical machine on mahogany veneered base containing a drawer on the lines of a late 18th-century toilet mirror. Accessories include pith figures, electrical whirl, discharger, etc. The prime conductor is missing.

Y 2 YORK
National Railway Museum

Leeman Road, York YO2 4XJ

0904 621261

The National Railway Museum results directly from the 1968 Transport Act when it was decided that the British

Railways Board should pass responsiblity for its historical relics (which by then included those at the British Transport Museum at Clapham, London, The Great Western Museum, Swindon, and the York Queen Street Museum) to the *Science Museum, London*. A site for the new museum was found at the old York North Motive Power Depot, which building was refurbished and the collections from Clapham and York united displayed there, opening in 1975. Since then adjacent properties have been acquired. The National Railway Museum is an outstation of the Science Museum.

The collections are extremely extensive, concentrating on railways in Great Britain. Nearly all exhibits lie outside the scope of this inventory, though there are a few scientific instruments. In addition to those listed, there are specialised items such as track gauges, pressure indicators and a recording speedometer.

1 Theodolite with box, telescope missing, signed: *Bancks London* (sic). Trade label in box signed: *Banks, No 111 Strand London, Optical and Mathematical Instrument Maker to his Royal Highness the Prince of Wales*. (78.35.5)

2 Theodolite with box. Signed on compass: *Gilbert & Wright London*, and on arm: *T.O. Blackett, Newcastle*. (78.35.6)

3 Surveyor's level with box. Signed: *Stanley Gt Turnstile Holborn London WC* and *9544*, and *S.E.R. Locomotive Department*.

4 Arithmometer, signed: *Thomas de Colmar...44 rue de Chateaudun, Paris* and *no. 1565*. Used on the South Eastern Railway, *c*1860. (78/31/169)

5 Fuller's circular slide rule. (2217/63)

6 Large wooden protractor, graduated in single degrees. Signed: *T. Dakin*. Used in setting out the Settle and Carlisle line. (102/56)

7 Timetable slide rule, wood, brass and ivory, 16½in. long. Signed: *Dring & Fage... 56 Stamford Street. London*. (75/17/98)

8 Slide rule for calculating freight charges on the Great Eastern Railway lines between Littleport and Dereham, and Wisbeach (sic) and Watling Row. (75/17/102)

Y 3 YORK
Vickers Instruments

Vickers plc, Haxby Road, York YO3 7SD

0904 24112

The firm of Vickers, successors of Cooke, Troughton and Simms, owns a number of instruments which relate to the succession of instrument makers – John Worgan, Thomas Wright, Benjamin Cole and Edward Troughton – which led up to the firm of Troughton and Simms, later Cooke, Troughton and Simms. In addition there are a number of later 19th- and 20th-century instruments, as well as several interesting books on surveying and astronomical instruments. There are also catalogues of Simms and T Cooke and Sons. An interesting account of

the origins of the firm has been published: TAYLOR/WILSON/MAXWELL (1960). Regrettably, it does not refer to any of the instruments owned by the firm. Some material formerly retained by the firm is now to be found at the *Royal Museum of Scotland (Chambers Street), Edinburgh*.

1 Reflecting telescope of brass on pillar and tripod stand. Signed: *Adams Mathematl Instrut Maker to his Majesty Fleet Street London*.

2 Universal equinoctial sundial, circular on adjustable screw feet. Brass, partly silvered. Signed: *Troughton & Simms London*. Second quarter 19th century.

3 Octant. Ebony with brass arm, radius 19in. Vernier. Signed on arm. *Troughton London*.

4 Octant. Ebony frame, wood and brass arm, vernier, radius 17in. Signed: *Gregory & Wright London, c1785*.

5 Reflecting circle. Brass, radius 10in. Signed: *Troughton London 162*. Early 19th century.

6 Reflecting circle and stand. Brass, radius 10in. Signed: *Troughton & Simms, London No. 234. c1830*. (fig 20)

7 Sextant. Ebony frame, ivory degree arc, brass arm, radius 8in. Signed on ivory plate: *Troughton London*.

8 Circumferentor. Brass, radius 11in. Signed: *Benjan Cole Maker Bull Alley, Lombard Street London*.

9 Surveyor's level. Brass, incorporating a compass mounted over the telescope and bubble level. Signed: *Troughton London*. Early 19th century.
Lit: REES (1819) 'Surveying' plate 5

10 Surveyor's level. Signed: *Troughton & Simms, London*. Second quarter 19th century.

11 Theodolite. Brass with silver scale. Signed: *Troughton & Simms London*.

12 Theodolite. Brass with silver scale. Signed: *Troughton & Simms London. c1830*

13 Compass. In mahogany box with lid. Engraved paper compass card, surrounded by silvered brass degree ring. Signed: (on compass card) *John Worgan Londini fecit 1696*.

14 Compound microscope. Signed: *Troughton & Simms London*. Possibly 1840s.

15 Compound microscope. Signed: *Troughton & Simms 100*. Possibly 1840s.

16-17 Compound microscopes, similar type. Signed: *Baker 244 High Holborn London*. One with presentation date 1864.

Y 4 YORK
Yorkshire Museum

Museum Gardens, York YO1 2DR

0904 29745/6

Extensive natural history collections, together with archaeology, geology, ceramics, coins and medieval architectural material. The Museum contains a few scientific instruments. The most important of its possessions in this field was the universal equatorial instrument constructed by Abraham Sharp, at one time assistant to John Flamsteed, the first Astronomer Royal. This was presented to the Yorkshire Philosophical Society in 1835, and has been transferred to the *National Maritime Museum, Greenwich.*

1 Miniature transit theodolite signed: *Robinson, Devonshire Street, Portland Plce London.* Silver scale, brass, partly lacquered black. Horizontal circle with three verniers and tangent screw. The vertical circle is supported by a column mounted in the centre of the horizontal circle. It is equipped with a telescope and a bubble level. Both circles are equipped with magnifiers for reading the verniers. Diameter of horizontal circle approx 4in, *c*1830.

2 Matthew Loft form of Culpeper microscope. Ebonised octagonal box, lignum vitae mounts to tube,

continuous curved brass legs, revolving object holder, black ray-skin tube. Brass snout. Objective and eyepiece, including turned eyepiece-holder missing.

3 Cuff-type microscope. Unsigned. Rectangular base with drawer. Case and eyepiece missing.

4 Smith & Beck 'Popular' microscope. Signed: *R. & J. Beck 8605 London and Philadelphia.*

5 Smith & Beck 'Star' microscope. Signed: *R. & J. Beck London 18165, c*1890.

6 'Paragon' binocular microscope. Signed: *J. Swift & Son 61 Tottenham Court Road, London.*

7 Compound monocular microscope. Signed: *J. Swift & Son 21 Tottenham Court Road, London.*

8 Mortar, bronze, date 1308, by William de Towthorpe for the Infirmary of St Mary's Abbey in York.
Lit: HEMMING (1929); CRELLIN/HUTTON (1973) p274

Eire

Eire 1 BIRR
Birr Observatory and Museum Trust

The Estate Office, Rosse Row, Birr, County Offaly

(010 353) 509 20023

Birr Castle is the seat of the Earls of Rosse. The third and fourth Earls took an avid interest in astronomy, and most of the instruments which survive relate to their work in this field over the period 1840 to 1916.

In 1845 William Parsons, the third Earl (1800-1876), completed building a giant Newtonian reflecting telescope with an aperture of six feet. This, the 'Leviathan of Parsonstown', remained the largest in the world until the Mount Wilson (USA) 'Hooker' reflector of 1918. The meridian walls which supported it and the 56ft long wooden tube survive *in situ* (no. 1). One of the two specula cast is in the *Science Museum, London* (the other cannot be traced). Apart from the material listed below there are various tubes, lenses, etc. which are not enumerated. All have been catalogued in an unpublished typescript by JA Bennett of the *Whipple Museum, Cambridge* (*Lit:* BENNETT (1983)). In addition there are electrical collections and a substantial collection of early photographic equipment. The Castle grounds are open to visitors in the Summer months.
Lit: MOORE (1971) for a general account of astronomy at Birr Castle; BENNETT/HOSKIN (1981); CAMBRIDGE (1983). A number of instruments are described in the collected scientific papers of the 3rd Earl: PARSONS (1926). Numbers following the entries below refer to the Bennett Catalogue (above); 'Whipple' numbers refer to numbers in the catalogue of the exhibition held in

Cambridge in 1983 (above).

1 Reflecting telescope, Newtonian, 6ft aperture. Mirror mount, wooden tube and meridian stone walls survive in the grounds of Birr Castle. Completed 1845.
Lit: ROBINSON (1847); ROSSE (1862)

2 Refracting telescope. Brass, single-draw tubes, 2in aperture. (8)

3 Refracting telescope. Barrel bound with asbestos. Altazimuth mount on brass pillar support with large wooden tripod. 3¾in aperture. (34)

4 'Lunar heat' telescope. Designed by the 4th Earl. Length 24½in, diameter of barrel 26½in. (35)

5 Refracting telescope, used as a 'finder'. Brass barrel, 2 lens eyepiece. 2¾in aperture. (36) Whipple 40

6 Refracting telescope, used as a 'finder'. Brass barrel. 3in aperture (37) Whipple 41

7 Refracting telescope. Brass barrel, right-angle eyepiece. 2in aperture. (39)

8 Tube for Newtonian telescope (no optics). Metal tube made up of thin sheets, would accommodate mirror of about 8in aperture. (44)

9 Zoellner astrophotometer, signed: *H. Ausfeld. Gotha No. 17.* An oil lamp creates an 'artificial star' of variable brightness to compare with intensity of real star. *c*1860. (43) Whipple 46

10 Part of mounting for an equatorial telescope. (40) Whipple 43

11 Clockwork telescope drive, possibly for 10. (41) Whipple 44

12 Electrically-controlled driving clock for an equatorial telescope, *c*1884. (13) Whipple 45
Lit: PARSONS (1885)

13 Clockwork movement, dead-beat escapement. (53)

14 Driving clock movement, spring driven. (75)

15 Refracting telescope, single draw ('night telescope'). Signed: *Berge London late Ramsden*. An inscription in the 4th Earl's hand states that it was given to the 3rd Earl by Sir James South and suggests that it had been used by Admiral Beaufort at Trafalgar. Inscription dated 18 April 1892. (92)

16 Refracting telescope. Brass, signed: *Watson & Sons, 313 High Holborn, London* and numbered: *654*. Altazimuth mount to pillar on tripod stand. 2¾in aperture. (96)

17 Refracting telescope, single draw ('day and night telescope'). Wooden body covered with brown leather. 2¾in aperture. (93)

18-20 Elliptical flats (3) for Newtonian reflector, speculum metal, in brass mounts, length (max) 4, 4¾, 7½in. (9,10,14)

21 Glass disc, possibly mirror blank. Diameter 36in. (100)

22 Primary mirror for Gregorian reflector, speculum metal. Diameter 17¾in. (11)

23 Circular flat for Newtonian reflector, speculum metal, in metal case. Diameter 8¼in. (12)

24 Mirror, silver on glass, in wooden cell. 34in aperture. (87)

25 Mirror, silver on glass. Now mounted in circular wooden frame, but probably a searchlight mirror for the 4th Earl's lunar heat telescope. Aperture 25in. (88)

26 Cast iron disc, with pattern of grooves and central hole, possibly used for grinding. Diameter 8½in. (98)

27 Cast iron disc, with pattern of grooves, and rim. One of the grinders for a 3ft mirror. Diameter 36in. (99)
Lit: PARSONS (1926) p82, fig 9

28 Lens in brass tube with 5 aperture stops. Signed: *Howard Grubb, Dublin. Aplanatic. 8½ x 6½. 5281*. (2)

29 Lens mounted in brass tube (? telescopic eyepiece), signed: *Lerebours et Secretan a Paris*. (3)

30 Lens mounted in brass tube with sliding adjustment, signed: *Lerebours et Secretan a Paris*. (26)

31 Various mounted lenses (9), unsigned. (21, 22, 24, 25, 27-30, 56)

32 Convex lenses (2 examples) which are mounted to pivot about horizontal axis on adjustable brass pillar stand. Signed: *W. Ladd & Co. London*. (69, 70)

33 Lens in brass tube. Signed: *Bland & Long, 153 Fleet Street, London. 1593*. (31)

34 Lenses mounted in brass tube, fitted for rack and pinion adjustment. Signed: *J.H. Dallmeyer. London. No. 51482. 4.B. Patent*. (77)

35 Lens mounted in brass tube. Signed: *A. Ross. London. 4926*. (78)

36 Various eyepieces in brass tubes (7), unsigned. (6, 7, 18-20, 32, 57)

37 Eyepiece filters in brass mounts (3). (4, 16, 17)

38 Eyepiece moving wire micrometer (incomplete). Signed: *Grubb Dublin*. Used with the 6ft telescope to measure angles and apparent angular distances and diameters. (15) Whipple 39
Lit: PARSONS (1880-81) p2

39 Transit instrument, oxidised brass. Signed: *Troughton London*. Diameter of circle 6¾in. (42)

40 Portable altazimuth instrument, oxidised brass, vertical circle with silvered scale. Signed: *Robinson, 38 Devonshire Street, Portland Place, London*. Diameter of vertical circle 3½in. (66) Whipple 50

41 Sextant, signed: *Elliott Brothers, 30 Strand, London*. (65) Whipple 51

42 Portable transit instrument, signed: *Troughton & Simms, London*. (38) Whipple 42

43 Astronomical spectroscope, two-prism type to design of William Huggins. Signed: *John Browning, London*. Spectroscopic work at Birr started in the early 1860s. (48) Whipple 47
Lit: BROWNING (1888) p24

44 Direct vision spectroscope, signed: *John Browning, London*. (49) Whipple 49

45 Direct vision spectroscope, signed: *Yeates & Son, Dublin*. (47) Whipple 48

46 Spectroscope collimator (72)

47 Dumpy level, brass, in fitted wooden box with trade label signed: *Buckley, Dublin*. (50)

48 Quadrant, brass, fitted with bubble level. Inscribed: *Lord Oxmanton. 1840*. This setting quadrant was probably used by William Parsons (before he succeeded to the title Earl of Rosse) with the 3ft reflecting telescope. (51) Whipple 38

49 Pantograph, brass, in fitted wooden box. Signed: *Troughton & Simms. London*. (46)

50 Robinson anemometer, signed: *Casella...London* and *No.383*. (45)

51 Anemometer gauge with 2 dials in oxidised brass casing. (73)

52 Divided brass circle, signed *Apps... 433 Strand, London*. Diameter 7¾in. (52)

53 Plotting protractor, brass, in fitted mahogany box. Signed: *Troughton & Simms, London*. (67)

54 Camera lucida, brass. (68) Whipple 55

55 Planimeter, Amsler's, German silver. Printed instructions, wooden case. Signed: *Elliott Bros. London*. (79)

56 Thermometer, mercury in glass, graduations in fahrenheit and centigrade. Signed: *R & G Knight, Foster Lane, London*. (80)

57 Bi-metallic thermometer, recording type. Signed:

Negretti & Zambra. London. (82)

58 Barometer, cistern type, with thermometer. Signed: *Yeates & Son, 2 Grafton St. Dublin.* (91)

59 Chemical balance in glazed wooden case with drawer containing fitted box of weights. Signed: *L. Oertling, London.* Beam length 13½in. (94)

60 Set of weights in fitted box. Signed: *F.E. Becker & Co.* (95)

61 Crookes' radiometer (97)

62 Astronomical regulator clock, mercury compensation pendulum, dead-beat escapement. Signed: *James Shearer, Devonshire Street, Queen Square, London.* (89)

63 Astronomical regulator clock. Signed: *Arnold & Dent, Strand, London* and 368. (90)

64 Compound microscope with fitted wooden box. Signed: *A. Ross. London No. 509* with eyepiece signed: *Dollond London.*

65 Compound microscope with fitted wooden box. Signed: *E. Davis, Leeds.* (64) Whipple 54

66 Compound microscope (incomplete). Signed: *Newton & Co. London.* (81)

67 Microscope slides in wooden cabinet of 19 drawers. (63) Whipple 53

Eire 2 CARLOW
College Library, Carlow Regional Technical College

Kilkenny Road, Carlow

(010 353) 503 31324

The physicist John Tyndall (1820-1893) was born at Leighlin Bridge, near Carlow. His papers and printed works have been collected by the Carlow County Library, while the Regional Technical College (founded 1970) has acquired a few artefacts associated with him. Other material with Tyndall associations is in the *Royal Institution, London.*

1 Sound rattle, wood, constructed by Tyndall. On loan from the *Royal Institution, London.*

2 Wet and dry bulb hygrometer, signed: *J. Hicks Hattons Gdns.*

3 Set of drawing instruments in case, including T-square, dividers, compasses, etc. Said to have been the property of Tyndall.

4 Infusion tube containing urine. Prepared by Tyndall.

Eire 3 CASTLEKNOCK
Dunsink Observatory

Castleknock, Dublin 15

(010 353) 13 87911

The teaching of astronomy at Trinity College, Dublin, was established following a bequest made in 1774 by Provost Francis Andrews: Dunsink Observatory, five miles to the north west of the college, was built and Henry Ussher was installed as the first professor in 1783. He ordered the two clocks by John Arnold (nos. 1 and 2), and a transit instrument and astronomical circle by Jesse Ramsden. Until recently the circle survived, but it was damaged by fire in 1977 and now only fragments remain (*Lit:* KING (1955) pp168-9; WAYMAN (1971) p126). Other instruments belonging to the early equipment of the observatory are known from an entry in the Visitation Book of the University for 2 May 1791, which records a series of instruments delivered to John Brinkley (Ussher's successor). The list includes a theodolite and portable barometer by Nairne and Blunt, a Robinson thermometer, two rain and evaporation gauges, and a clock 'with flat diamond suspension'.

During the 19th century the Observatory achieved international importance, William Rowan Hamilton (of quaternion fame) holding the chair of astronomy from 1827 to 1865. During the early 20th century the Observatory declined, though it was revived in 1940 when it was incorporated into the newly founded Dublin Institute for Advanced Studies.
Lit: WAYMAN (1987)

1-2 Astronomical regulators. Month movements, dead-beat escapement, five bar zinc and steel composition pendulums. Mahogany case. Signed: *John Arnold & Son London.* Delivered 1787.
Lit: WAYMAN (1971) pp275-282

3 Sidereal clock, signed: *E. Dent & Co London 61 Strand & 34 Royal Exchange.* Regulated by transit circle observations. Set up in 1874.

4 Telescope objective by Cauchoix, Paris. Diameter 9¾in, 19ft focus. Originally purchased by James South for his observatory at Campden Hill, Kensington, and mounted by Troughton and Simms. The telescope proved unsatisfactory, and South presented it to Dunsink. It is currently used on the 12in Grubb refractor (no. 5 below).
Lit: BALL (1940); KING (1955) p236

5 South's equatorial refractor, 12in aperture. Displayed at the Dublin Exhibition of 1863, installed by Thomas Grubb at Dunsink in 1868.

6 Double star micrometer for the Grubb telescope (no. 5) by Pistor and Martin of Berlin.

7 Micrometer by Grubb, 1875, for use in the Grubb telescope installed in 1868 (no. 5).

8 Telescope mechanism by Grubb, for use with 15in telescope installed in 1890.

9 Divided object glass micrometer, signed: *Dollond London*, early 19th century.

10 Refracting telescope on wooden tripod stand. Signed: *Utzschneider und Fraunhofer in Munchen*. Constructed between 1819 and 1826. Damaged in the 1977 fire.

11 Refracting telescope, 5in aperture, wooden tripod stand. Signed: *Plossl in Wien*. Fire-damaged.

12 Refractory telescope, 3in aperture, with terrestrial and astronomical eyepieces; folding tripod stand. Signed: *Troughton & Simms London* and also engraved: *Telford Premium 1874*. Awarded by Institution of Civil Engineers to John McCarthy Meadows. Never used at Dunsink.

13 Sextant, signed: *G. Whitbread London*.

14 Orrery, comprising planetarium, tellurium and lunarium parts, brass, on pillar and claw stand. Signed: *W & S Jones Fecerunt 30 Lower Holborn London*. Early 19th century.

Eire 4 CORK
Experimental Physics Department, University College, Cork

College Road, Cork

(010 353) 21 276871

Queen's College, Cork, Queen's College, Galway and Queen's College, Belfast, were all incorporated in 1845. The first two, with University College, Dublin, now form the National University of Ireland. Queen's College, Cork, was opened to students in November 1849. From the beginning, science was taught and a medical school established. George Frederick Shaw held the chair of natural philosophy from 1849 until 1855 and John England from 1855 to 1894. Thereafter the chair was renamed experimental physics, the first holder of which was William Bergin, from 1895 to 1931. There survive in the department the remains of what was probably an extensive collection of mid- to late 19th-century demonstration apparatus. In all the three University Colleges, instruments signed by Yeates of Dublin are prominent; there are also Paris-made items.

The observatory, within the college grounds, was built with funds supplied by a benefactor, WH Crawford. He also supplied the equatorial telescope and meridian circle. The former, constructed by Howard Grubb, won a gold medal at the Paris Exhibition of 1878. It was used by John England to observe the transit of Venus on 6 December 1882. The instruments are not currently displayed.

COLLEGE

1 Sextant, signed on ivory label: *J. Buckley Dublin*, in oak box with trade label: *T. Bennett Optician...Mathematical & Philosophical Instruments 124 Patrick Street Cork*.

2 Small refracting telescope on brass stand, signed: *J.C. Hofmann Rue de Buci 3, Paris*.

3 Small telescope, with intervening rotatable optical system (? polarising), on brass stand, signed: *Yeates & Son Dublin*.

4 Mounted conical glass prism on brass stand, signed: *Lerebours & Secretan Paris*.

5 Norremberg demonstration polariscope, signed: *Yeates & Son, Dublin*.

6 Polarising apparatus consisting of a blackened adjustable mirror, a microscope and interposing filters, on a stand, signed: *Lerebours et Secretan Paris*.

7 Binocular microscope in box, unsigned but with signed accessories by various makers, including objectives signed: *E. Leitz Wetzlar* and *C Zeiss Jena*.

8 Travelling microscope, signed: *Becker, Hatton Wall, London*.

9 Spectroscope, simple Bunsen and Kirchhoff type, on a Steinheil-type stand. Unsigned.

10 Two glass prisms for liquids.

11 Beckman thermometer, signed: *Dr Peters u.Rost Berlin* and numbered: *1009*.

12 Minimum thermometer, signed: *J.J. Hicks 8, 9 & 10 Hatton Garden London* and numbered: *1256704*.

13 Marcet's steam apparatus (for experiments with high pressure steam), signed: *Watkins & Hill, Charing Cross, London*.

14 Set of four lever systems mounted on wooden base, with weights.

15 Model steam engine with two cylinders and crankshaft.

16 Small balance with solid brass beam, in glazed case, signed: *Beckers Sons Rotterdam*.

17 Disc electrical machine (disc 2ft diameter), signed: *Thos. Bennett Cork*.

18 Spark tube for variable spark length, tube 24½in long.

19 Henley's universal discharger

20 Wimshurst machine (disc 12in diameter), signed: *Dollond 35 Ludgate Hill London*.

21 Eight Crookes tubes of different kinds, mounted in a frame.

22 Small induction coil with label: Supplied by *T.H. Mason 5 & 6 Dame Street Dublin*.

23 Two bar magnets (2ft x 1in) in wooden case.

24 Two very large two-dimensional 'working' models of a simple and force pump on board 8in x 3ft, and a two cylinder air pump and gas jar, both models bearing plaques signed: *Watkins & Hill Charing Cross London*.

25 Model Harrison gridiron pendulum, constructed from wood (!).

26 Coloured chart, framed, showing coloured spectra of 9 elements, with title: *Spectraltafel nach Originalzeichnung von G. Kirchhoff und R. Bunsen*.

27 Goniometer, Borsch type. Diameter 11in.

OBSERVATORY

28 Equatorial telescope, aperture 8in, with clock drive, signed: *Grubb Dublin* and dated *1876*.

29 Chronograph, signed: *Grubb Dublin*.

30 Transit telescope, aperture 5in. Apparently unsigned.

31 Siderostatic telescope, unsigned.

Eire 5 DUBLIN
Chester Beatty Library and Gallery of Oriental Art

20 Shrewsbury Road, Dublin 4

(010 353) 1 2692386

Alfred Chester Beatty (1875-1968), an American with Irish ancestry, was a mining consultant whose enterprises flourished and who collected avidly. His major interests were Islamic and oriental cultures, though European material was also acquired. The collections were brought to Dublin in 1950. Apart from manuscripts with scientific bearings, the collection includes a group of Turkish astrolabe-quadrants and a single astrolabe.

1 Astrolabe-quadrant, painted wood, Turkish, signed: *Ahmad ash-Sharabatli*. Dated AH 1230 (1814/5 AD). (Astr. no. 1)

2 Horary quadrant, painted wood, Turkish, unsigned. 19th century. (Astr. no. 2)

3 Quadrant, Turkish, with monogram 'J'. Dated AH 1218 (AD 1803/4). (Astr. no. 3)

4 Astrolabe-quadrant, painted wood, possibly Syro-Egyptian, signed: *Abu-t-Tahir Muhammad*. Dated AH 775 (AD 1373/4) (Astr. no. 4)

5 Astrolabe-quadrant, Turkish, signed: *al-Hajj Hasan al-Mu'arrif*. Dated AH 1118 (AD 1706/7). (Astr. no. 5)

6 Sundial, Turkish, square form, incorporating small English-made compass. The reverse is a horizontal dial for different latitudes (compass missing, on this side). Signed: *al-Amir Ridwan at-Tawil*. Dated AH 1201 (AD 1786/7). (Astr. no. 7)

7 Double quadrant, Turkish, mounted on semi-circular piece of wood, for two latitudes. Signed: *Ahmad ash-Sharabatli*. Dated AH 1171 (AD 1757/8). (Astr. no. 8)

8 Astrolabe, Persian, brass, with six plates for eleven latitudes and a table of horizons. Signed: *Abd al-A'imma*, and inscribed as being designed in the time of Shah Sultan Husayn. Dated AH 1120 (AD 1708/9).

Eire 6 DUBLIN
Dublin Civic Museum

58 South William Street, Dublin 2

(010 353) 1 6794260

Sited in the old City Assembly House (erected between 1765 and 1771) the Dublin Civic Museum, founded in 1953, deals with the history of Dublin City and County from medieval to modern times. The Museum houses the Old Dublin Society and its associated reference facilities.

1 Circumferentor, very incomplete, signed: *Spear Dublin*. (223)

2 Level with compass, in case, signed: *Buckley Sackville St Dublin*.

3 Level, signed: *Ed. Springer Berlin 7957*. c1900.

4 Horizontal dial, brass, originally circular cut down to octagonal shape, 5in across. Signed: *William King Dublin Fecit*.

5 Horizontal 'moon dial', slate, oval 15¾in by Michael Lyons for use at latitude 53°10' (Powerscourt House, Enniskerry). The outer scale shows the equation of time, the inner scale an adjustment for 'lunar time'. Inscribed: *Construd Oct. 14 1844 by Mr Lyons*. Came from Professor Bayley Butler's roof garden at Glenlion, Baily.

Eire 7 DUBLIN
Egestorff Collection

25 Wellington Place, Dublin 4

(010 353) 1 689325

A private collection which contains a considerable number of instruments by Dublin makers and retailers. Emphasis is on surveying instruments, microscopes, telescopes and sundials, but there are also a number of fine early lathes, in particular an ornamental turning lathe of about 1800 by Holzapffel and Deyerlein, and also clockmakers' instruments. There is a small collection of late 19th-century cameras. The Smith Collection of electrical and pneumatic apparatus is also displayed (items 94-109). The collection is open to the public (adults only) on the first Wednesday of each month from 7 to 9 pm.
Lit: EGESTORFF (1984)

MICROSCOPES

1 Screw-barrel microscope. Brass body, ivory handle. In leather box with compass microscope. English, early 18th century.

2 Brass Culpeper-type microscope, in pyramidal box. Rack focusing. Signed on the stage: *J. Lynch & Son Dublin*. The trade card of James Lynch and Son is pasted inside the box, c1810.

3 Cuff microscope, in pyramidal box. Signed on the stage: *J. Cuff Londini Inv. et fecit.*, with printed tract dated 1744.

4 Cuff-type microscope. Rack focusing. Pyramidal box. Unsigned. 18th century.

5 Compound microscope: Jones 'Most Improved' type. In mahogany box complete with accessories. Signed: *Dollond London*. Early 19th century.

6 Compound microscope. Jones 'Most Improved' type, equipped with revolving objective carrier with six objectives. Signed: *Clarke Manufacturer Sackville Street Dublin*. Early 19th century.

7 Cary-type microscope. Signed: *Cary London. c*1825.

8 Cary-type microscope. Unsigned. Early 19th century.

9 'Opaque solar microscope', Benjamin Martin type. Signed: *Dollond London*. Recorded as having been made for Lord Provost Lawson of Edinburgh in 1860.

10 Solar microscope, signed: *Ross*. Mid-19th century.

11 Compound microscope. Cylindrical tube. Pillar screws into centre of short side of box. Signed: *T. Harris & Son London. c*1820.

12 Compound microscope: Powell & Lealand No. 3. Signed: *Powell & Lealand 4 Seymour Place Euston Square London*. Constructed between 1846 and 1857.

13 Compound microscope. Triangular pillar supported between trunnions. Signed: *Baker 244 High Holborn London. c*1860.

14 Binocular microscope. Mechanical stage with rotating movements. Signed: *Watson & Son 313 High Holborn London. No. 1848*.

15 Binocular microscope, signed: *Henry Crouch, London*. Late 19th century.

16 Microscopist's oil lamp, late 19th century.

TELESCOPES AND REFLECTING CIRCLES

17 Small refracting telescope. Closed length 4½in. Removable pillar and tripod stand. Wheel of four powers. Optional dark glass. Signed: *Dollond London*.

18 Refracting telescope on pillar and claw stand. 2¾in aperture. Unsigned. 19th century.

19 Gregorian reflecting telescope, 2½in aperture. Pillar and claw stand. Black leather covered tube. Unsigned. 18th century.

20 Gregorian reflecting telescope on pillar and claw stand. Brass tube. 2.5in mirror. Signed: *G. Adams London*.

21 Gregorian reflecting telescope. Box mounted. Signed: *James Short London 130/831 = 12*. Constructed *c*1755.

22 Gregorian reflecting telescope. Pillar and claw stand, elevating and traversing mechanism. 4in aperture. Signed: *W. & S. Jones 30 Holborn London. The Specula by John Cuthbert 1844*.

23 Gregorian reflecting telescope in pillar and claw stand. 6in aperture. Signed: *J. Davis Edinburgh*.

24 Gregorian reflecting telescope. 6in aperture. Finder telescope. Pillar and claw stand, the barrel supported on trunnions. Optional screw motion to horizontal and vertical motions. Signed: *J. Watson London*. Early 19th century.

25 Three-draw refracting telescope. Green ray-skin covered outer tube, *c*1800.

26 Refracting telescope, signed: *Tulley and Sons Islington London*. First half of 19th century.

27 Refracting telescope, signed: *H. Hunt Cork*. 19th century.

28 Reflecting circle, 11½in diameter. Signed: *Troughton & Simms London* and numbered *341*. 19th century.

ORRERIES, ETC.

29 Armillary sphere. Diameter of meridian ring 17½in. Brass, on wooden tripod stand. Painted horizon plate. Signed: *Lynch & Son, Dublin, c*1810.

30 Orrery. Brass on pillar and claw stand. Signed: *W. & S. Jones, Holborn, London*. With tellurium, lunarium and planetarium attachments, planets restored. *c*1800.

31 Orrery. Brass mechanical parts, wooden base plate covered with engraved paper. Signed: *Newton & Co., Opticians to H.M. the Queen, 3 Fleet St. London*.

32 Pair of globes, diameter 18in, terrestrial and celestial, on wooden claw foot stands with compasses. Signed: *London, Smith & Son, 63 Charing Cross*.

33 Terrestrial globe, 12in diameter, signed: *G & J Cary, 96 St James Street Jany 1 1840*.

ASTROLABES

34 Astrolabe. Brass. Diameter 6in. Rete with 12 star pointers. Seven plates include a table of horizons and plates for latitudes from 17 – 50°. Made by Balhumal, Lahore, *c*1850.

SUNDIALS, ETC.

35 Horizontal dial, for latitude of 48°. Octagonal in black leather case. Unsigned, 18th century.

36 Horizontal dial in turned brass box with lid, for latitude of Dublin. Signed: *Edwd Clarke Lowr Sackville St. Dublin*. Early 19th century.

37 Floating magnetic dial in turned ebony box. Signed: *C. Essex & Co. London, c*1820s.

38 Diptych dial, with 2 scaphe dials incorporated. Ivory. Maker's mark two crowns (? P. Reinman, Nurnberg) and dated 1624.

39 Analemmatic dial, signed: *Antoni Thomson fecit. c*1650.

40 Universal equinoctial dial. Roughly octagonal base with engraved rococo decoration. Brass. Unsigned but probably made in Augsburg. 18th century.

41 Universal equinoctial dial. Signed: *Joh. Schretteger Augsburg*.
Lit: BOBINGER (1966) p354, fig 166

42 Mechanical equinoctial dial with minute dial. Unsigned. *c*1800.

43 Mechanical equinoctial dial, signed: *T. Mason Dublin*, c1800.

44 Horizontal dial. Brass, split gnomon for 45 – 50°. Unsigned, c1750.

45 Large equinoctial dial, unsigned, 19th century.

46 Equinoctial ring dial. Brass. Diameter 8⅜₆in. Handwritten instructions for use in box. Signed: *Saunders Dublin*.

47 Equinoctial ring dial. Brass. Signed: *Seward Dublin*. Diameter 7½in.

48 Equinoctial ring dial. Brass. Diameter 6⅝₆in. Unsigned, 18th century.

49 Equinoctial ring dial. Silver. Diameter 3in. Signed: *George Parker Dublin*. First half of 18th century.

50 Inclining dial. Signed: *J. Webb 192 Tottenham Court Rd., London*, c1775.

51 Inclining dial. Silvered brass. Circular. Signed: *Melling & Co. 29 S. Castle Street Liverpool*.

52 Pillar dial, paper on wood, for 48° latitude. Signed: *AParis Henry Robert, Horologer au Palais Royal, no 164*.

53 Astronomical compendium. 4in square. Gilt brass. Combines equinoctial sundial, solar-lunar conversion volvelle and aspectrarium, compass with eight part windrose, astrolabe and geometric square (graduated quadrant and shadow-square). French, late 16th century.

54 Nocturnal. Boxwood. English, c1700. Unsigned.

55 Three sand glasses

NAVIGATION INSTRUMENTS

56 Backstaff. Rosewood and boxwood. One original sight. Unsigned. ? Early 18th century.

57 Octant. Ebony frame. Radius 14in. Floral decoration and military trophies engraved on the brass arm. Unsigned, but label in box signed: *Mrs Janet Taylor 104 Minories London*. Post 1845.

58 Octant. Ebony frame, ivory scale. Radius 13½in, signed: *B. Condy Philadelphia 1775 [?] for Captain C.T. Bethel* (partly erased).

59 Octant. Ebony frame. 9 11/16in radius. Unsigned. Early 19th century.

60 Miniature sextant in leather case. Radius 4in. Brass frame. Signed: *Spear Capel St, Dublin*. Early 19th century.

61 Sextant. Brass frame. Radius 7⅝in. Unsigned. Early 19th century.

62 Sextant. Ebony frame. Radius 10in. Signed: *J.W. Norie & Co. London*. c1820.

SURVEYING AND MINING INSTRUMENTS

63 Graphometer, 3½in semicircular limb. Brass. Signed: *Butterfield AParis*. c1700.

64 Holland Circle, 7¼in. diameter. Brass. Two pairs of fixed sights and an alidade with sights. Signed: *A. Odelem*

in Brauns, early 18th century.

65 Surveyor's compass. Brass, in mahogany box 6⅛ x 6¼in. Signed: *J. Ronchetti Optician, Manchester*.

66 Circumferentor. Long arms for sights. Complete with tripod. Signed: *Walker & Son 16½ Temple Barr Dublin*, c1800.

67 Circumferentor. Signed: *Spear College Green Dublin*. c1830.

68 Circumferentor. With tube sights hinged for taking altitude readings against a graduated quadrant. Signed: *J. White 95 Buchanan St. Glasgow*. c1860.

69 Circumferentor. Hinged sights, one provided with additional sights and a scale for altitude measurements. Signed: *James White Glasgow*. Second half 19th century.

70 Surveyor's level. Four screw. Signed on compass: *Troughton London*. c1790.

71 Surveyor's level. Signed: *Spencer & Son, 19 Grafton Street, Dublin*, c1870.

72 Theodolite with 8in diameter circles. Brass. Unsigned. c1800.

73 Theodolite. 7in. Signed: *Adams, London*. c1800.

74 Theodolite. 5in. Four screw. Signed: *Thos. Grubb Dublin*. c1850.

75 Theodolite, Everest style, 5in. Signed: *Troughton & Simms London*.

76 Clinometer, 2in radius quadrant arm. Signed: *Watkins & Hill, Charing Cross*, early 19th century.

77 Miner's dial with circular degree scale and sights for measuring elevation. Signed: *Braham Bristol*. Mid-19th century.

78 Lean's dial (Miner's theodolite). Signed: *E.T. Newton, St. Day Cornwall*. (fig 91)

79 Universal surveying instrument. Signed: *Schmalcalder's Patent 82 Strand London*. c1825.

80 Water level by JJ Hicks, acquired new in 1893 (accompanying MS notebook starts entries on 8 December 1893). With tripod and prismatic compass by *Troughton and Simms*.

DRAUGHTSMEN'S AND MATHEMATICAL INSTRUMENTS

81 Gunner's callipers. Brass. Signed: *Wm. Deane fecit*, c1700.

82 Gunter's scales, boxwood. Two examples, lengths 1 and 2ft. Unsigned.

83 Slide rule. Signed: *W. & S. Jones 30 Holborn London*, c1800.

84 Circular protractor. Brass. 8in diameter. Signed: *Spear & Co. Dublin*. c1825.

85 Sectors (4). Two ivory and unsigned; one boxwood and signed: *Jacob & Halse London*; one brass and signed *J. Sisson, London*.

86 Brass sector, radius 12in. Signed: *Thos. Heath fecit No. 37*, c1725.

87 Pantographs (2). One signed: *J & W Watkins Charing Cross*, early 19th century; the other signed *W & A Smith*.

88 Case of drawing instruments in shark-skin case, instruments inscribed: *Henry Tryon*, sector signed: *W. Elliott, 268 High Holborn, London*.

89 Circular protractor, inscribed owner's initials: *JBG*, 19th century.

90 Dividers, brass, steel points. 18th century.

91 Plane table, alidade and trough compass, signed: *Cary, London*.

92 Guinea balance, signed: *T Houghton Maker Farnworth near Warrington Lancashire*. c1800.

93 Money scales in wooden box with label: *James Warren... Skinner Row, Dublin*.

AIR PUMPS AND ACCESSORIES (SMITH LOAN COLLECTION)

94 Air pump, double barrel, table model. Signed on ivory label: *W. Fraser London*.

95 Two bell jars, one wide and low in form, the other tall and narrow.

96 Gas jar with brass cap and single threaded outlet. Another with brass cap and stopcock.

97 Small Magdeburg hemispheres.

98 Brass fountain experiment, on wooden base.

ELECTRICAL MACHINES AND ACCESSORIES (SMITH LOAN COLLECTION)

99 Plate electrical machine, 30in diameter plate, late 18th century.

100 Cylinder electrical machine, signed: *Nairne*, incomplete.

101 Henley discharger.

102 Thunder house.

103 Flickering jar and flickering tube.

104 Electric egg.

105 Swirling discharge.

106 Gold leaf electroscope.

107 Tangent galvanometer.

108 Galvanometer signed: *Baird & Tatlock (London) Ltd*.

109 Eudiometer, Cavendish type.

Eire 8 DUBLIN
National Botanic Gardens

Glasnevin, Dublin 9

(010 353) 1 374388

The Botanic Gardens were established in 1795 on their present site under the control of the Royal Dublin Society, aided by grants from the Irish Parliament. In 1877 the Gardens passed into state control and today are administered by the Department of Agriculture.

In 1970 the Herbarium of the *National Museum of Ireland, Dublin*, transferred to the Botanic Gardens. Items 4-9 below are instruments transferred at that time. Apart from the first item, all microscopes date from the end of the 19th century or early 20th century.

1 Compound microscope, Powell and Lealand portable model with folding tripod foot, in fitted box. Signed: *Powell & Lealand 170 Euston Road London* (address from 1863 onwards). Pencilled notes on box lid signed: *W.R. McNab March 4th 1880*. McNab was professor of botany at the Royal College of Science, Dublin, and Scientific Superintendent of the Botanic Gardens.

2 Compound microscope, signed: *E Leitz Wetzlar No 85173*, in box with supplier's label: *F.E. Becker & Co London*.

3 Dissecting microscope, signed: *E Leitz Wetzler New-York 30E. 23d. Str*.

4 Compound microscope, label in box signed: *E. Leitz Wetzlar* and dated 3.11.11 (i.e. 1911). Engraved on microscope: *No.113 Dublin Museum (Botany)*.

5 Compound microscope, in box, signed: *E. Leitz Wetzlar and No. 163776*.

6 Compound microscope, in box, signed: *E. Leitz Wetzlar No. 85486*. Plaque with supplier's name: *F.E. Becker & Co... 83-87 Hatton Wall London EC*.

7 Compound microscope, in box, signed: *E Leitz Wetzlar No. 29838*. Label in box signed: *J Robinson & Sons 65 Grafton Street Dublin*. Engraved microscope: *No. 116 Dublin Museum (Botany)*.

8 Dissecting microscope, signed: *E Leitz Wetzlar* and engraved: *No 71 Dublin Museum*. Label in box dated 1905.

9 Dissecting microscope, signed *Carl Zeiss Jena*, and numbered: *1640*. In box with supplier's label: *C. Baker 243 & 244, High Holborn London*.

Eire 9 DUBLIN
National Museum of Ireland

Kildare Street, Dublin 2

(010 353) 1 618811

The National Museum's collections can be traced to those formed by the Royal Dublin Society and the Royal Irish Academy. The RDS established a museum in 1733 and formed large collections in the decorative arts and natural history. The arts collections were transferred following the Dublin Science and Art Museum Act of 1877 which established the National Museum. The Royal Irish Academy passed over its collection of Irish antiquities in 1890.

The architect of the National Museum was Thomas Deane; the building was opened on 29 August 1890. There are important collections of archaeological material and antiquities, decorative arts, folk-life, geology and zoology. Scientific instruments are few in number, though several Irish makers and retailers are represented. A number of garden and wall sundials of stone, chiefly of the 18th and 19th century, have not been listed.

1   Astrolabe. Gilt copper, octagonal, width 7⅛in. With 3 plates (diameter 5¾in) for latitudes 36, 39, 42, 45, 48 and 51°. Signed in shadow square: *Erasmus Habermeli* and engraved at the back of the compass holder on the bracket: *Francisci de Padoanis Foroliuiensis Mediciae D*, with arms consisting of three roses with leaves and barbs. Late 16th century. (fig 8)
Lit: GUNTHER (1932) no. 279, Illd; ZINNER (1956) p338; ECKHARDT (1976-77) p40, no. 69, Illd.

2   Mechanical equinoctial dial (diameter 3¼in) in fish-skin case. On reverse, latitudes of various towns, including several Irish ones. Signed: *Seward Dublin. c1800.* (578-1910)

3   Compass, English, paper windrose, diameter 2¼in. Turned brass box and lid. (1039-1891)

4   Inclining dial, brass. Signed: *Thos Harris & Son London.* diameter 4½in. Early 19th century. (438-1909)

5   Equinoctial sundial. Brass, partly gilt. Octagonal. The dial is unsigned, but an applied silver disc in a contemporary leather box with silk lining is engraved with various latitudes and the initials: *LTM* (Ludwig Theodor Muller of Augsburg). 18th century. (113-1901)

6   Sector, brass, 6½in length. Signed: *N. Bion AParis.* Early 18th century.

7   Sundial, with small compass, inscribed: *J. Baum & Co Birmingham* and *Patent 1875.* (86-1971)

8   Horizontal sundial, brass, octagonal (gnomon missing). Signed: *Jn Fawcett Maker Dublin.* (14-1975)

9   Horizontal sundial, brass, octagonal, 9¼in across. Signed: *Mason Dublin.* (7-1977)

10   Graphometer, 4in radius. Typical French type. Signed: *Butterfield AParis.* (273-1894)

11   Circumferentor, arms and compass needle missing. Signed: *Seward Dublin. c1800.* (84-1940(15))

12   Circumferentor, arms missing. Signed: *Buckley Dublin. c1850.* (2-1964)

13   Level in mahogany box, with trade label signed: *Spencer Dublin.*

14   Octant, ebony, with ivory plaque signed: *Yeates Dublin.* Radius 10½in. (9-1976)

15   Waywiser. Oak wheel (15¾in radius), mahogany handle. Signed: *Jonathan Sisson Londini Fecit.* (59-1933)

16   Botanic microscope (Gould type) in red leather case.

17   Chondrometer in mahogany box, with printed instructions. Signed: *R. Spear 27 College Green Dublin.* Mid-19th century. (67-1934)

18   Glass rod from medical director, light green glass, about 2ft long. Said to be of Dublin manufacture. (33-1956)

Eire 10 DUBLIN
Royal College of Surgeons of Ireland

123 St Stephen's Green, Dublin 2

(010 353) 1 780200

The Dublin Society of Surgeons was founded in 1780, largely as a result of the efforts of Sylvester O'Halloran, and it received its Royal Charter four years later. In 1789 teaching facilities were made available, with the object of the founders, as stated in the Charter, being 'to establish a liberal and extensive system of Surgical Education'. The earliest part of the current building, which still houses a medical school, was started in 1825.

Most of the instruments which come within the scope of the inventory are microscopes. There is, in addition, a collection of surgical instruments. Plans to display these instruments have been shelved pending the restoration of the old museum building.
Lit: WIDDESS (1949)

1   Microscope, Cuff type, signed: *Geo. Adams in Fleet Street London,* mid-18th century.

2   Microscope, Jones type with tripod foot, signed: *Dollond London.* Owner's name engraved on box lid: *DFS 1844.*

3   Microscope, Cary type (Gould pattern) with folding tripod, unsigned.

4   Microscope, Culpeper type in wooden box with trade label: *Edwd Clarke 18 Lower Sackville St Dublin.* Early 19th century.

5   Microscope, large compound, signed: *A. Ross London 522.* Second quarter of the 19th century.

6   Microscope, large compound, signed: *Ross, London, 3450.* Third quarter of the 19th century.

7   Microscope, large, compound Ross style, signed: *Baker 244 High Holborn London.* Two objectives signed: *Gunlach Berlin.*

8   Microscope, Society of Arts Educational pattern, signed: *Armstrong Bro 88 Deansgate Manchester.*

9   Microscope, Education pattern, signed: *J.H. Steward 456 Strand.*

10   Microscope, Educational pattern, Ross style, signed: *J.H. Steward 66 Strand.*

11   Microscope, signed: *Smith & Beck 1816.* Constructed in 1857.

12 Microscope, Wenham type, polarising, signed: *Henry Crouch 3167*.

13 Microscope, Leitz type, signed: *J Robinson & Sons Grafton Street* and *Presented by Professor Barley, 1960*.

14 Microscope, wood and card, Nuremberg tripod type, early 19th century.

15 Sykes hydrometer, with label on box lid: *M. Jordi & Co. Dublin*, thermometer signed: *Mason & Son, Opticians, Dublin*.

16 Thermometer, early clinical type with bent tube, signed: *Crichton. c1840*.

Eire 11 DUBLIN
Royal Dublin Society

Ballsbridge, Dublin 4

(010 353) 1 680645

The RDS was founded on 25 June 1731 as the Dublin Society for improving Husbandry, Manufactures and other Useful Arts; the word 'Sciences' was added at a subsequent meeting. Later in the century sums were voted to provide premiums for agricultural and manufacturing schemes. In 1795 William Higgins was appointed professor of chemistry and mineralogy, and two years later a chemical laboratory was established where instruction was given. A large laboratory and lecture room was completed in 1810, and Humphry Davy gave courses of lectures. Robert Kane became lecturer in natural philosophy in 1834. In 1836, evening scientific meetings were inaugurated for the communication of scientific discoveries. Instruments and apparatus were acquired for the laboratories and for public instruction.

The Royal Dublin Society once held the central position in the scientific activities of Ireland. Its range of activities has varied in a complex way which can be appreciated with reference to the work cited below.
Lit: MEENAN/CLARKE (1981)

1 Double burning lens on stand, diameter of lenses 14½ and 6in. Probably supplied by Parker of Fleet Street, London. Formerly the property of the chemist Richard Kirwan (1733-1812) and purchased by the RDS from his executors in December 1812.

2 Air pump, double-barrelled, floor standing. Signed: *J. Lynch & Son, Dublin.* ? *c1800*.

3 Condensing lens, signed *Yeates & Son, Dublin* and marked *Laboratory Royal Dublin Society*.

4 Wenham's binocular microscope with box. Signed: *Smith, Beck & Beck 3095. c1850s*.

5 Simple microscope with box. Signed: *M. Pillischer No. 66. c1860*.

6 Microscope signed *A. Ross, London*. An objective is dated 1856.

7 Projection microscope. Signed *Wright & Newton, made by Newton & Co., 3 Fleet Street, London*. Late 19th century.

8 Lens in brass mount, signed *B' Grubb, Dublin, 3712*.

9 Heliostat, signed: *Yeates & Son, Dublin*. Late 19th century.

10 Polarising apparatus, signed: *J. Spencer & Son, 19 Grafton St., Dublin*.

11 Direct vision spectroscope, signed: *Adam Hilger Ltd* and *No. J.10.301/20215*. In box with supplier's name: *T. Mason 5 Dame Street, Dublin*.

12 Spectroscope, on stand with brass telescope and collimator, unsigned. Moving needle (? provision for blackened paper recording). Second half 19th century

13 Multi-prism spectrometer, Duboscq pattern, unsigned. ? *c1880*.

14 Polarising apparatus, signed: *J. Spencer & Son 19 Grafton St., Dublin*.

15 Polarimeter, Norremberg pattern, signed: *Yeates & Son, Dublin*.

16 Polarimeter, signed *P. Harris & Co Limited Birmingham* and *G.C. James* (owner?).

17 Optical bench, brass, mounted on wooden stand incorporating fitted drawer with accessories. Signed: *J. Duboscq...a Paris*.

18 Pair of hollow prisms (for liquids) in box.

19 Stand with lens and pair of prisms.

20 Reflecting telescope, 5in aperture, signed: *Dollond London* (optics missing).

21 Chemical balance, long beam, signed *Oertling London*. Late 19th century.

22 Chemical balance, long beam, signed: *Ladd & Oertling, London*, with retailer's name: *Spencer & Son, 19 Grafton St., Dublin. c1870*.

23 Hydrostatic balance, with box. Signed: *F. Sartorius, Gottingen*.

24 Apothecary scales, signed: *James Robinson, Optician and Philosophical Artist, 65 Grafton Street, Dublin*.

25 Miner's safety lamp, Davy design, signed: *Newman, London, c1825*.

26 Mortars and pestles (4): one ceramic, marked: *Wedgwood Best Composition*, two agate, one iron.

27 Redwood viscometer, numbered: *No. 163*.

28 Abel flash-point apparatus, signed: *Yeates & Son Dublin 1886* and numbered: *465*.

29 Wollaston-type goniometer, signed *Cary, London*. Early 19th century.

30 Set of six nickel crucibles, signed: *F.W. & J.* and marked: *Reinnickel*.

31 Eudiometer, Cavendish pattern, on brass stand.

32 Moissan arc furnace, signed: *Ducretet a Paris*.

33 A silver and copper flask

34 Glass hydrometers (3) signed: *Yeates & Son Dublin.*

35 Conical measuring vessel, glass, graduated ¼, ½, 1, 2, 3 naggins.

36 Gas cylinder, glass, graduated in cubic inches to 12.

37 Set of drawing instruments, signed: *W.H. Harling.*

38 Wheatstone bridge, signed: *Telegraph Works, Silvertown, London. No. 1417.*

39 Induction coil, signed: *Apps 433 Strand London Patd 1881 – 264 No.1133.*

40 Voltmeter in wooden casing, signed: *Electrical Power Storage Co Limited No. 519 London.*

41 Air-speed meter, signed: *Yeates & Son, Dublin. No. 1594.*

42 Mercury regulator, signed: *J.R. Ryan & Co. 13 College Green, Dublin.* Presented to the RDS on 1 May 1834.

43 Standard yard, signed: *Troughton & Simms London and Mr Baily's Metal No. 15.* Cast in 1845.

44 Model time ball (?), sphere, free to run up and down vertical rod, with electromagnetic mechanism in base. Signed: *Yeates & Son Dublin.*

Eire 12 DUBLIN
Chemistry Laboratory, Trinity College

Trinity College, Dublin 2

(010 353) 1 772941

Trinity College was founded in 1591 as a constituent college of the University of Dublin (though no other colleges were subsequently established). Medicine was the first scientific subject to be taught. In 1711 a medical school was founded, with appointments in surgery, medicine, botany and chemistry. The first holder of a chair in chemistry was Robert Perceval (1756-1839) in 1785. Nothing survives from this early period (though a lamp furnace described by Perceval in 1791 is to be found in the *Royal Museum of Scotland, (Chambers Street), Edinburgh).*

The present laboratory was built in the period 1875-87. It contains a number of items connected with James Emerson Reynolds (1844-1920) who was professor at Trinity College between 1875 and 1903, and with Sydney Young (1857-1937), professor from 1904 to 1928. *Lit:* COCKER (1978)

1 Chemical balance, with heavy brass frame beam supported by steel knife-edge on an agate plane. Signed: *Grubb Dublin.* Constructed by Thomas Grubb, this balance was awarded a silver medal by the Royal Dublin Society in 1884.
Lit: DUBLIN (1931), C.29

2 Chemical balance, long beam, signed *Oertling London.* Belonged to Emerson Reynolds. Late 19th century.

3 Atometer, designed by Emerson Reynolds for comparison of atomic heats, *c*1880.

Lit: REYNOLDS (1881) p58; DUBLIN (1931), C.28

4 Tumbling aspirator ('chemical hygrometer') consisting of two 2100cc cylindrical glass graduated vessels which pivot about a brass frame through which gases can be introduced, on wooden stand with 20 holes (? for recording number of inversions). Label on base signed: Dancer Optician Manchester. Possibly connected with James Apjohn (1796-1886), professor from 1850 to 1875.

5 Spectrometer, signed: *John Browning 138 Strand London WC2,* numbered *7389.*

6 Saccharometer, signed: *Franz Schmidt & Haensch Berlin,* numbered *11436.*

7 Eudiometer, Cavendish type, signed on tap: *Griffin London.*

8 Two glass vessels with brass caps with taps, for collecting gases over water. Mid-19th century.

9 Glass distillation apparatus, including tubulated retort and quilled receivers. First half 19th century.

10 Glass distillation columns, designed and constructed by Sydney Young, *c*1890-1922.

11 Apparatus constructed by Sydney Young (of type first described by Thomas Andrews) for determining properties of liquids and vapours under pressure, and hence thermodynamical relations. *c*1890.

Eire 13 DUBLIN
Physical Laboratory, Trinity College

Trinity College, Dublin 2

(010 353) 1 772941

The first professor of natural and experimental philosophy at Trinity College was Richard Helsham, appointed in 1724. During the following century there was a succession of eminent physicists who held the chair: Bartholomew Lloyd, his son Humphrey Lloyd (1800-1881) who built a magnetical observatory in the grounds of Trinity in 1837, and John Hewitt Jellett (1817-1888) who studied optical properties of solutions. The Department of Physics moved into its present Physical Laboratory building in 1909. The surviving apparatus, mainly of the 19th century, reflects the research and teaching activities of the department.

1 Lodestone. Brass mounted, finely decorated and inscribed: *The Gift of his Excely Thomas Lord Wyndham Baron of Finlas Lord Chancellor and one of the Lord Justices of Ireland to Trinity College near Dublin. c1735.* (fig 69)

2-4 Large concave mirror (17½in diameter) and two small mirrors on mahogany tripod stands. Late 18th century.

5 Concave mirror, copper (19½in diameter), used for focusing radiant heat.

6 Small air-pump, double-barrel table model, signed: *Cary London.*

7 Air-pump, double-barrel, table model. The platform on which the plate for the receiver would have been mounted is set on two columns at the back of the pump and two sets of twin columns at the front. Signed on ivory label: *Newman & Son 122 Regent St. London.*

8 Air-pump, single-barrel, table model. Brass barrel at oblique angle to rectangular brass plate. First half 19th century.

9 Gregorian reflecting telescope on pillar and claw stand aperture 2½in. Unsigned, 18th century.

10 Instrument for accurate plotting of star-fields from photographs. Believed to have been used at *Birr Castle, Birr* by Lord Rosse, though acquired by the department from *Dunsink Observatory, Castleknock.* Signed: *Troughton & Simms London.*

11 Compass on gimbals, signed *Buckley Dublin* and *TCD.*

12 Anamorphic mirror, cylindrical type, and board mounted with distorted image. Also, five further boards of different pattern.

13 Jellet's saccharimeter, signed: *Spencer & Sons Dublin. Lit:* DUBLIN (1931), item P.26

14 Polarimeter, signed: *Reichert.*

15 Direct vision spectroscope, signed: *John Browning 63 Strand London,* second half of 19th century.

16 Gas microscope in box, signed: *Newton & Co 5 Fleet Street London.*

17 Projection microscope, signed: *Sold by Ross London.*

18 Dip circle, signed: *Henry Barrow & Co 26 Oxendon St London,* mid-19th century.

19 Dip circle, signed: *Robinson & Barrow 26 Oxendon St London,* mid-19th century.

20 Galvanometer, signed: *Thomas Jones, 4 Rupert Street, London,* early 19th century.

21 Siren, signed: *F. Kerby 12 Spanns Buildings St Pancras London.*

22 Double siren of French manufacture, though signed: *Yeates & Son Dublin.*

23 Set of Helmholz resonators

24 Samples of the chemical elements in glass tubes, contained in leather case, with trade label: *Yeates & Son Dublin.* Very nearly complete. Must be post-1861 as thallium is included: possibly *c*1870.

25 Kelvin's electric balance, signed: *J. White Glasgow* and *311.*

Eire 14 DUBLIN
School of Engineering, Trinity College

Trinity College, Dublin 2

(010 353) 1 772941

The School of Engineering was founded in 1841 when John MacNeill was appointed professor. It is older than all other engineering schools in the British Isles except for that at Glasgow University. During the 19th century its major reputation was for railway engineering. A collection of surveying and drawing instruments has recently been brought together, deriving from both within the School and from outside sources. It may be viewed on application. An important collection of 19th-century steam engine models, formerly in the School, is now housed in the Parsons Building nearby.

1 Hadley octant, 13in radius, signed: *For The Revnd Mr Charles Jones 1752* (the son of Lewis Jones, sub-Auditor General of Dublin).

2 Sextant, 16in radius, signed: *J. Bennett London. c*1800.

3 Box sextant, signed: *Troughton & Simms.*

4 Box sextant, signed: *Robinson, London,* mid-19th century.

5 Box sextant, signed: *Yeates & Son, Dublin.*

6 Circumferentor, unsigned, *c*1800.

7 Circumferentor, signed *T. & J. Mason Dublin,* arms missing. *c*1820.

8 Surveyor's level (Y level), signed: *Horatio Yeates, 12 Wicklow St Dublin.*

9 Surveyor's level (Y level), 16in, signed: *Troughton & Simms, London. c*1850.

10 Transit theodolite, 6in, signed: *Troughton & Simms, London.*

11 Theodolite, 7in, signed: *Spear Dublin, c*1800.

12 Theodolite, 5in, signed: *W. Elliott 268 High Holborn London.*

13 Artificial horizon, *c*1840.

14 Burel reflecting level, *c*1880.

15 Various ivory drawing scales, signed: *Spear, Moore, Spencer, Yeates* and *Troughton & Simms.*

16 Circular protractor, brass, signed: *Yeates 2 Grafton St Dublin, c*1850.

17 Circular protractor, brass, signed: *Troughton & Simms, London.*

18 Parallel rule, ebony, signed: *Spencer & Son Dublin, c*1850.

19 Plane table, alidade and accessories, signed: *Stanley, London, c*1880.

20 Beam compass in box, with trade label signed: *Yeates & Son Optician to the University, Dublin.*

21 Proportional compasses, *c*1840.

22 Eidograph, signed: *Dunn Edinr, c*1840.

23 Computing scales, as arranged by Major General Hannyngton, signed: *Aston & Mander 25 Old Compton St London WC, c*1880.

24 Solid geometrical models, wood, signed: *Yeates & Son Dublin.*

25 Equipment for calibration and testing of spirit bubbles, with supplier's label: *J. Spencer & Son...19 Grafton Street Dublin.*

Eire 15 DUBLIN
Department of Physics, University College

Belfield, Stillorgan Road, Dublin 4

(010 353) 1 693244

University College was founded in 1851 as a result of the efforts of John Henry Newman. Science teaching was developed in the 1870s by Mgr Gerald Molloy (1834-1906) who had studied under Professor Nicholas Callan at *St Patrick's College, Maynooth*, where he later became Professor of Theology. Much of the extensive collection of physics demonstration apparatus must have been acquired by Molloy. In 1873 he was appointed Vice-Rector of the Catholic University in Dublin, and ten years later, Rector. He also occupied the position of Professor of Natural Philosophy and established a reputation as a gifted lecturer in popular science. His apparatus, mostly electrical, but also relating to the early telegraph and telephone, is divided between University College and Maynooth. A perhaps surprising amount of demonstration apparatus for sound experiments survives at both institutions.

Numbers following entries refer to a typescript inventory of the collection compiled in 1984.

1 Concave diffraction grating on speculum metal, in wooden box, ruled by Henry Augustus Rowland in 1897. 6 by 1⅞in ruled surface with 14438 lines to the inch. The grating belonged originally to the Royal University of Ireland and was housed in a Rowland mounting dismantled in 1932. Scratched on the edge: *Ruled on Prof. Rowland's engine Johns Hopkins University Baltimore, Md, USA 1894.*

2 Diffraction grating on glass, 1 x ⅞in, in leather case. 17300 lines to the inch. Said to have been constructed by ? Chapman who was connected with the Naval Observatory at Arlington, Virginia, USA, and plate prepared by JT Brashear, Allegheny City, Pa, USA, 1894. Name of Professor DL Waire scratched in corner.

3 Echelon grating, said to have been constructed by Adam Hilger in 1898 for Thomas Preston, and intended for his work on the Zeeman effect. Preston died before he could use it.

4 Electromagnet with which Preston discovered the anomalous Zeeman effect.

5 Large electromagnet, constructed for Gerald Molloy, *c*1870.

6 Gregorian reflecting telescope on pillar and claw stand. Complete except for lens caps. Aperture 2½in, tube length 14in. Signed: *Sterrop Maker.*

7 Gregorian reflecting telescope on pillar and tripod stand. Aperture 4in, tube length 25in, the barrel mounted on trunnions. Signed: *Harris 47 Holborn London.*

8 Achromatic telescope on pillar and tripod stand. Aperture 3in. Signed: *Yeates and Son, Dublin*, mid-19th century.

9 Heliostat, overall height 19in. Signed: *Yeates & Son Dublin.*

10 Heliostat, signed: *Spencer & Son Dublin.* (M4)

11 Hollow prism for liquids, signed: *Pixii Pere & Fils Rue de Grenelles German 13 (a Paris).* (L25)

12 Concave mirror on stand, for focusing radiant heat, copper.

13 Melloni 'optical' bench (for study of radiant heat) with accessories. (T16)

14 Large spouting can, with orifices of different cross-sections, used to demonstrate pressures at different heads of water. (H1)

15 Model gear wheel and screw, brass. (M25)

16 Double cone and inclined plane, wood.

17 Manometric flame interference apparatus, signed: *Rudolph Koenig a Paris.* (A11)

18 Box of tuning forks, signed: *RK.* (A17)

19 Double siren signed: *Rudolph Koenig a Paris.* (A12)

20 Siren with revolution counter, signed: *Harvey & Peak London W* and *Yeates & Son Dublin.* (A13)

21 Savart disc (rotatable toothed disc for producing a sound when card rubs against it). (A23)

22 Chladni plate on tripod stand, signed: *Baird & Tatlock London and Glasgow.* (A7)

23 Helmholz resonators with rotating mirror, to analyse sound by 'manometric flames'. (A28)

24 Air thermometer, signed: *Ducretet et Cie 21 Rue des Ursulines a Paris.* (T18)

25 Dip circle, signed: *Troughton & Simms London.*

26 Magnetic needle in glazed case, signed: *Spencer & Son 19 Grafton St Dublin.*

27 Two small and one large plate electrical machines.

28 Plate electrical machine driven by an electric motor, signed: *Leybold's Nachfolger AG Koln-Rhein* and by supplier: *T.H. Mason, 5 & 6 Dame Street, Dublin.* (B9)

29 Two Wimshurst machines, one signed *Philip Harris & Co Ltd* (B11), the other (disc broken): *Yeates & Son Dublin.*

30 Wimshurst machine of cylinder variety, unsigned, *c*1890. (B1)

31 Number of Leyden jars in various forms. (B118-126)

32 Insulating stool (B110)

33 Voltaic pile, demonstration type, late 19th century. (B80)

34 Faraday rotating wire, annular trough of mercury.

35 Voltameter, glass, on stand.

Eire 16 GALWAY
Physics Department, University College

Galway

(010 353) 91 24411

University College, Galway, was first opened to students as Queen's College in October 1849. (It is now a constituent of the National University of Ireland.) Physics was taught from the beginning, Morgan W Crofton being appointed to the chair. Between 1852 and 1857 the professor was George Johnstone Stoney, particularly remembered for introducing the term 'electrine' in 1874, which was modified to 'electron' in 1891; he calculated an approximate value for its charge. Stoney was succeeded by Arthur Hill Curtis.

A substantial quantity of demonstration apparatus for the second half of the 19th century survives. It is not displayed. Nearly 30% of the items listed in a printed inventory of 1902 can be traced. There is also some early 20th-century material preserved which is not listed here. *Lit:* GALWAY (1902)

MECHANICS, DYNAMICS, PNEUMATICS, ETC.

1 System of model levers on base.

2 Geometrical rhomboid, separable into three pieces.

3 Set of brass frictionless gears mounted on base.

4 Rolling double cone, wood.

5 Trolley and inclined plane, wood.

6 Part of centrifugal forces machine (?), signed: *W.M. Stiles, London.*

7 Whirling machine, incomplete, signed: *Yeates & Son Dublin.*

8-9 Sphere and cone, with internal point of suspension (on mounted needle), for precession demonstration (?).

10 Three glass vessels mounted on tank, to show that water finds a common level.

11-13 Three syringe pumps, one signed: *F.E. Becker & Co London.*

14 Sprengel pump

15 Glass pressure vessel mounted within brass frame.

16 Vacuum fountain apparatus

17 Magdeburg hemispheres

LIGHT

18 Refracting telescope, with stand, aperture 3in. Signed: *J. Brown 76 St Vincent St Glasgow.*

19 Observatory telescope, on wheeled stand, aperture 4in. Unsigned.

20 Split telescope on stand, with intervening polarising device (?), signed: *Yeates & Son Dublin.*

21 Lenses mounted on brass stands, signed: *Yeates & Son, Dublin.*

22 Plain mirror on wooden stand, diameter 5in.

23 Convex mirror on wooden stand, diameter 8in.

24 Optical bench, signed: *Elliott Bros. Strand London.*

25 Simple spectroscope, collimator and telescope (brass) attached to wooden box. Prism can be manipulated from outside with a lever device.

26-28 Three simple teaching spectroscopes, signed: *John Browning 63 Strand London; Harvey & Peak London;* and *Griffin London.*

29 Multi-prism spectrometer, signed: *Max Kohl Chemnitz.*

30 Spectroscope for attachment to microscope, in box, signed: *John Browning London.* With printed instructions dated January 1877.

31 Adjustable rectangular plain black mirror on stand, for polarisation experiments.

32 Model eye, can be dismantled, papier mache.

33 Spherometer, in box, signed: *Griffin London.*

34 Conical anamorphic mirror, in box.

35 Machine for rotating card discs for optical effects, with various cards, signed: *Elliott Bros. 30 Strand London.*

36 Heliostat, signed: *Yeates & Son Dublin.*

HEAT

37 Differential air thermometer on wooden stand, signed: *Elliott Bros 119 Strand London.*

38 Differential air thermometer, with Leslie's cube interposed between bulbs (to demonstrate different radiant properties of black and white surface), signed: *Harvey & Peak London.*

39 Demonstration pyrometer on stand.

40 Model Harrison gridiron pendulum, wood.

41 Mechanical equivalent of heat apparatus, signed: *W.G. Pye & Co. Cambridge.*

42 Hypsometer on folding brass stand.

43 s'Gravesande ball and torus apparatus, signed: *Elliott 30 Strand London.*

44 Hope's apparatus (for determining temperature of maximum density of water), with attached label signed: *Griffin Kingsway London W.*

45 Davy lamp

46 Iron retort (used for distilling mercury).

47 Iron pressure cooker (Papin's digester).

MAGNETISM AND ELECTRICITY

48 Compass on gimbals, in box, signed: *Elliott Brothers Strand London.*

49 Dip circle, signed: *Elliott Bros London.*

50 Model earth, wound with electromagnet, and marked: *N* and *S.* Signed: *Elliott Bros 30 Strand London.*

51 Delezenne circle, or earth inductor, for demonstrating electrical currents induced by terrestrial magnetism, signed: *Yeates & Son Dublin*.

52 Plate electrical machine, diameter 18in, signed: *W & S Jones No 30 Holborn London*.

53 Wooden stand of large plate electrical machine, diameter 30in, signed: *Elliott Bros 30 Strand London*.

54 Wimshurst machine, diameter 11in, unsigned.

55 Wimshurst machine in glazed case, diameter 14in.

56 Part of Wimshurst machine on oval base, signed: *J Robert Voss Berlin Agent F E Becker & Co London WC*.

57 Electrophorus, diameter 5¼in.

58-59 Leyden jars

60 Leyden jar on wooden stand.

61 Cylindrical conductor, brass.

62 Cylindrical conductor, painted black, signed: *Elliott Bros 30 Strand London*.

63 Cylindrical conductor, brass, on two glass insulating legs, signed: *Elliott Bros 30 Stand London*.

64 Spherical conductor, copper.

65, 66 Biot's apparatus (two separable insulated hemispheres for showing distribution of static electricity).

67 Flickering plate, 12¼ x 7½in.

68 Electric bells on stand.

69 Electric bells (2) mounted in wooden frame, base 14½ x 4½in, height 25in, signed: *Elliott Bros 30 Strand London*.

70 Electric whirl (revolving points).

71 Gamut of bells (combination of bells and revolving points).

72-73 Electric pistols.

74 Faraday cage.

75 Electric egg.

76 Spark tube.

77 Water bucket (when charged, water discharges through a siphon from the bucket).

78 Insulated table, 16 x 14in.

79 Cuthbertson's electrometer, signed: *W & S Jones 30 Holborn London*.

80 Cuthbertson's electrometer, signed: *Elliott Bros 30 Strand London*.

81 Four bichromate cells.

82-83 Single and double Barlow's wheels.

84 Electromagnet with a variety of contacts and platform with weights.

85 Induction coil, base 27¼ x 13½in, signed: *L Miller's (Patent) Jointless Section Coil 66 Hatton Garden London*.

86 Induction coil, base 29 x 16½in, signed: *Harvey & Peak London W*.

87 Model dynamo, horseshoe magnets, on wooden stand.

88 Gramme's magneto-electric machine, horseshoe magnets, geared system for coil rotation, metal stand, signed: *Gramme inv... Breguet ft No 157*.

89 Revolving coil on stand, rotated by pulley wheel, signed: *Max Kohl Chemnitz*.

90 Electrometer, signed: *Elliott Bros. 5, Charing Cross & 56 Strand London*.

91 Quadrant electrometer in brass cage, signed: *Griffin London*.

92 Galvanometer with asymmetric coil, signed: *Elliott Brothers 30 Strand*.

93 Twenty-seven Crookes' tubes of different types.

SOUND

94 Kundt's tube (evacuable glass tube with plungers), length 4ft 6in.

95 Musical flames apparatus (two tubes on hollow base).

96 Set of Helmholz resonators and revolving mirror.

97 Model organ system (two pipes), signed: *Made by Yeates & Son Dublin*.

98-99 Two sirens, one signed: *Max Kohl Chemnitz*, the other *Yeates & Son Dublin*.

100 Large disc siren, signed: *Harvey & Peak London*.

101 Electromagnet-operated tuning fork, signed: *Yeates & Son Dublin*.

102 Chladni's plate apparatus

103 Metronome, signed: *Enregistrant par l'Air & l'Electricite system de Ch Verdin 6 Rue Rollis Paris 1886*.

MISCELLANEOUS

104-105 Terrestrial and celestial globes, 18in diameter, on wooden stands with barley-twist legs, overall height 3ft 10in. Signed: *Cary's New Terrestrial Globe...with Corrections and Additions to 1818*, and *Cary's New Celestial Globe...1 March 1816*.

106 Terrestrial globe, 12in diameter, on short stand. Signed: *Malby's Terrestrial Globe...Hodges & Smith Agents Grafton Street Dublin...Jany 1858*.

107 Flint-lock mounted on wooden block.

108 Chemical balance in glazed case. Frame-type beam, length 12in. Signed: *J Robinson & Sons 65 Grafton Street Dublin*. (? Made by Oertling.)

109-110 Two identical chemical student balances in glazed cases, solid beams, 7in long. Signed: *Beckers Sons Rotterdam*.

111 Chemical balance in glazed case, solid beam 7in long. Signed: *Manufactured in Germany especially for F.E. Becker & Co 33,35 & 37 Hatton Wall London*. Date pencilled on drawer by cabinet maker: 16.11.95 (i.e. 1895).

112 Demonstration beam balance on stand, with scales for pointer and beam end.

113 Case for two Twaddle hydrometers, both missing, but thermometer and jar survive.

114 Eight Guyton hydrometers

115 Wollaston goniometer, in box, signed: *Elliott Bros. London.*

116 Anemometer, signed: *L. Casella London* and numbered *351.*

Eire 17 MAYNOOTH
St Patrick's College (Maynooth College)

Maynooth, County Kildare

(010 353) 1 6285222

The College is a Pontifical University, a major seminary for the training of priests, and a Recognised College of the National University of Ireland. It was founded in 1795 and was of scientific significance in the 1830s because of the activities of its Professor of Natural Philosophy, the Revd Nicholas Joseph Callan (1799-1864). Callan was appointed in 1826 and held his appointment until his death. He worked in the field of electromagnetism and was a pioneer in high tension electricity (he invented the induction coil in 1836), and discovered the principle of self-excitation in dynamo-electric machines. He had plans to use electrically propelled coaches on the Dublin – Kingstown (Dun Laoghaire) Railway. Considerable quantities of apparatus used by Callan are preserved in the College Museum.

The Museum also contains a large amount of physics apparatus used for teaching. Much of this is signed by Irish makers and retailers; some is undoubtedly of foreign import but signed by Dublin dealers. There are considerable quantities of instruments bearing the name of the Dublin firm of Yeates, including electrical equipment, thermometers, hydrometers, and demonstration apparatus. A certain amount of material is associated with Mgr Gerald Molloy (see *University College, Dublin, Department of Physics*). In addition to those in the inventory, there are instruments connected with Marconi's transmission of radio signals across the Irish Sea, AW Conway's work on the wireless, and some early telegraph and telephone equipment.

Lit: For Callan's scientific work and material, see McLAUGHLIN (1964 and 1965); CASEY (1965); HEATHCOTE (1965); SHIERS (1971). A catalogue in typescript, MAYNOOTH – ST PATRICK'S COLLEGE (1955), is referred to in the listing below by the numbers in brackets which follow the entry.

MATERIAL ASSOCIATED WITH NICHOLAS CALLAN

1 Large horseshoe magnet. Constructed in 1834 by the local blacksmith, James Briody. Measures 13 ft from end to end. Secondary coil missing. (1611)
Lit: McLAUGHLIN (1964) p29; CASEY (1965) p558

2-11 Electromagnets and coils constructed and used by Callan. (1600-609)

12-13 Two induction coils. One (612) has a primary coil 11½ in long and 6in in diameter, the secondary coil has a diameter of 20in and a width of 4½in The second and larger coil (613) has a primary measuring 40in with a diameter of 7in, while the secondary consists of three coils, each of 21in diameter and 4in thick. The coils were exhibited in London in 1837 and 1851.
Lit: CASEY (1965) pp558-60, figs 2 and 3

14 Repeater or rapid contact breaker. Constructed by Callan in 1836 from a clock escapement with a crank and three cups of mercury. (610)
Lit: McLAUGHLIN (1964) plate 4 and fig 6 on p27

15 Point-plate valve. When the plate is negative a discharge occurs more rapidly than when it is positively charged.

16-18 Three electromagnetic engines. The frame of one was cast at the foundry of Yeates and Son of Dublin and bears the firm's name. Callan's experiments commenced in 1836. (620-622)

19 Wollaston cells. The remains of Callan's first large battery (50 out of the original 280) used in 1836 for the giant electromagnet. (623-673)

20 Poggendorff bichromate cell. (673)

21 Daniell cells, seven, 1836. (674-680)

22 Grove cell, 1839. (681)

23 Bunsen cells, eight, 1843. (682-689)

24 Callan cast iron cells, three, consisting of zinc, dilute nitro-sulphuric acid, porous pot, concentrated nitro-sulphuric acid and cast iron, which served as positive element and container.

25 Maynooth battery. The battery was constructed from 577 improved Callan cells. These three cells are signed: *Maynooth Battery F.M. Clarke Maker 488 Strand London.* (693-695)
Lit: McLAUGHLIN (1964) p18, plate 4; CASEY (1965) pp560-1, fig 4

26 Sine galvanometer designed by Callan in 1854 for measuring large currents of electricity.
Lit: CALLAN (1854)

27 Container and positive element of Callan's single fluid cell. Designed for illuminating purposes such as lighthouses and street lighting. (697-698)
Lit: CALLAN (1855)

28 Arc lamps, hand-feed and clockwork type, for pushing forward the points of the carbons. (699-703)

29 Lime light apparatus and oxy-hydrogen burners. (704-707)

OTHER INSTRUMENTS AND APPARATUS

Microscopes

30 Culpeper-type microscope. Matthew Loft form with continuous curved legs. Black ray-skin outer tube, inner tube vellum covered. Lignum vitae eyepiece and snout, also octagonal base. Brass screw fittings for objectives. Pyramidal oak case. *c*1740. (800)

31 Two dissecting microscopes, unsigned.

32 Microscope signed: *Nachet*, late 19th century.

33 Microscope with camera lucida attachment, in box. Signed: *Mon.E.Hart. & A. Prazmouski sucr Rue Bonaparte I, Paris. c*1880.

Surveying

34 Circumferentor. Brass, with cover, dial diameter 5¼ in, needle missing. Brass lugs at side of compass with screw threads for attachment of sight arms (missing). Signed: *Johannes Lewis Dublini Fecit anno Domini 1688.* (801.1)

35 Circumferentor. Brass dial and cover, dial diameter 5¼in, distance between sights 16in. Signed: *Spicer Dublin.* (801.2)

36 Circumferentor. Brass dial and cover, dial diameter 6¼in, sight arms missing. Signed: *Seacomb Mason No. 8 Arran Quay Dublin.* (801.3)

37 Circumferentor. Screw attachment of sight arms. Two bubble levels in base of compass. Dial diameter 6in. Signed: *Walker & Son No. 17 Temple Barr Dublin.*

38 Circumferentor, signed: *Spear & Clarke.*

39 Theodolite, signed: *F.W. Breithaupt & Sohn in Cassel 1876.*

40 Theodolite, diameter of horizontal scale 8¼in. Sighting telescope with bubble level carried by arm moving over vertical semicircle toothed internally. Signed: *Tho. Heath Londoni fecit. c*1740.

41 Transit theodolite, signed: *J. Hughes London.*

42 Celestial globe, signed: *New Celestial Globe* and *Pubd by J.& W. Cary Strand.*

43 Terrestrial and celestial globes, terrestrial (no stand), signed: *J.& W. Cary Strand March 13th 1808*; celestial (with wooden stand) signed: *Made and sold by J.& W. Cary No 181 Strand...1799.* (709-710)

44 Octant, signed: *Elliot Bros. 47 Charing Cross London.*

45 Octant. Ebony frame, radius 12in. Vernier, locking screw. Signed: *Yeates Dublin.*

46 Nautical compass mounted in gimbals in wooden box. Signed: *Mason Essex Bridge Dublin.* First half 19th century. (874)

47 Variation compass, signed: *Yeates & Son Dublin.*

48 Cannon sundial on tripod, face inscribed with cities and latitudes including Dublin, Glasgow, Paris and St Petersburg.

Heat

49 Heliostat (827)

50 'Dial thermometer' (expanding or contracting mercury causes a mechanical effect to move pointed over a dial), signed: *Made by Yeates & Son Dublin.*

51 Harrison gridiron pendulum, for demonstrations.

52 Demonstration thermopile

Pneumatics

53 Air pump, double-barrel, floor-standing type, table base, elevated plate. Signed: *Newman London*, early 19th century.

54 Air syringe, single-barrel, obliquely mounted cylinder, elevated plate on mahogany base. Unsigned, first half 19th century.

55 Glass tube for guinea and feather apparatus. (713)

Chemistry

56 Goniometer on tripod stand, Borsch type, signed *R. Fuess Berlin.* Transferred from University College, Dublin, Geology Department (as are nos. 57 and 58).

58 Goniometer, Wollaston type, signed: *Yeates & Son Dublin.*

59 Goniometer, Wollaston type, signed: *Spencer & Son 19 Grafton St Dublin.* (872.0)

60 Goniometer on tripod stand, Borsch type, signed: *Elliot Brothers 30 Strand.*

61 Polariscope, Norremberg type, signed: *Spencer & Son Dublin.*

62 Polarimeter signed: *Mon Jules Duboscq Ph Phellin Paris.*

63 Box of six glass Twaddle hydrometers.

64 Sikes hydrometer signed: *Yeates & Sons Dublin.*

65 Nicholson hydrometer, white metal.

66 Very large spectrometer, signed: *Adam Hilger London* and used by Walter Noel Hartley (1846-1913) at the Royal College of Science, Dublin, in his research on the spectra of the chemical elements. (Another spectrometer used by Hartley for absorption spectra, constructed by Meacher of London and Yeates and Son of Dublin, is in the *Science Museum, London.*) Presented by *University College, Dublin.*

67 Eudiometer, Cavendish type.

Sound

68 Siren mounted on wooden box, signed: *Yeates & Son Dublin.*

69 Electrically-maintained siren

70 Cardboard disc siren, signed: *Yeates & Son Dublin.* (829)

71 Manometer flame interference apparatus, signed: *Rudolph Koenig a Paris.*

Electricity

72 Cylinder electrical machine, unsigned. Mid-19th century.

73 Winter electrical machine, 36in diameter plate. Maker's name plate missing. *c*1830

74 Carre electrical machine, *c*1870.

75 Wimshurst electrical machine, signed: *Yeates & Son Dublin.*

76 Large Wimshurst electrical machine. Presented in 1958 by *Department of Physics, University College, Dublin.*

77 Holz electrical machine.

78 Voltaic cannon. (825)

79 Electric chimes. (836)

80 Apparatus for producing water by exploding oxygen and hydrogen together. (7222-3)

81 Insulated conductors and stools. (715-8)

82 Nobili astatic galvanometer. (721)

83 Variable condenser, signed: *Made by Yeates & Son Dublin.*

84 Sine and tangent galvanometers, signed: *Yeates & Son, c*1900. (843,4)

85 Clarke's magneto electric machine (806)

Miscellaneous

86 Spinner to demonstrate the difference between static and dynamical equilibrium where the position is governed by both gravitational and centrifugal forces. Mahogany, turned wooden wheel with hand knob. Brass spinning head at end of horizontal arm connected with handwheel by cords running over pulleys.

87 Tyndall's apparatus (machine for boiling water by friction pad rotating against cylinder), signed: *George Prescott & Co. 8 South King Street Dublin.* (714)

88 Demonstration water turbine, signed: *Yeates & Son Dublin.*

89 Small Brinell hardness tester, signed: *Yeates & Son Dublin.*

90 Linear dividing engine, with attached label: *5000-1 inch 196.8523-1mm* and *Yeates & Son Dublin.*

91 Episcope, signed: *Yeates & Son Dublin.*

92 Parts of seismograph used at Rathfarnham by Professor JP O'Reilly of the Royal College of Science, Dublin, *c*1880.

Bibliography

ALLODI (1962) — F Allodi, *I microscopi Culpeper di Norimberga*, Florence, 1962.

AMSTERDAM, RIJKSMUSEUM (1975) — Rijksmuseum Amsterdam, *Anamorfosen spel met Perspectief*, Amsterdam, 1975.

ANDERSON (1972) — RGW Anderson, *The Mariner's Astrolabe*, Royal Scottish Museum, Edinburgh, 1972.

ANDERSON (1978) — RGW Anderson, *The Playfair Collection and the Teaching of Chemistry at the University of Edinburgh 1713-1858*, Royal Scottish Museum, Edinburgh, 1978.

ANDERSON (1982) — RGW Anderson, *Science in India*, Science Museum, London, 1982.

ANDERSON (1983) — RGW Anderson, 'Early Islamic Chemical Glass', *Chemistry in Britain*, **19**, 822-3, 1983.

ANDERSON/ SIMPSON (1976) — RGW Anderson and ADC Simpson, *Edinburgh and Medicine: A Commemorative Catalogue*, Royal Scottish Museum, Edinburgh, 1976.

ANDRADE (1929) — EN da C Andrade, 'The Air Pump, Past and Present', *Proceedings of the Royal Institution of Great Britain*, **26**, 114-126, 1929.

ANDRADE (1957) — EN da C Andrade, 'The Early History of the Vacuum Pump', *Endeavour*, **16**, 29-35, 1957.

ANDREWS (1981-82) — JH Andrews, 'The Copying of Engineering Drawings and Documents', *Transactions of the Newcomen Society*, **53**, 1-15, 1981-82.

ANON (1800) — Anon, *Archaeologia*, **13**, 410 and plate XXVII, 1800.

ANON (1954) — Anon, 'Elliott Instruments in the Museum of the Cavendish Laboratory', *Elliott Journal*, **2**, 128-130, 1954.

ARMSTRONG (1840) — WG Armstrong, 'On the Electricity of Effluent Steam', *Philosophical Magazine* 3rd series, **17**, 452-457, 1840.

ARMSTRONG (1843) — WG Armstrong, 'An Account of a Hydro-electric Machine', *Philosophical Magazine* 3rd series 194-202, 1843.

ATKIN/ MARGESON (1985) — M Atkin and S Margeson, *Life on a Medieval Street: Excavations on Alms Lane Norwich, 1976*, Norwich Survey, Norwich, 1985.

BAILLIE (1939) — GH Baillie, *Guide to the Museum of the Worshipful Company of Clockmakers in London*, Worshipful Company of Clockmakers, London, 1939.

BALL (1940) — R Ball, 'Extracts from the Diary of Sir Robert Ball', *Observatory*, **63**, 197-206, 1940.

BANFIELD (1976) — E Banfield, *Antique Barometers*, Hereford, 1976.

BARCLAY (1937) — A Barclay, *Handbook of the Collection Illustrating Pure Chemistry Parts I and II*, Science Museum, London, 1937.

BARNARD (1916) — FP Barnard, *The Casting-Counter and the Counting-Board*, Oxford, 1916.

BARNES (1915-16) — CL Barnes, 'The Society's House', *Memoirs and Proceedings of the Manchester Literary and Philosophical Society*, **60**, ii-viii, 1915-16.

BARTY-KING (1986) — H Barty-King, *Eyes Right: the Story of Dollond and Aitchison*, London, 1986.

BAXANDALL (1975) — D Baxandall, *Calculating Machines and Instruments*, 2nd edn, revised by J Pugh, Science Museum, London, 1975.

BEDARIDA/ MADDISON (1970) — F Bedarida and F Maddison, *Cinq siecles de l'art francais des instruments scientifiques XVe-XIXe siecle*, Maison Francaise, Oxford, 1970.

BEESON (1962) — CFC Beeson, *Clock-making in Oxfordshire 1400-1850*, Monograph 2, Antiquarian Horological Society, 1962.

BEEVERS (1958) — SB Beevers, 'The John Gershom Partington Collection of Time Measurement Instruments', *The Connoisseur Year Book 1958*, London, 1958.

BEEVERS (1971) — SB Beevers, 'The John Gershom Partington Collection of Time Measurement Instruments', *The Connoisseur*, **176**, 93-105, 1971.

BELL (1981) — AS Bell, ed., *The Scottish Antiquarian Tradition*, Edinburgh, 1981.

BELLAIGUE (1974) — G de Bellaigue, *Furniture, Clocks and Gilt Bronzes*, Fribourg, 1974; in the series *The James A. de Rothschild Collection at Waddesdon Manor*, ed. A Blunt.

BENNETT (1983A) — JA Bennett, *Science at the Great Exhibition*, Whipple Museum of the History of Science, Cambridge, 1983.

BENNETT (1983B) — JA Bennett, *Whipple Museum of the History of Science, Catalogue 3: Astronomy and Navigation*, Cambridge, 1983.

BENNETT (1983)   JA Bennett, 'A Catalogue of Scientific Instruments at Birr Castle, Co. Offaby, Ireland', typescript at Whipple Museum of the History of Science, Cambridge, 1983.

BENNETT (1984A)   JA Bennett, *The Celebrated Phaenomena of Colours*, Cambridge, 1984.

BENNETT (1984B)   JA Bennett, *The Whipple Museum of the History of Science, Catalogue 5: Spectroscopes, Prisms and Gratings*, Cambridge, 1984.

BENNETT (1987)   JA Bennett, *The Divided Circle: a History of Instruments for Astronomy, Navigation and Surveying*, Oxford, 1987.

BENNETT (1990)   JA Bennett, *Church, State and Astronomy in Ireland: 200 Years of Armagh Observatory*, Armagh, 1990.

BENNETT/ HOSKIN (1981)   JA Bennett and M Hoskin, 'The Rosse Papers and Instruments', *Journal of the History of Astronomy*, **12**, 216-229, 1981.

BENYON (1958)   T Benyon, *Howell Harris, Reformer and Soldier 1714-73*, Caernarvon, 1958.

BERMAN (1977)   M Berman, *Social Change and Scientific Organisation, the Royal Institution 1799-1844*, London, 1977.

BION (1972)   N Bion, *The construction and principal uses of Mathematical instruments*, translated E. Stone, second English edition, London, 1758; facsimile edition London, 1972.

BIRMINGHAM (1936)   Assay Office, Birmingham, *The Assay Office at Birmingham*, Birmingham, 1936.

BIRMINGHAM (1940)   W & T Avery Ltd, *Avery Historical Museum*, London, 1940; typescript catalogue, copies held at the Museum and at the Science Museum Library, London.

BLACKMORE (1976)   HL Blackmore, *The Armouries of the Tower of London*, London, 1976.

BLACKWOOD (1970)   B Blackwood, 'The Origin and Development of the Pitt Rivers Museum', *Occasional Papers on Technology*, **77**, 7-16, 1970.

BLAIR (1964)   C Blair, 'A Royal Compass-dial', *The Connoisseur*, **157**, 246-8, 1964.

BLAIR (1974)   C Blair, *Arms, Armour and Base Metal Work*, Fribourg, 1974; in the series *The James A. de Rothschild Collection at Waddesdon Manor*, ed. A Blunt.

BOBINGER (1954)   M Bobinger, *Christop Schissler der Altere und der Jungere*, Schwabische Geschichtsquellen und Forschungen, Band 5, Augsburg, 1954.

BOBINGER (1966)   M Bobinger, *Alt-Augsburger Kompassmacher; Sonnen-, Mond- und Sternuhren, Astronomische und Mathematische Gerate Raderuhren*, Abhandlungen zur Geschichte der Stadt Augsburg Band 16, Augsburg, 1966.

BOBINGER (1969)   M Bobinger, *Kunstuhrmacher in alten Augsburg*, Abhandlungen zur Geschichte der Stadt Augsburg Band 18, Augsburg, 1966.

BOOTH (1964)   AD Booth, 'A Drum-type Microscope made by John Davis', *Journal of the Royal Microscopical Society* series III, **82**, 293-5, 1964.

BRACEGIRDLE (1977)   B Bracegirdle, 'J.J. Lister and the Establishment of Histology', *Medical History*, **21**, 187-191, 1977.

BRACEGIRDLE (1978)   B Bracegirdle, *A History of Microtechnique*, London, 1978.

BRACEGIRDLE (1981)   B Bracegirdle, *The Wellcome Museum of the History of Medicine: a part of the Science Museum*, Science Museum, London, 1981.

BRACEGIRDLE (1983)   B Bracegirdle, ed., *Beads of Glass: Leeuwenhoek and the Early Microscope*, Science Museum, London, 1983.

BRADBURY (1967)   S Bradbury, *The Evolution of the Microscope*, Oxford, 1967.

BRADBURY (1968)   S Bradbury, 'The Development of the Reflecting Microscope', *Microscopy, Journal of the Quekett Microscopical Club*, **31**, 1-19, 1968.

BRADBURY/ TURNER (1967)   S Bradbury and GL'E Turner, *Historical Aspects of Microscope*, Cambridge, 1967; originally published in *Proceedings of the Royal Microscopical Society*, **21**, 1-227, 1967.

BRADFORD (1963)   Bolling Hall Museum, Bradford, *Abraham Sharp, Mathematician and Astronomer 1653-1742*, Bradford, 1963.

BRAGG (1959)   L Bragg, 'Treasures in the Collections of the Royal Institution', *Proceedings of the Royal Institution*, **37**, 259-75, 1959.

BROADBENT (1949)   LH Broadbent, *The Avery Business 1730-1918*, Birmingham, 1949.

BROOK (1900-1)   AJS Brook, 'Notice of a Bracket Timepiece ... and also of Three Timepieces in the University Library St Andrews', *Proceedings of the Society of Antiquaries of Scotland*, **35**, 418-430, 1900-1.

BROWN (1970)   HM Brown, *Cornish Clocks and Clockmakers*, 2nd edn, Newton Abbot, 1970.

BROWN (1982A)   O Brown, *The Whipple Museum of the History of Science, Catalogue 1: Surveying*, Cambridge, 1982.

BROWN (1982B) O Brown, *The Whipple Museum of the History of Science, Catalogue 2: Balances and Weights*, Cambridge, 1982.

BROWN (1984) O Brown, *The Whipple Museum of the History of Science, Catalogue 4: Spheres, Globes and Orreries*, Cambridge, 1984.

BROWN (1986) O Brown, *The Whipple Museum of the History of Science, Catalogue 7: Microscopes*, Cambridge, 1986.

BROWNING J Browning, *How to Work with the Spectroscope*, London, 1878.

BRUCK (1972) HA Bruck, *The Royal Observatory Edinburgh 1822-1972*, Edinburgh, 1972.

BRUNT (1956) D Brunt, 'The Centenary of the Meteorological Office', *Science Progress*, **44**, 193-207, 1956.

BRYDEN (1968) DJ Bryden, *James Short and his Telescopes*, Royal Scottish Museum, Edinburgh, 1968.

BRYDEN (1970) DJ Bryden, 'The Jamaican Observatories of Colin Campbell FRS and Alexander Macfarlane FRS', *Notes and Records of the Royal Society of London*, **24**, 261-272, 1970.

BRYDEN (1972A) DJ Bryden, 'George Brown, Author of the Rotula', *Annals of Science*, **28**, 1-29, 1972.

BRYDEN (1972B) DJ Bryden, *Scottish Scientific Instrument Makers*, Royal Scottish Museum, Edinburgh, 1972.

BRYDEN (1972C) DJ Bryden, 'Three Edinburgh Microscope Makers', *The Book of the Old Edinburgh Club*, **33**, 165-176, 1972.

BRYDEN (1972D) DJ Bryden, 'Balthazar Knie, a Provincial Barometer Maker', *The Connoisseur*, **179**, 172-175, 1972.

BRYDEN (1973) DJ Bryden, 'A Didactic Introduction to Arithmetic, Sir Charles Cotterell's Instrument for Arithmeticke of 1667', *History of Education*, **2**, 5-18, 41-2, 1973.

BRYDEN (1975) DJ Bryden, 'The Whipple Museum, Cambridge', *The Connoisseur*, **188**, 166-123, 1975.

BRYDEN (1976) DJ Bryden, 'Scotland's Earliest Surviving Calculating Device: Robert Davenport's Circles of Proportion of c.1650', *Scottish Historical Review*, **55**, 54-60, 1976.

BRYDEN (1978) DJ Bryden, *Selected Exhibits in the Whipple Museum of History of Science*, Whipple Museum of the History of Science, Cambridge, 1978.

BRYDEN (1985) DJ Bryden, 'The Arithmetical Jewell or Jewell of Arithmetick', *Quarto*, **23**, 9-14, Abbot Hall Art Gallery, Kendal, 1985.

BRYDEN (1988) DJ Bryden, *The Whipple Museum of the History of Science, Catalogue 6: Sundials and related instruments*, Cambridge, 1988.

BRYDEN (1990) DJ Bryden, 'The Edinburgh Observatory 1736-1811: the Story of a Failure', *Annals of Science*, **47**, 445-74, 1990.

BUCHANAN (1982) PD Buchanan, 'Quantitative Measurement and the Design of the Chemical Balance 1750-c1900', unpublished PhD Thesis, University of London, 1982.

BURNETT (1982) J Burnett, 'The Giustiniani Medicine Chest', *Medical History*, **26**, 325-33, 1982.

CAJORI (1909) F Cajori, *A History of the Logarithmic Slide Rule*, New York, 1909.

CAJORI (1920) F Cajori, 'On the History of Gunter's Scale and the Slide Rule during the 17th Century', *University of California Publications in Mathematics*, **1**, 181-209, 1920.

CALLAN (1854) NJ Callan, 'On the Results of a Series of Experiments', *Philosophical Magazine* 4th series, **7**, 87-9, 1854.

CALLAN (1855) NJ Callan, 'On a New Single Fluid Galvanic Battery', *Philosophical Magazine* 4th series, **9**, 260-272, 1855.

CALLANDER (1910) JG Callander, 'Notices of (1) a Seventeenth-Century Sundial from Wigtownshire; ... ', *Proceedings of the Society of Antiquaries of Scotland*, **44**, 169-80, 1910.

CALTHORP JE Calthorp, 'James Ferguson the Astronomer and the Ferguson Relics', *Journal of Scientific Instruments*, **11**, 145-150, 1932.

CALVERT (1967) HR Calvert, *Scientific Trade Cards in the Science Museum Collection*, Science Museum, London, 1967.

CAMBRIDGE (1936) Cambridge Philosophical Society, *Catalogue of a Loan Exhibition of Historic Apparatus in Cambridge*, Cambridge, 1936.

CAMBRIDGE (1944) History of Science Lectures Committee, Cambridge, *An Exhibition of Historic Scientific Instruments and Books*, Cambridge, 1944.

CAMBRIDGE (1949) History of Science Lectures Committee, Cambridge, *A Guide to the Historic Scientific Instruments in the Whipple Museum of the History of Science*, Cambridge, 1949.

CAMBRIDGE (1966) Cavendish Laboratory, Cambridge, *Museum of the Cavendish Laboratory: An Outline Guide to Exhibits*, Cambridge, 1966.

CAMPBELL (1980) AM Campbell, *Catalogue of the Collection of Historical Scientific Instruments in the University of Strathclyde*, Glasgow, 1980.

CASEY (1965) — MT Casey, 'Nicholas Callan, Inventor of the Induction Coil', *School Science Review*, **160**, 557-562, 1965.

CHALDECOTT (1949) — JA Chaldecott, *Outline Guide to the King George III Collection*, Science Museum, London, 1949.

CHALDECOTT (1951) — JA Chaldecott, *Handbook of the King George III Collection of Scientific Instruments*, Science Museum, London, 1951.

CHALDECOTT (1954) — JA Chaldecott, *Heat and Cold: Part I Historical Review; Part II Descriptive Catalogue*, Science Museum, London, 1954.

CHALDECOTT (1969) — JA Chaldecott, 'Cromwell Mortimer FRS and the Invention of the Metalline Thermometer for Measuring High Temperatures', *Notes and Records of the Royal Society of London*, **24**, 113-135, 1969.

CHALDECOTT (1976) — JA Chaldecott, *Temperature Measurement and Control: Part I Historical Review; Part II Descriptive Catalogue*, Science Museum, London, 1976.

CHALDECOTT (1981) — JA Chaldecott, 'Wedgwood's Ceramic Wares for Commercial Use, Production and Supply from 1779 to 1794', *Ambix*, **28**, 184-205, 1981.

CHEW (1968) — VK Chew, *Physics for Princes*, Science Museum, London, 1968.

CHEW (1981) — VK Chew, *Talking Machines*, 2nd edn, Science Museum, London, 1981.

CHILD (1940) — E Child, *The Tools of the Chemist*, New York, 1940.

CITTERT (1947) — PH van Cittert, 'Proportionalpassers', *Netherlandsch tijdschrift voor natuurkunde*, **13**, 1-22, 1947.

CLARKE/MORRISON-LOW/SIMPSON (1989) — TN Clarke, AD Morrison-Low and ADC Simpson, *Brass and Glass: Scientific Instrument Making Workshops in Scotland*, Edinburgh, 1989.

CLAY (1951) — RS Clay, 'The Whipple Museum at Cambridge', *Journal of Scientific Instruments*, **28**, 286-7, 1951.

CLAY/COURT (1928) — RS Clay and TH Court, *The History of the Microscope*, London, 1928.

CLUTTON/DANIELS (1975) — C Clutton and G Daniels, *Clocks and Watches in the Collection of the Worshipful Company of Clockmakers*, London, 1975.

COCKER (1978) — W Cocker, 'A History of the University Chemistry Laboratory, Trinity College, Dublin: 1711-1946', *Hermathena*, **124**, 58-76, 1978.

COCKS (1980) — AS Cocks, *The Victoria and Albert Museum: The Making of the Collection*, London, 1980.

CONGREVE (1819) — W Congreve, *A description of sights, or instruments for pointing guns*, London, 1819.

CONNOR (1987) — RD Connor, *The Weights and Measures of England*, London, 1987.

COTTER (1968) — CH Cotter, *A History of Nautical Astronomy*, London, 1968.

COTTER (1983) — CH Cotter, *A History of the Navigator's Sextant*, Glasgow, 1983.

COUSINS (1969) — FW Cousins, *Sundials: a Simplified Approach by means of the Equatorial Dial*, London, 1969.

CRELLIN (1969) — JK Crellin, *Medical Ceramics*, Wellcome Institute of the History of Medicine, London, 1969.

CRELLIN (1972) — JK Crellin, 'Pharmaceutical History and its Sources in the Wellcome Collection, IV: Tiles, Pills and Boluses', *Medical History*, **16**, 81-5, 1972.

CRELLIN/HUTTON (1973) — JK Crellin and DA Hutton, 'Pharmaceutical History and its Sources in the Wellcome Collection, V: Communication and English Bell Metal Mortars, 1300-1850', *Medical History*, **17**, 266-287, 1973.

CRELLIN/SCOTT (1969) — JK Crellin and JR Scott, 'Pharmaceutical History and its Sources in the Wellcome Collection, II: Drug Weighing in Britain, c1700-1900', *Medical History*, **13**, 51-67, 1969.

CRELLIN/SCOTT (1970) — JK Crellin and JR Scott, 'Pharmaceutical History and its Sources in the Wellcome Collection, III: Fluid Medicines, Prescription Reform and Posology, 1700-1900', *Medical History*, **14**, 132-153, 1970.

CRELLIN/SCOTT (1972) — JK Crellin and JR Scott, *Glass and British Pharmacy; a survey and guide to the Wellcome Collection of British Glass*, Wellcome Institute for the History of Medicine, London, 1972.

CRICHTON (1803) — James Crichton, 'On the Freezing Point of Tin and the Boiling Point of Mercury; with a Description of a Self Registering Thermometer', *Philosophical Magazine*, **15**, 147-8, 1803.

CROM (1970) — TR Crom, *Horological Wheel Cutting Engines 1700-1900*, Gainsville, 1970.

CROSTHWAITE (1877-8) — JF Crosthwaite, 'Peter Crosthwaite, the Founder of Crosthwaite's Museum, Keswick', *Transactions of the Cumberland Association of Literature and Science*, Pt III, 151-164, 1877-8.

DANIELL (1951) — JA Daniell, 'The Making of Clocks and Watches in Leicestershire and Rutland', *Transactions of the Leicestershire Archaeological Society*, **27**, 1-36, 1951.

DANIELL (1975)   JA Daniell, *Leicester Clockmakers*, Leicestershire Museums, Art Galleries and Records Service, Leicester, 1975.

DAUMAS (1972)   M Daumas, *Scientific Instruments of the 17th & 18th C. and their Makers*; translation by M Holbrook of the Paris 1953 edition, London, 1972.

DEER (1977)   LA Deer, 'Italian Anatomical Waxes in the Wellcome Collection: The Missing Link', *Rivista di Storia delle Scienze Mediche e Naturali*, **20**, 281-98, 1977.

DERBY (n.d.)   Museum and Art Gallery Guide, Derby, n.d.

DESTOMBES (1969)   M Destombes, 'Deux astrolabes nautiques inedit de J.et A. de Goes, Lisbonne', *Agrupamento de Estudos de Cartographia Antiga, Seccao de Coimbra, serie separatas*, **22**, 1969.

DIBNER (1957)   B Dibner, *Early electrical machines*, Norwalk, 1957.

DICKINSON (1956)   HW Dickinson, 'A Brief History of Draughtsmen's Instruments', *Transaction of the Newcomen Society*, **27**, 73-84, 1956.

DICKINSON/ JENKINS (1927)   HW Dickinson and R Jenkins, *James Watt and the Steam Engine*, London, 1927.

DIGBY (1973)   S Digby, 'The Bhugola of Ksema Karna: a Dated Piece of Sixteenth Century Metalware', *Art and Archaeology Research Papers*, **4**, 10-13, 1973.

DISNEY/HILL/ WATSON-BAKER (1928)   AR Disney, CF Hill and WE Watson-Baker, eds, *The origin and development of the Microscope as illustrated by a catalogue of the instruments and accessories in the collections of the Royal Microscopical Society*, Royal Microscopical Society, London, 1928.

DRAKE (1977)   S Drake, 'Tartaglia's Squadra and Galileo's Compasso', *Annali dell' Istituto e Museo di Storia della Scienza di Firenze*, **2**, 35-44, 1977.

DRAKE (1978)   S Drake, *Galileo Galilei, Operations of the Geometric and Military Compass, 1606*, National Museum of History and Technology, Washington DC, 1978.

DREYER (1883)   JLE Dreyer, *An historical account of the Armagh Observatory*, Armagh, 1883.

DUBLIN – DUBLIN ROYAL SOCIETY (1931)   Dublin Royal Society, *Royal Dublin Society Bicentenary Celebration: Official Handbook and Catalogue of the Museum*, Dublin, 1931.

ECKHARDT (1976)   W Eckhardt, 'Erasmus Habermal – zur Biographie des Instrumentenmachers Kaiser Rudolphs', two papers reprinted from *Jahrbuch der Hamburger Kunstsammlungen*, **21**, 55-92, 1976.

ECKHARDT (1977)   W Eckhardt, 'Erasmus und Jusuo Habermal – Kunstgeschichtliche Ammerkungen zu den Werken der beiden Instrumentenmacher', *Jahrbuch der Hamburger Kunstsammlungen*, **22**, 13-74, 1977.

EDINBURGH (1954)   Edinburgh, The Royal Scottish Museum, *The Royal Scottish Museum 1854-1954*, The Royal Scottish Museum, Edinburgh, 1954.

EDINBURGH (1959)   Edinburgh, The Royal Scottish Museum, *The St Andrews University Astrolabes*, The Royal Scottish Museum, Edinburgh, 1959.

EGGESTORFF (1984)   P Eggestorff, *The Eggestorff Collection: An Abridged Catalogue*, Dublin, 1984.

ERSKINE (1787)   DS Erskine (Earl of Buchan), *An Account of the Life, Writings and Inventions of John Napier of Merchiston*, p. 41, Perth, 1787.

EVANS (1956)   J Evans, *A history of the Society of Antiquaries*, Oxford, 1956.

FALCONER (1980)   IJ Falconer, *Apparatus from the Cavendish Museum, Cambridge*, University of Cambridge, Department of Physics, Cambridge, 1980.

FARADAY (1859)   M Faraday, *Experimental Researches in Electricity II*, London, 1859.

FARRAR (1963)   K Farrar, 'A Note on a Eudiometer Supposed to have Belonged to Henry Cavendish', *British Journal for the History of Science*, **1**, 375-380, 1963.

FARRAR (1968)   K Farrar, 'Dalton's Scientific Apparatus' in DSL Cardwell, ed., *John Dalton and the Progress of Science*, pp. 159-186, Manchester, 1968.

FELDHAUS (1959)   FM Feldhaus, *Geschichte des Technischen Zeichnens*, 2nd edn, revised by E Schruff, Wilhelmshaven, 1959.

FERGUSON (1767)   J Ferguson, *Tables and Tracts*, London, 1767.

FERGUSON (1920)   W B Ferguson, ed., *The Photographic Researches of Ferdinand Hurter and Vero C Driffield*, Royal Photographic Society, London, 1920.

FIELD/WRIGHT   JV Field and MT Wright, *Early Gearing: Geared Mechanisms in the Ancient and Medieval World*, Science Museum, London, 1985.

FIELD/WRIGHT (1985B) JV Field and MT Wright with DR Hill, *Byzantine and Arabic Mathematical Gearing*, Science Museum, London, 1985; reprinted from *Annals of Science*, **42**, 87-163, 1985.

FINDLAY (1935) A Findlay, *The Teaching of Chemistry in the University of Aberdeen*, Aberdeen, 1935.

FOLLETT (1978) DH Follett, *The Rise of the Science Museum under Henry Lyons*, Science Museum, London, 1978.

FORBES (1916) G Forbes, *David Gill, Man and Astronomer*, London, 1916.

GALWAY (1902) Queen's College, Galway, *Catalogue of Apparatus in the Physics Department of Queen's College, Galway*, Dublin, 1902.

GARVAN (1964) ANB Garvan, 'Slide Rule and Sector: a Study in Technology and Society', *Proceedings of the 10th International Congress of the History of Science, Ithaca, 1962*, **2**, 397-400, Paris, 1964.

GAVINE (n.d.) D Gavine, 'Notes on the History of Astronomy' unpublished MS, Department of Natural Philosophy, University of Aberdeen, 197-.

GAVINE (n.d.) D Gavine, 'The Astronomical Observatories of the Aberdeen Universities' unpublished MS, Department of Natural Philosophy, University of Aberdeen, 197-.

GEDDES (1955) AEM Geddes, 'The Development of the Study and Practice of Astronomy at Aberdeen', *Weather*, **10**, 385, 1955.

GERNSHEIM (1969) H and A Gernsheim, *The History of Photography*, London, 1969.

GIBBS/ HENDERSON/ PRICE (1973) SL Gibbs, JA Henderson and DJ de S Price, *A Computerized Checklist of Astrolabes*, New Haven, 1973.

GILBERT (1965) KR Gilbert, *The Portsmouth Block-Making Machinery*, Science Museum, London, 1965.

GILBERT (1966) KR Gilbert, *The Machine Tool Collection*, Science Museum, London, 1966.

GILBERT (1975) KR Gilbert, *Early Machine Tools*, Science Museum, London, 1975.

GILL (1913) D Gill, *A History of the Royal Observatory, Cape of Good Hope*, London, 1913.

GLASGOW (1967) Glasgow Museum and Art Galleries, *Fulton's Orrery*, Glasgow, 1967.

GLASGOW (1980) Collins Exhibition Hall, University of Strathclyde, *Catalogue of the Collection of Historical Scientific Instruments in the University of Strathclyde*, Glasgow, 1980.

GOODISON (1977) N Goodison, *English Barometers 1680-1860: a history of domestic barometers and their makers*, 2nd revised edn, Woodbridge, 1977.

GOUK (1988) P Gouk, *The Ivory Sundials of Nuremberg 1500-1700*, Cambridge, 1988.

GOULD (1935) RT Gould, 'John Harrison and his Timekeepers', *Mariner's Mirror*, **21**, 115-239, 1935.

GREEN/LLOYD (1970) G Green and JT Lloyd, *Kelvin's instruments and the Kelvin Museum*, University of Glasgow, Glasgow, 1970.

GREENAWAY (1951) F Greenaway, *A Short History of the Science Museum*, Science Museum, London, 1951.

GREW (1681) N Grew, *Museum Regalis Societatis, or a catalogue and description of the natural and artificial rarities belonging to the Royal Society and preserved at Gresham College*, London, 1681.

GUNTHER (1967) AE Gunther, 'Robert T. Gunther: a Pioneer in the History of Science, 1869-1940', *Early Science in Oxford*, Vol. XV, ed. RWT Gunther, Oxford, 1967.

GUNTHER (1975) AE Gunther, *A Century of Zoology at the British Museum through the Lives of Two Keepers, 1815-1914*, Folkestone, 1975.

GUNTHER (1980) AE Gunther, *The Founders of Science at the British Museum 1753-1900*, Halesworth, Suffolk, 1980.

GUNTHER, (1920-23) RWT Gunther, *Early Science in Oxford Vol. 1: Part 1, Chemistry; Part 2, Mathematics; Parts 3 & 4, Physics and Surveying; Vol. 2: Astronomy*, Oxford, 1920-23.

GUNTHER (1927) RWT Gunther, 'The Great Astrolabe and other Scientific Instruments of Humphrey Cole', *Archaeologia*, **76**, 273-317, 1927.

GUNTHER (1932) RWT Gunther, *The Astrolabes of the World*, Oxford, 1932.

GUNTHER (1935) RWT Gunther, *Handbook of the Museum of the History of Science in the Old Ashmolean Building*, Museum of the History of Science, Oxford, 1935.

GUNTHER (1936) RWT Gunther, 'The Newly Found Astrolabe of Queen Elizabeth', *Illustrated London News*, **189**, 738-9, 1936.

GUNTHER (1937A) RWT Gunther, *Early Science in Cambridge*, Oxford, 1937.

GUNTHER (1937B) RWT Gunther, 'The First Observatory Instruments of the Savilian Professors at Oxford', *The Observatory*, **60**, 190-197, 1937.

GUNTHER (1939) RWT Gunther, *The Old Ashmolean Building and its Historical Scientific Instruments. Reports of the Committee of Management of the Lewis Evans Collection, afterwards the Museum of the History of Science*, Museum of History of Science, Oxford, 1939.

HACKMANN (1973) WD Hackmann, 'John and Jonathan Cuthbertson, the Invention and Development of the 18th Century Plate Electrical Machine', Communication 142, Rijksmuseum voor de Geschiedenis der Natuurwetenschappen, Leyden, 1973.

HACKMANN (1978A) WD Hackmann, *Electricity from glass: the history of the frictional electrical machine 1600-1850*, Alphen aan den Rijn, 1978.

HACKMANN (1978B) WD Hackmann, 'Eighteenth Century Electrostatic Measuring Devices', *Annali dell' Instituto e Museo di Storia della Scienza di Firenze*, **3**, 3-58, 1978.

HALL (1951) AR Hall, 'Whipple Museum of the History of Science, Cambridge', *Nature*, **167**, 878-9, 1951.

HALL (1970) I Hall, *William Constable as Patron*, Ferens Art Gallery, Hull, 1970.

HALL (1982) I Hall, 'Range of a Dilettante, William Constable and Burton Constable', *Country Life*, 22 April, 1982.

HAMBLY (1982) M Hambly, *Drawing Instruments, their History, Purpose and Use for Architectural Drawing*, Royal Institute of British Architects, London, 1982.

HAMER (1924) SH Hamer, 'Long Case Clocks and some Notable Clock Makers', Reports 45-46, Papers of the Halifax Antiquarian Society, Halifax, 1924.

HAMMOND (1981) JH Hammond, *The Camera Obscura: A Chronicle*, London, 1981.

HAMMOND/ AUSTIN (1987) JH Hammond and J Austin, *The Camera Lucida in Art and Science*, Bristol, 1987.

HARRIS (1827) WS Harris, 'On ... Metallic Substances as Conductors of Electricity', *Philosophical Transactions*, **117**, 18-24, 1827.

HARRISON (1912) WA Harrison, 'The R.E. Museum', *The Royal Engineers Journal*, **15**, 345-54, 1912.

HARTNER (1932) W Hartner, 'The Principle and Use of the Astrolabe' in AU Pope, ed., *A Survey of Persian Art from Prehistoric Times to the Present*, **6**, 1397-1402, 1932.

HATTON/ FLOWETT (1963-4) AP Hatton and JW Flowett, 'Clegg's Model of a Watt Beam Engine', *Memoirs and Proceedings of the Manchester Literary and Philosophical Society*, **106**, 104-7, 1963-4.

HAYWARD (1950) JF Hayward, 'The Celestial Globes of Georg Roll and Johannes Reinhold, *The Connoisseur*, **125**, 220 167-72, 1950.

HAYWARD (1969) JF Hayward, *English Watches*, 2nd edn, Victoria and Albert Museums, London, 1969.

HEATHCOTE (1965) NH de V Heathcote, 'N.J. Callan, Inventor of the Induction Coil', *Annals of Science*, **21**, 145-167, 1965.

HEILBRON (1979) JL Heilbron, *Electricity in the 17th and 18th Centuries*, Berkeley, 1979.

HEMMING (1929) AG Hemming, 'Dated English Bell-metal Mortars', *Connoisseur*, **83**, 158-166, 1929.

HENDERSON (1867) E Henderson, *Life of James Ferguson FRS*, Edinburgh, 1867.

HENRION (1618) D Henrion, *Usage du Compass de Proportion*, Paris, 1618.

HERBERT (1967) AP Herbert, *Sundials Old and New*, London, 1967.

HEYWOOD (1954) H Heywood, 'A Brief Description of the 18th C. Scientific Instruments owned by the Most Honourable, the Marquis of Bath', *Bulletin of the British Society for the History of Science*, **1**, 246, 1954.

HEYWOOD (n.d.) H Heywood, 'An Illustrated Catalogue of Scientific Instruments owned by the Most Honourable, the Marquis of Bath', unpublished MS, copies at Longleat and the Museum of History of Science, Oxford, n.d.

HIGGINS (1953) K Higgins, 'The Classification of Sundials', *Annals of Science*, **9**, 342-358, 1953.

HILL (1971) CR Hill, *Catalogue 1: Chemical Apparatus*, Museum of the History of Science, Oxford, 1971.

HILL/DREY (1980) CR Hill and REA Drey, *Catalogue 3: Drug Jars*, Museum of the History of Science, Oxford, 1980.

HILL/PAGET-TOMLINSON (1958) HO Hill and EW Paget-Tomlinson, *Instruments of Navigation: A Catalogue of Instruments at the National Maritime Museum with Notes upon their Use*, National Maritime Museum, London, 1958.

HOLBROOK (1974) M Holbrook, *A Girdle about the Earth: Astronomical and Geographical Discovery, 1490-1630*, Holburne of Menstrie Museum, Bath, 1974.

HOPKINSON (1980) T Hopkinson, *Treasures of the Royal Photographic Society 1839-1919*, London, 1980.

HOPPEN (1976)   KT Hoppen, 'The Early Royal Society', *British Journal for the History of Science*, **9**, 1-24 and 243-273, 1976.

HOWSE (1975)   D Howse, *Greenwich Observatory 3: The Buildings and Instruments*, London, 1975.

HOWSE (1968-70)   D Howse, 'Captain Cook's Marine Timekeepers: The Arnold Chronometers', *Antiquarian Horology*, **6**, 276-280, 1968-70.

HUGGINS (1906)   W Huggins, *The Royal Society, or Science in the State and in the Schools*, **21**, 1218-131, London, 1906.

INSLEY (1982)   J Insley, 'Court, Crisp and Clay – Some Notes on Collectors and Collections of Antique Microscopes', *Microscopy*, **34**, 345-353, 376, 1982.

JAY (1980)   JP Jay, 'The Museum of Meteorological Instruments', *Meteorological Magazine*, **109**, 246-248, 1980.

JENEMANN (1977)   HR Jenemann, 'Eine kurze Entwicklungsgeschichte der Wissenschaflichen Waage', *Festschrift zum 125 Jahrigen Jubilaum der Firma Gebr. Bosch*, Jungingen, 1977.

JENEMANN (1979)   HR Jenemann, *Die Waage des Chemikers*, Frankfurt, 1979.

JENKIN (1925)   CF Jenkin, *The Astrolabe, its Construction and Use*, Oxford, 1925.

JONES (1871)   HB Jones, *The Royal Institution: its Founders and its First Professors*, London, 1871.

JONES (1904)   F Jones, 'The Collection of Apparatus made by Dalton, now in the Possession of the Society', *Memoirs of the Manchester Literary and Philosophical Society*, **48**, 1-5, 1904.

JONES (1980)   D Jones, *Toy with the Idea*, Norfolk Museums Service, Norwich, 1980.

JONES (1990)   M Jones, *Fake? The Art of Deception*, London, 1990.

JOSTEN (1955)   CH Josten, *A catalogue of scientific instruments from the 13th to the 19th centuries from the collection of J.A. Billmeir CBE, exhibited by the Museum of the History of Science, Oxford*, 2nd edn, London, 1955.

JOULE (1884-87)   JP Joule, 'The Scientific Papers of James Prescott Joule', *Scientific Papers, Vol. 1, 1840*, pp. 27-42, G3.62, London, 1884-87.

KAESTLIN (1970)   JP Kaestlin, *Catalogue of the Museum of Artillery in the Rotunda at Woolwich: Part 1, Ordnance*, 2nd revised edn, Museum of Art, Woolwich, 1970.

KENDAL (1973)   Kendal, *Borough Museum Handbook*, Kendal, 1973.

KIELY (1947)   ER Kiely, *Surveying Instruments: their History and Classroom Use*, 19th Year Book, United States National Council of Teachers of Mathematics, New York, 1947.

KING (1955)   HC King, *The History of the Telescope*, London, 1955.

KING (1973)   R King, *Michael Faraday of the Royal Institution*, Royal Institution of Great Britain, London, 1973.

KING/ MILLBURN (1978)   HC King and JR Millburn, *Geared to the Stars: the Evolution of Planetariums, Orreries and Astronomical Clocks*, Toronto, 1978.

KIRBY (1761)   J Kirby, *The Perspective of Architecture*, London, 1761.

KNIGHT (1990)   R A Knight, 'Carpenter's Rule from the Mary Rose', *Tools and Trades*, **6**, 43-55, 1990.

KNOBEL (1909)   EB Knobel, 'On a Chinese Planisphere', *Monthly Notices of the Royal Astronomical Society*, **69**, 436, 1909.

KUHNELT (1962)   H Kuhnelt, 'Alte Tiroler Bergbauinstaumerle', *Tiroler Heimat*, **26**, 113-120, 1962.

KYLE (1970)   D Kyle, 'The Story of Two Stethoscopes', *Journal of the Royal College of General Practitioners*, **20**, 302-6, 1970.

LANCASTER-JONES (1925)   E Lancaster-Jones, *Catalogue of the Collection in the Science Museum, South Kensington: Geodesy and Surveying*, Science Museum, London, 1925.

LAVEN/van CITTERT-EYMERS (1967)   WS Laven and JG van Cittert-Eymers, *Electrostatic Instruments in the Utrecht University Museum*, Utrecht, 1967.

LAW (1969)   RJ Law, *James Watt and the Separate Condenser*, Science Museum, London, 1969.

LAWRENCE (1978)   CJ Lawrence, 'The Marey Sphygmograph', *Medical History*, **22**, 196-200, 1978.

LAWRENCE (1979A)   CJ Lawrence, 'The Dudgeon Sphygmograph and its Descendants', *Medical History*, **23**, 96-101, 1979.

LAWRENCE (1979B)   CJ Lawrence, 'Early Sphygmomanometers', *Medical History*, **23**, 474-478, 1979.

LECKY (1875-6)   RJ Lecky, 'Report on the Condition of some of the Instruments Belonging to the Royal Astronomical Society', *Monthly Notices of the Royal Astronomical Society*, **36**, 126 and 132, 1875-6.

LEEDS (1974) — Leeds City Museum, *The Bicentenary of the Discovery of Oxygen by Joseph Priestley*, Leeds Leisure Services Department, 1974.

LEYBOURN (1672) — W Leybourn, *Panorganon, or a Universal Instrument*, London, 1672.

LINDSAY (1969) — EM Lindsay, 'The Astronomical Instruments of H.M. King George III presented to Armagh Observatory', *Irish Astronomical Journal*, **9**, 57-68, 1969.

LIVEING/DEWAR (1879) — GD Liveing and J Dewar, 'Note on a Direct Vision Spectroscope', *Proceedings of the Royal Society of London*, **28**, 242, 1879.

LLOYD (1944) — HA Lloyd, 'Five Hundred Years of Precision Timekeeping', *Horological Journal*, **86**, 338-355, 1944.

LLOYD (1958) — HA Lloyd, *Some Outstanding Clocks over Seven Hundred Years*, London, 1958.

LLOYD (1969) — JT Lloyd, 'Item ane Shipe Skin: An Account of Early Experimentation in the Natural Philosophy Department', *The College Courant: Journal of the Glasgow University Graduates Association*, **21**, no. 43 5-9, 1969.

LOCKYER/FRANKLAND (1869, 1870) — JN Lockyer and E Frankland, 'Researches on Gaseous Spectra in Relation to the Physical Constitution of the Sun', *Proceedings of the Royal Society of London*, **17**, 288-91 453-4, 1869; **18**, 79-80, 1870.

LONDON (1876) — London – Science and Art Department of the Committee of Council on Education, *Catalogue of the Special Loan Collection of Scientific Apparatus at the South Kensington Museum*, London, 1876.

LONDON (1899) — London – Royal Institution of Great Britain, *The Spottiswoode Collection of Physical Apparatus*, London, 1899.

LONDON (1900) — London – Board of Education, *Catalogue of the Science Collections for teaching and research in the Victoria and Albert Museum; Physiography, Part II, Meteorology including Terrestrial Magnetism*, London, 1900.

LONDON (1905) — London – Board of Education, *Catalogue of the Science Collections for teaching and research in the Victoria and Albert Museum; Part II, Physics*, 2nd edn, London, 1905.

LONDON (1920) — London – Hertford House, *Illustrated Catalogue of Furniture, Marbles ... and objects of art ... in the Wallace Collection*, 6th edn, Wallace Collection, London, 1920.

LONDON (1922) — London – Board of Education, *Catalogue of the Collections in the Science Museum ... Meteorology*, Science Museum, London, 1922.

LONDON (1949) — London – Worshipful Company of Clockmakers, *Catalogue of the Museum of the Worshipful Company of Clockmakers of London*, 3rd edn, London, 1949.

LONDON (1957) — London – Science Museum, *The Science Museum: The First Hundred Years*, London, 1957.

LONDON (1966) — London – Wellcome Historical Medical Museum and Library, *Chinese Medicine: An Exhibition Illustrating the Traditional System of Medicine of the Chinese People*, London, 1966.

LONDON (1970) — London – London Museum, *Glass in London*, London, 1970.

LONDON (1971) — London – National Maritime Museum, *An Inventory of the Navigation and Astronomy Collections in the National Maritime Museum Greenwich*, London, 1971.

LONDON (1975) — London, *Survey of London, Vol. XXVIII: The Museums Area of South Kensington and Westminster*, London 1975.

LONDON (1976A) — London – British Museum, *British Museum Guide*, London, 1976.

LONDON (1976B) — London – National Trust, *The Fox Talbot Museum, Lacock*, London, 1976.

LONDON (1982) — London, National Trust, *Petworth House*, London, 1982.

LONDON (1985) — London, The Iveagh Bequest, *John Joseph Merlin, the Ingenious Mechanic*, London, 1985.

LOWERY (1930) — H Lowery, 'The Joule Collection in the College of Technology, Manchester', *Journal of Scientific Instruments*, **7**, 375, 1930.

MADDISON (1957) — FR Maddison, *A supplement to a catalogue of scientific instruments in the collection of J.A. Billmeir Esq CBE, exhibited by the Museum of the History of Science, Oxford*, London, 1957.

MADDISON (1958) — FR Maddison, 'An Eighteenth-Century Orrery by Thomas Heath, and some Earlier Orreries', *The Connoisseur*, **11**, 163-7, 1958.

MADDISON (1962) — FR Maddison, 'A 15th Century Islamic Spherical Astrolabe', *Physis*, **IV**, 101-9, 1962.

MADDISON (1963) — FR Maddison, 'Early Astronomical and Mathematical Instruments: a Brief Survey of Sources and Modern Studies', *History of Science*, **2**, 17-50, 1963.

MADDISON (1966) FR Maddison, 'Hugo Hett and the Roias Astrolabe Projection', *Agrupamento de Estudos de Cartografia Antiga, (Seccao de Coimbra)*, **12**, 1966.

MADDISON (1969) FR Maddison, 'Medieval Scientific Instruments and the Development of Navigational Instruments in the XVth and XVIth Centuries', *Agrupamento de Estudos de Cartografia Antiga, (Seccao de Coimbra)*, **30**, serie separatas, 1969.

MADDISON/ TURNER (1973) FR Maddison and AJ Turner, *Catalogue 2: Watches*, Museum of the History of Science, Oxford, 1973.

MANN (1962) J Mann, *Wallace Collection Catalogues: European Arms and Armour I and II*, Wallace Collection, London, 1962.

MARTIN (1961) T Martin, *The Royal Institution*, 3rd edn, Royal Institution of Great Britain, London, 1961.

MARTIN (1932-36) T Martin, ed., *Faraday's Diary I-VIII*, London, 1932-36.

MAY (1973) WE May, *A History of Marine Navigation*, Henley on Thames, 1973.

MAYER (1956) LA Mayer, *Islamic Astrolabists and their Work*, Geneva, 1956.

MAYNOOTH (1955) Maynooth – St Patrick's College Museum, *Third Tostal Display, Souvenir Catalogue*, Maynooth, 1955.

McCANN (1983) A McCann, 'A Private Laboratory at Petworth House, Sussex, in the Late Eighteenth Century', *Annals of Science*, **40**, 635-655, 1983.

McCONNELL (1980) A McConnell, *Geomagnetic Instruments before 1900*, London, 1980.

McCONNELL (1981) A McConnell, *Historical Instruments in Oceanography*, Science Museum, London, 1981.

McCONNELL (1983) A McConnell, *No Sea Too Deep: the History of Oceanographic Instruments*, Bristol, 1983.

McCONNELL (1986A) A McConnell, *Geophysics and Geomagnetism: Catalogue of the Science Museum Collection*, London, 1986.

McCONNELL (1986B) A McConnell, 'The Scientific Life of William Scoresby Jnr with a Catalogue of his Instruments in the Whitby Museum', *Annals of Science*, **43**, 257-286, 1986.

McDONNELL (1963) J McDonnell, *A History of Helmsley, Rievaulx and District*, York, 1963.

MacGREGOR (1983) A MacGregor, ed., *Tradescant's Rarities: Essays on the Foundation of the Ashmolean Museum 1683, with a Catalogue of the Surviving Early Collections*, Oxford, 1983.

McKIE (1965-6) JN McKie, 'Gideon Scott 1765-1844', *Journal of the British Astronomical Association*, **76**, 53-8, 1965-6.

McLAUGHLIN (1964) PJ McLaughlin, 'Some Irish Contemporaries of Faraday and Henry', *Proceedings of the Royal Irish Academy*, **64**, 24-35, 1964.

McLAUGHLIN (1965) PJ McLaughlin, *Nicholas Callan, Priest Scientist*, Dublin, 1965.

MEEHAN/ CLARKE (1981) J Meehan and D Clarke, *The Royal Dublin Society 1731-1981*, Dublin, 1981.

MENZIES-CAMPBELL (1966) J Menzies-Campbell, *Catalogue of the Menzies-Campbell Collection*, Royal College of Surgeons, Edinburgh, 1966.

MEYRICK (1979) R Meyrick, *The John Gershom Parkington Memorial Collection of Time Measurement Instruments*, Bury St Edmunds, 1979.

MICHEL (1956) H Michel, 'Boussoles de Mines des XVIe et XVIIe siecles', *Ciel et Terre*, 72nd year, nos 11-12, 1-15, 1956.

MICHEL (1966) H Michel, *Instruments des sciences dans l'art et l'histoire*, Rhode-Saint-Genese, 1966.

MICHEL (1967) H Michel, *Traite de l'astrolabe*, Paris, 1947; reprinted with addenda and corrigenda, Paris, 1967.

MIDDLETON (1964) WEK Middleton, *The History of the Barometer*, Baltimore, 1964.

MIDDLETON (1966) WEK Middleton, *A History of the Thermometer and its Use in Meteorology*, Baltimore, 1966.

MIDDLETON (1969) WEK Middleton, *The Invention of the Meteorological Instruments*, Baltimore, 1969.

MILLBURN (1972-73) JR Millburn, 'Benjamin Martin and the Development of the Orrery', *British Journal for the History of Science*, **6**, 378-399, 1972-73.

MILLER (1974) EJ Miller, *That Noble Cabinet*, London, 1974.

MILLS/TURVEY (1979) AA Mills and PJ Turvey, 'Newton's Telescope: An Examination of the Reflection Telescope Attributed to Sir Isaac Newton in the Possession of the Royal Society', *Notes and Records of the Royal Society*, **33**, 133-155, 1978.

MITCHELL (1975) AA Mitchell, 'The BOA Museum', *The Optician*, **170**, no 4397 4-7; no 4399 33-4; no 4401 33-5, 1975.

MITFORD (1951) RLSB Mitford, *The Society of Antiquaries of London: Notes on its History and Possessions*, London, 1951.

MOORE (1967) PA Moore, *Armagh Observatory 1790-1967*, Armagh Observatory, Armagh, 1967.

MOORE (1971) P Moore, *The Astronomy of Birr Castle*, London, 1971.

MOOREHOUSE (1972) S Moorehouse, 'Medieval Distilling-Apparatus of Glass and Pottery', *Medieval Archaeology*, **16**, 79-121, 1972.

MORRISON-LOW (1984) AD Morrison-Low, 'Scientific Apparatus Associated with Sir David Brewster: An Illustrated Catalogue', *'Martyr of Science': Sir David Brewster 1781-1868*, eds AD Morrison-Low and JRR Christie, pp. 82-103, Edinburgh, 1984.

MUIR (1950) J Muir, *John Anderson, Pioneer of Technical Education and the College he Founded*, Glasgow, 1950.

NATIONAL PHYSICAL LABORATORY (1967) National Physical Laboratory, *NPL Historical Exhibition of Standards and Measuring Equipment, Catalogue of Exhibits*, Teddington, 1967.

NATIONAL PHYSICAL LAB (1983) National Physical Laboratory, *Brief Guide to the National Physical Laboratory Museum*, London, 1983.

NEDOLUHA (1957-9) A Nedoluha, 'Kulturgeschichte des Technischen Zeichnens', *Blatter fur Technikgeschichte*, **19-21**, 1957-9.

NEEDHAM (1959) J Needham, *Science and Civilization in China, Vol. 3*, Cambridge, 1959.

NEEDHAM/LU (1966) J Needham and G-D Lu, 'A Korean Astronomical Screen of the mid-18th C from the Royal Palace of the Yi Dynasty', *Physis*, **8**, 137-162, 1966.

NEWCASTLE-UPON-TYNE (1950) Municipal Museum of Science and Industry, Newcastle-upon-Tyne, *Catalogue of the Museum*, Newcastle-upon-Tyne, 1950.

NEWSTEAD (1931) R Newstead, 'Excavations on the Site of the New Telephone Exchange, St John St, Chester', *Journal of the Chester and North Wales Architectural, Archaeological and Historical Society*, **33**, new series, 9-31, 1931.

NOLTE (1922) F Nolte, 'Die Armillarsphare', *Abhandlungen zur Gerschichte der Naturwisserschafter und der Medezin* 2, Erlangen, 1922.

NORTH (1969) JD North, 'A Post-Copernican Equatorium', *Physis*, **11**, 418-457, 1969.

NORTH (1974) JD North, 'The Astrolabe', *Scientific American*, **230**, no. 1, 96-106, 1974.

NUTTALL (1973A) RH Nuttall, *The Arthur Frank Loan Collection of Early Scientific Instruments*, Glasgow, 1973.

NUTTALL (1973B) RH Nuttall, 'C.R. Goring, J.J. Lister and the Achromatic Microscope', *Microscopy*, **32**, 253-261, 1973.

NUTTALL (1979) RH Nuttall, *Microscopes from the Frank Collection 1800-1860*, Jersey, 1979.

OWEN (1969) J Owen, 'Catalogue of Items in the Museum', unpublished MS, Department of Physics, University College of Aberystwyth, 1969.

OXFORD (1969) Oxford – Museum of the History of Science, *A Brief Guide to the Museum of the History of Science, Oxford*, Oxford, 1969.

PALMER/SAHIAR (1971) FW Palmer and AB Sahiar, *Microscopes to the End of the 19th C*, Science Museum, London, 1971.

PARSONS (1862) W Parsons, 3rd Earl of Rosse, 'On the Construction of Specula of Six-Feet Aperture', *Philosophical Transactions*, **151**, 681-707, 1862.

PARSONS (1880-81) L Parsons, 4th Earl of Rosse, 'Observations of Nebulae and Clusters of Stars Made with the Six-Foot and Three-Foot Reflectors at Birr Castle, from the Year 1848 up to the Year 1878', *Scientific Transactions of the Royal Dublin Society*, **2**, 1-178, 1880-81.

PARSONS (1885) L Parsons, 4th Earl of Rosse, 'On an Electric Control for an Equatorial Clock-Movement', *Report of the Fifty-Fourth Meeting of the British Association for the Advancement of Science 1884*, London, 1885.

PARSONS (1926) C Parsons, ed., *The Scientific Papers of William Parsons, third Earl of Rosse, 1800-1867*, London, 1926.

PEATE (1975) IC Peate, *Clock and Watch Makers in Wales*, 3rd revised edn, National Museum of Wales, Welsh Folk Museum, Cardiff, 1975.

PEDERSEN/PHIL (1974) O Pedersen and M Phil, *Early Physics and Astronomy*, London, 1974.

PINDER-WILSON (1975) R Pinder-Wilson, 'The Malcolm Celestial Globe', *The Classical Tradition: British Museum Yearbook I*, pp. 83-101, London, 1975.

PINTO (1969) EH Pinto, *Treen and Other Wooden Bygones*, London, 1969.

PLENDERLEITH (1959) RW Plenderleith, 'An Old Scottish Yard and Ell Measure', *Scottish Studies*, **3**, 105-6, 1959.

PLENDERLEITH (1960) RW Plenderleith, 'Discovery of an Old Astrolabe', *Scottish Geographical Magazine*, **76**, 25, 1960.

PLYMOUTH (n.d.) City Art Gallery, Plymouth, *Cottonian Collection Catalogue*, Plymouth, n.d.

POULLE (1969) E Poulle, 'Les instruments astronomiques du Moyen Age', *Le Ruban Rouge*, **32**, 1967; reprinted as *Selected Offprints of the Museum of the History of Science, Oxford* no. 7, 1969.

PRICE (1952)   DJ de S Price, 'The Early Observatory Instruments of Trinity College Cambridge', *Annals of Science*, **8**, 1-12, 1952.

PRICE (1954)   DJ de S Price, 'A Collection of Armillary Spheres and other Antique Scientific Instruments', *Annals of Science*, **10**, 152-87, 1954.

PRICE (1955)   DJ de S Price, 'An International Checklist of Astrolabes', *Archives International d'Histoire des Sciences*, **8**, 243-263 and 361-381, 1955.

PRICE (1956A)   DJ de S Price, 'Two Mariner's Astrolabes', *Journal of the Institute of Navigation*, **9**, 338, 1956.

PRICE (1956B)   DJ de S Price, 'Mariner's Astrolabes: A Recent Scottish Find', *Scottish Geographical Magazine*, **72**, 24-5, 1956.

PRICE (1959A)   DJ de S Price, 'On the Origin of Clockwork, Perpetual Motion Devices, and the Compass', *United States National Museum Bulletin*, **218**, 81-112, 1959.

PRICE (1959B)   DJ de S Price, 'The First Scientific Instrument of the Renaissance', *Physis*, **1**, 26-30, 1959.

PRICE (1959C)   DJ de S Price, 'An Ancient Greek Computer', *Scientific American*, **200**, 128-30, 1959.

PRICE (1975)   DJ de S Price, 'Gears from the Greeks: the Antikythera Mechanism – a Calendar Computer from ca. 80 BC', *Translations of the American Philosophical Society*, new series, **64**, pt 7, 1975.

PRICE (1982)   DJ de S Price, review of WARD (1981), *Journal of the History of Astronomy*, **13**, 129-131, 1982.

PULLAN (1970)   JM Pullan, *The History of the Abacus*, 2nd edn, London, 1970.

PYATT (1983)   E Pyatt, *The National Physical Laboratory: A History*, Bristol, 1983.

READ (1893)   CH Read, 'Notes on a Planispheric Astrolabe given to the Society by the Rev. I.G. Lloyd, FSA', *Proceedings of the Society of Antiquaries*, **14**, 362-364, 1893.

READ (1935)   J Read, 'An Old Balance', *Chemistry and Industry*, **13**, 1028, 1935.

READ (1953)   J Read, 'Schools of Chemistry in Great Britain and Ireland; I, the United College of St Salvator and St Leonard in the University of St. Andrew's', *Journal of the Royal Institute of Chemistry*, **77**, 8-18, 1953.

REES (1819)   A Rees, *The Cyclopaedia or Universal Dictionary of Arts, Sciences and Literature*, London, 1819.

REID (1982)   JS Reid, 'The Castlehill Observatory, Aberdeen', *Journal for the History of Astronomy*, **13**, 84-96, 1982.

REID (1983)   JS Reid, 'Patrick Copland (1748-1822)', unpublished MLitt. thesis, University of Aberdeen, 1983.

REPSOLD (1908) JA Repsold, *Zur Geschichte der astronomischen Messwerkzeuge von Purbach bis Reichenbach 1450-1830*, Leipzig, 1908.

REPSOLD (1914) JA Repsold, *Zur Geschichte der Astronomischen Messwerkzeuge von 1830 bis um 1900*, Leipzig, 1914.

RICHESON (1966) AW Richeson, *English Hand Measuring to 1800: Instruments and Practices*, Cambridge, Mass., 1966.

RIEKHER (1957) R Riekher, *Fernrohe und ihre Meister*, Berlin, 1957.

ROBBINS (1929-30) F Robbins, 'The Radcliffe Observatory, Oxford', *Journal of the British Astronomical Association*, **40**, 310-326, 1929-30.

ROBINSON (n.d.) DM Robinson, 'Early Navigational Instruments in the N.E. of England', unpublished MS catalogue, Marine and Technical College, South Shields, n.d.

ROBINSON (1829) TR Robinson, *Astronomical Observations made at the Armagh Observatory*, London, 1829.

ROBINSON (1833) TR Robinson, 'On the Dependence of a Clock's Rate on the Height of the Barometer', *Memoirs of the Royal Astronomical Society*, **5**, 125-6, 1833.

ROBINSON (1836) TR Robinson, 'Description of the Mural Circle of the Armagh Observatory', *Memoirs of the Royal Astronomical Society*, **9**, 17-33, 1836.

ROBINSON (1847) TR Robinson, 'On Lord Rosse's Telescope', *Proceedings of the Royal Irish Academy*, **3**, 114-133, 1847.

ROBINSON (1850) TR Robinson, 'Description of an Improved Anemometer for Registering the Direction of the Wind and Space which it Traverses in given Intervals of Time', *Transactions of the Royal Irish Academy*, **22**, 155-78, 1850.

ROBINSON (1859) TR Robinson, *Places of 5,345 Stars, observed from 1828 to 1854 ... (Second Armagh Catalogue of 3,300 Stars...)*, Dublin, 1859.

ROCHE (1981)   JJ Roche, 'The Radius Astronomicus in England', *Annals of Science*, **38**, 1-32, 1981.

ROHR (1970)   RRJ Rohr, *Sundials: History, Theory and Practice*, Toronto, 1970; English edition of *Les cadrans solaires: traite de gnomonique theorique et applique*, Paris, 1965.

ROOSEBOOM (1956)   M Rooseboom, *Microscopium*, Communication No 95, Nederlandsch Historisch Natuurwetenschappelijk Museum, Leyden, 1956.

ROSE (1968)   PL Rose, 'The Origins of the Proportional Compass from Mordente to Galileo', *Physis*, **10**, 53-69, 1968.

ROYAL PHOTOGRAPHIC SOCIETY (1953)   (1953) *The Centenary of the Royal Photographic Society 1853-1953*, London, 1953.

RULE (1982)   M Rule, *The Mary Rose: The Excavation and Raising of Henry VIII's Flagship*, London, 1982.

RULE (1989)   M Rule, *The Mary Rose: A Guide*, London, 1989.

RYAN (1972)   WF Ryan, 'John Tradescant's Russian Abacus', *Oxford Slavonic Papers* **V** new series, 83-88, 1972.

ST ANDREWS   St Andrews – University of St Andrews, University Muniments SM110 MB F37

SANDERS (1960)   L Sanders, *A Short History of Weighing*, Birmingham, 1960.

SAUNDERS (1984)   HN Saunders, *All the Astrolabes*, Oxford, 1984.

SAVAGE-SMITH (1985)   E Savage-Smith, *Islamicate Celestial Globes: Their History, Construction and Use*, Smithsonian Studies in History and Technology 46, Washington DC, 1985.

SCARISBRICK (1898)   J Scarisbrick, *Revenue Series 3: Spirit Assaying*, 2nd edn, Wolverhampton, 1898.

SCHNEIDER (1970)   I Schneider, *Der Proportionzirkel: ein universelles Analogrechen-instrument der Vergangenheit Deutsches Museum, Abhandlungen und Berichte 38 Heft 2*, 1970.

SCORESBY (1821)   W Scoresby, 'Description of a Magnetometer being a New Instrument for Measuring Magnetic Attractions and Finding the Dip of the Needle', *Transactions of the Royal Society of Edinburgh*, **9**, 243-258, 1821.

SCORESBY (1831, 1832)   W Scoresby, 'On the Uniform Permeability of all Known Substances to Magnetic Influence', *Edinburgh New Philosophical Journal*, **12**, 319-339, 1831; **13**, 97-132, 1832.

SCORESBY/ JOULE (1846)   W Scoresby and JP Joule, 'On the Mechanical Powers of Electro-magnetism', *Philosophical Magazine*, **28**, 448-55, 1846.

SCOTT (1899-1905)   DD Scott *et al.*, 'The Evolution of Mine Surveying Instruments', *Transactions of the American Institution of Mining Engineers*, **28**, 679-745, 1899; **29**, 931-1015, 1900; **30**, 783-837, 1901; **31**, 25-112, 716-747, 884-913, 921-935, 1902; **33**, 1035-1037, 1904; **35**, 322-326, 1905.

SHAPIN (1984)   S Shapin, 'Pump and Circumstance: Robert Boyle's Literary Technology', *Social Studies of Science*, **14**, 481-520, 1984.

SHAPIN/ SCHAFFER (1985)   S Shapin and S Schaffer, *Leviathan and the Air Pump: Hobbes, Boyle and the Experimental Life*, Princeton, 1985.

SHAW (1936)   H Shaw, *Applied Geophysics, a Brief Survey of the Development of Apparatus and Methods*, 3rd edn, Science Museum, London, 1936.

SHIERS (1971)   G Shiers, 'The Induction Coil', *Scientific American*, **224**, no. 5 80-7, 1971.

SHORTT (1968)   H Shortt, 'A Thirteenth-Century 'Steelyard' Balance from Huish', *Wiltshire Archaeological and Natural History Magazine*, **63**, 66-71, 1968.

SIMCOCK (1985)   AV Simcock, ed., *Robert T Gunther and the Old Ashmolean*, Museum of the History of Science, Oxford, 1985.

SIMPSON (1984)   ADC Simpson, 'Newton's Telescope and the Cataloguing of the Royal Society's Repository', *Notes and Records of the Royal Society*, **38**, 187-214, 1984.

SKINNER (1967)   FG Skinner, *Weights and Measures*, Science Museum, London, 1967.

SKINNER (1986)   GM Skinner, 'Sir Henry Wellcome's Museum for the Science of History', *Medical History*, **30**, 383-418, 1986.

SPILLER (1968)   B Spiller, ed., *Cowper: Poetry and Prose*, London, 1968.

STEARN (1981)   WT Stearn, *The Natural History Museum at South Kensington*, London, 1981.

STENUIT (1971)   R Stenuit, *Les Tresors de L'Armada*, Paris, 1971.

STEVENS (1870)   ET Stevens *et al.*, eds, *Catalogue of the Salisbury and South Wilts Museum*, revised edn, Salisbury, 1870.

STEVENSON (1921)   EL Stevenson, *Terrestrial and Celestial Globes*, New Haven, 1921.

STIMPSON (1988)   A Stimpson, *The Mariner's Astrolabe: A Survey of Known Surviving Sea Astrolabes*, Utrecht, 1988.

STIMPSON/ DANIEL (1977)   AN Stimpson and C Daniel, *The Cross Staff: Historical Development and Modern Use*, London, 1977.

STOCK (1969) — JT Stock, *Development of the Chemical Balance*, Science Museum, London, 1969.

STOCK (1973) — JT Stock, 'Weighed in the Balance', *Analytical Chemistry*, **45**, 974A-980A, 1973.

STOCK/BRYDEN, (1972) — JT Stock and DJ Bryden, 'A Robinson Balance by Adie and Son Edinburgh', *Technology and Culture*, **13**, 44-54, 1972.

STOCK/VAUGHAN (1983) — JT Stock and D Vaughan, *The Development of Instruments to Measure Electric Current*, Science Museum, London, 1983.

STROUP (1981) — A Stroup, 'Christiaan Huygens and the Development of the Air Pump', *Janus*, **68**, 129-158, 1981.

SUTCLIFFE/MITCHELL/CHITTELL (1932) — JH Sutcliffe, M Mitchell and E Chittell, *British Optical Association Library and Museum Catalogue*, no. 4, London, 1932.

SWINBANK (1982) — P Swinbank, 'Experimental Science in the University of Glasgow at the time of Joseph Black', *Joseph Black 1728-1799*, ed. ADC Simpson, p. 26, Royal Scottish Museum, Edinburgh, 1982.

SYMONDS (1951) — RW Symonds, *Thomas Tompion, his Life and Work*, London, 1951.

TAIT (1968) — H Tait, *Clocks in the British Museum*, Trustees of the British Museum, London, 1968.

TAIT (1983) — H Tait, *Clocks and Watches*, British Museum Publications, London, 1983.

TANGYE (1905) — R Tangye, *The Cromwellian Collection of Manuscripts, Miniatures, Medals, etc., in the Possession of Sir Richard Tangye*, London, 1905.

TATE (1930) — FGH Tate, *Alcoholometry*, London, 1930.

TAYLOR (1980) — RS Taylor, 'Swan's First Electric Light at Cragside', *National Trust Studies 1981*, ed. G Jackson-Stops, pp. 27-34, London, 1980.

TAYLOR/RICHEY (1962) — EGT Taylor and MW Richey, *The Geometrical Seaman*, London, 1962.

TAYLOR/WILSON/MAXWELL (1960) — EW Taylor, JS Wilson and PDS Maxwell, *At the Sign of the Orrery*, 2nd edn, York, 1960.

THODAY (1971) — AG Thoday, *Astronomy 2: Astronomical Telescopes*, Science Museum, London, 1971.

THODAY (1976) — AG Thoday, *A List of the Apparatus of Sir Francis Galton 1822-1911*, Galton Laboratory, University College, London, 1976.

THOMAS (1964) — DB Thomas, *The First Negatives*, Science Museum, London, 1964.

THOMAS (1966) — DB Thomas, *Cameras*, Science Museum, London, 1966.

THOMAS (1969A) — DB Thomas, *The First Colour Motion Pictures*, Science Museum, London, 1969.

THOMAS (1969B) — DB Thomas, *The Science Museum Photography Collection*, Science Museum, London, 1969.

THOMAS (1981) — DB Thomas, *The Science Museum Camera Collection, incorporating the Arthur Frank Collection*, Science Museum, London, 1981.

THOMPSON (1927) — CJS Thompson, 'An Historical Case of Surgical Instruments in the Possession of the Royal College of Physicians of London', *The Medical Press and Circular*, **124**, new series 344-6, 1927.

THOMPSON (1970) — FH Thompson, 'Dodecahedrons Again', *The Antiquaries Journal*, **50**, 93-6 and plate xx, 1970.

TRUCKELL (n.d.) — AE Truckell, *An account of the Camera Obscura and the Other Instruments Belonging to the Observatory*, Dumfries Museum, Dumfries, n.d.

TRURO (1966) — Truro – Royal Institution of Cornwall, *The County Museum Truro, A Guide*, Truro, 1966.

TURNBULL (1939) — HW Turnbull, ed., *James Gregory Memorial Volume*, Edinburgh, 1939.

TURNER (1966) — GL'E Turner, 'Decorative Tooling on 17th and 18th Century Microscopes and Telescopes', *Physis*, **8**, 99-128, 1966.

TURNER (1969) — GL'E Turner, 'The History of Optical Instruments: a Brief Survey of Sources and Modern Studies', *History of Science*, **8**, 53-93, 1969.

TURNER (1972) — GL'E Turner, 'Micrographia Historica: the Study of the History of the Microscope', *Proceedings of the Royal Microscopical Society*, **7**, 121-1149, 1972.

TURNER (1973A) — AJ Turner, *The Clockwork of the Heavens, an Exhibition of Astronomical Clocks, Watches and allied Scientific Instruments*, Asprey and Co. Ltd, London, 1973.

TURNER (1973B) — AJ Turner, 'Mathematical Instruments and the Education of Gentlemen', *Annals of Science*, **30**, 51-88, 1973.

TURNER (1979) — GL'E Turner, 'Johann Daniel von Berthold: A Clerical Craftsman and his Universal Ring-Dial', *Annali Dell' Istituto e Museo di Storia della Scienza di Firenze*, **4**, 15-20, 1979.

TURNER (1980) — H Turner, *Sir Henry Wellcome and his Collections*, London, 1980.

TURNER (1981A) AJ Turner, 'William Oughtred, Richard Delamain and the Horizontal Instrument in 17th Century England', *Annali Dell' Instituto e Museo di Storia della Scienza di Firenze*, **VI**, 49-125, 1981.

TURNER (1981B) GL'E Turner, *Collecting Microscopes*, London, 1981.

TURNER (1982) AJ Turner, 'The Making of an Instrument Catalogue', *Annals of Science*, **39**, 407-420, 1982.

TURNER (1985) GL'E Turner, 'Charles Whitwell's Addition c.1595 to a Thirteenth Century Quadrant', *Antiques Journal*, **LXV**, part II 454-5 and plates XCI - CI, 1985.

TURNER (1989) G L'E Turner, *The Great Age of the Microscope: Collection of the Royal Microscopical Society*, Bristol, 1989.

TURNER/ LEVERE (1973) GL'E Turner and TH Levere, 'Van Marum's Scientific Instruments in Teyler's Museum, Leyden', *Martinus van Marum: Life and Work* Vol. IV, ed. E Lefebvre and JG De Bruijn, Leyden 1873.

UNDERWOOD (1951A) EA Underwood, *Catalogue of an Exhibition Illustrating the History of Pharmacy*, Wellcome Historical Medical Museum, London, 1951.

UNDERWOOD (1951B) EA Underwood, *Catalogue of an Exhibition Illustrating Prehistoric Man in Sickness and in Health*, Wellcome Historical Medical Museum, London, 1951.

UNDERWOOD (1952) EA Underwood, *Catalogue of an Exhibition Illustrating the Medicine of the Aboriginal Peoples in the British Commonwealth*, Wellcome Historical Medical Museum, London, 1952.

VENN (1901, 1912) J Venn, *Biographical History of Gonville and Caius College*, Vol. III pp. 194-5, Cambridge, 1901; Vol. IV plates VII-XI, Cambridge, 1912.

VOICE (n.d.) E Voice, 'Catalogue of Scientific and Navigation Instruments in the Admiral's Room, Snowshill Manor', unpublished MS, Snowshill Manor, n.d.

WALLACE (1836) W Wallace, 'An Account of the Invention of the Pantograph', *Transactions of the Royal Society of Edinburgh*, **13**, 418-439, 1836.

WARD (1876-7) JC Ward, 'Jonathan Otley, the Geologist and Guide', *Transactions of the Cumberland Association of Literature and Science*, part II, pp. 125-169, 1876-7.

WARD (1955) FAB Ward, *Handbook of the Collection Illustrating Time Measurement: Part II, Descriptive Catalogue*, 3rd edn, Science Museum, London, 1955.

WARD (1963) FAB Ward, *Timekeepers, Clocks, Watches, Sundials, Sandglasses*, Science Museum, London, 1963.

WARD (1965-8) FAB Ward, 'A 14th Century Portable Sundial', *Antiquarian Horology*, **5**, 124-125, 1965-8.

WARD (1966) FAB Ward, *Descriptive Catalogue of the Collection Illustrating Time Measurement*, Science Museum, London, 1966.

WARD (1970) FAB Ward, *Handbook of the Collection Illustrating Time Measurement: Part I, Historical Review*, 4th edn, Science Museum, London, 1958; 3rd impression with amendments, 1970.

WARD (1972) FAB Ward, *Clocks and Watches 2: Spring-Driven*, Science Museum, London, 1972.

WARD (1973) FAB Ward, *Clocks and Watches 1: Weight Drive Clocks*, Science Museum, London, 1973.

WARD (1981) FAB Ward, *A Catalogue of European Scientific Instruments in the Department of Medieval and Later Antiquities of the British Museum*, British Museum, London, 1981.

WARTNABY (1957) J Wartnaby, *Seismology, a Brief Historical Survey and a Catalogue of Exhibits in the Seismology Section of the Science Museum*, Science Museum, London, 1957.

WARTNABY (1968) J Wartnaby, *Surveying Instruments and Methods*, Science Museum, London, 1968.

WATERS (1958) DW Waters, *The Art of Navigation in England in Elizabethan and early Stuart Times*, London, 1958.

WATERS (1964) DW Waters, 'Some Early Globes in England – a Provisional Hand List of the Terrestrial Globes and Armillary Spheres in the National Maritime Museum', *Der Globusfreund*, **13**, 70-87, 1964.

WATERS (1966) DW Waters, 'The Sea or Mariner's Astrolabe', *Agrupamento de Estudos de Cartografia Antiga (Seccao de Coimbra)* serie separatas, **15**, 1966.

WAYMAN (1971) PA Wayman, 'Henry Usher at Dunsink, 1783-1790'; 'The Visitation Book of Dunsink Observatory'; 'The Arnold Clocks', *Irish Astronomical Journal*, **10**, 121-9; 135-41; 275-82, 1971.

WAYMAN (1987) PA Wayman, *Dunsink Observatory, 1785 to 1985*, Dublin, 1987.

WEBSTER (1974) RS Webster, *The Astrolabe: Some Notes on its History, Construction and Use*, Lake Buff, 1974.

WEDGWOOD (1978)
Josiah Wedgwood and Sons Ltd, *Josiah Wedgwood: The Arts and Sciences United*, Barlaston, 1978.

WEISS (1975)
A Weiss, 'Zu den Anwendung smoglichkeiten des Pentagon-dodekaeders bei den Romern', *Archaeologisches Korrespondenzblatt*, **5**, 221024, 1975.

WESTCOTT (1932-33)
GF Westcott, *Handbook of the Collections Illustrating Pumping Machinery: Part 1 Historical Notes; Part 2 Descriptive Catalogue*, Science Museum, London, 1932-33.

WHEWELL (1838)
W Whewell, 'Description of a New Tide Gauge Constructed by Mr T.G. Blunt and Erected on the ... Bristol, 1837', *Philosphical Transactions*, **128**, 249-251, 1838.

WHIPPLE (1926)
RS Whipple, 'An Old Catalogue and what it Tells Us of the Scientific Instruments and Curios Collected by Queen Charlotte and King George III', *Proceedings of the Optical Convention, 1926*, Part 2 pp. 502-528, London, 1926.

WHITBY (1972)
Whitby – Literary and Philosophical Society, *Whitby Museum Guide*, Whitby, 1972.

WHITEHEAD (1981)
P Whitehead, *The British Museum (Natural History)*, London, 1981.

WIDDES (1949)
JDH Widdes, *A Dublin School of Medicine and Surgery – An Account of the Schools of Surgery: Royal College of Surgeons, Dublin 1789-1948*, Edinburgh, 1949.

WILLIAMS (1983)
MR Williams, 'From Napier to Lucas: the Use of Napier's Bones in Calculating Instruments', *Annals of the History of Computing*, **5**, 279-296, 1983.

WILSON (1849)
G Wilson, 'On the Early History of the Air-pump in Great Britain', *Edinburgh New Philosophical Journal*, **46**, 330-355, 1849.

WILSON (1935)
DK Wilson, *A History of Mathematical Teaching in Scotland*, London, 1935.

WOLLASTON (1807)
WH Wollaston, 'Description of the Camera Lucida', *Philosophical Magazine*, **27**, 343-347, 1807.

WOLLASTON (1814)
WH Wollaston, 'A Synoptic Scale of Chemical Equivalents', *Philosphical Transactions*, **104**, 1-22, 1814.

WOOD (1972)
C Wood, 'The Astronomical Clocks at the Observatory, Catton Hill, Edinburgh', *Antiquarian Horology*, **8**, 55-61, 1972.

WOODBURY (1958)
RS Woodbury, *History of the Gear-Cutting Machine*, Cambridge, Mass., 1958.

WOODWARD (1966)
MH Woodward, *One at London*, London, 1966.

WRAY (1983A)
EM Wray, ed., *Catalogue of Apparatus in the Museum of the Natural Philosphy Class in the United Colleges of St Salvator and St Leonard, St Andrews, prepared by W. Swan Ll.D. May 1880*, St Andrews, 1983.

WRAY (1983B)
EM Wray, *Historical Scientific Instruments from the Collection of the Department of Physics, University of St Andrews: A Guide to Selected Exhibits*, St Andrews, 1983.

YONGE (1968)
EL Yonge, *A Catalogue of Early Globes made prior to 1850 and Conserved in the United States*, American Geographical Library Series, Vol. 6, New York, 1968.

YOUNG (1975)
RM Young, *Teaching Toys*, 2nd edn, Castle Museum, Norwich, 1975.

ZINNER (1956)
E Zinner, *Deutsche und Niederlandische Astronomische Instrumente des 11-18 Jahrhunderts*, Munchen, 1956.

ZUCK (1978)
D Zuck, 'Dr Nooth and His Apparatus', *British Journal of Anaesthesia*, **50**, 393-401, 1978.

Indexes

Two indexes are provided, one of names and the other of instruments. In the former are included not only the names of instrument-makers, but also scientists, astronomers, surveyors, etc, with whom the instruments are associated. In the case of instrument-makers their place of professional activity is also included. In some cases the name on an instrument refers to an owner rather than a maker. Though it is sometimes difficult to decide on the status of a name, where there is a likelihood that the name is not that of a maker, the indication (prov), i.e. provenance, is shown.

It should be noted that these indexes only refer to items and makers actually listed in the main body of the text. Since the six largest collections do not have their contents listed in detail, it follows that the majority of items in these collections will not appear in the index. Thus, always, in any search for material, information regarding specific instruments or makers in these six collections must be sought directly from the museums concerned.

Index of names

Index of instruments